C0-AUG-202

CADCAM IN EDUCATION AND TRAINING

Proceedings of the CAD ED 83 Conference

Edited by Dr Paul Arthur

Kogan
Page

First published in 1984 by Kogan Page Ltd
120 Pentonville Road, London N1

Copyright © The CADCAM Association 1984
All rights reserved

British Library Cataloguing in Publication Data
CADCAM in education and training
  1. Design, Industrial—Data processing
  2. Manufacturing processes—Data processing
  3. CAD/CAM systems
  I. Arthur, Paul 19—
650.2'028'54   TS171.4

  ISBN 0-85038-808-2

Photosetting by Typesetters (Birmingham) Ltd.,
Smethwick, Warley, West Midlands

Printed in Great Britain by Anchor Press
and bound by Wm. Brendon & Son, both of
Tiptree, Essex.

670.427
C11

Contents

Part 1:
Introducing CADCAM

Editor's note:

All contributing authors were requested to supply figures of a quality suitable for reproduction – many failed to do so, and in such cases the figures have been omitted.

Readers wishing to view any of these missing figures should approach either the author or the CADCAM Association who will be happy to supply photocopies.

1. Welcoming address

H H Rosenbrock

The first thing I would like to do is of course welcome you to UMIST. It is a very great pleasure to be host here to this meeting. The subject is one in which UMIST has had an interest for many years and it is a subject which is at present only at the very beginnings of its development. What we have seen so far, I believe, is comparable to the development of the motor car up to say 1900, and the developments which are going to take place are going to have a very great impact. Technological institutions like UMIST feel that they need to be deeply involved in this kind of activity, so welcome to UMIST and I hope that your conference will prove an interesting, profitable and enjoyable one for all of you.

My own interest in CAD goes back to the 1960s when it was very young. The first contact I had was in 1963–4 when I happened to spend a year at MIT when the sketch pad programme was just terminating. My own technical work was in a very specialized area of CAD, that is to say the computer-aided design of control systems. We didn't realize it at the time but looking back on it, it was somewhat different in its development from the major development which has taken place over the last five years or so in the wider application of CAD to engineering design. We were concerned with designing control systems of the abstract things: they don't give rise by themselves to graphical output pictures and the problem was to develop a theory which gave rise to the pictures, and that's where a lot of the work went. Then of course there were all the problems of the role which the user should be given in the system. My own particular interest in the social effects stemmed from the strong feeling that I got at that time, that in giving to the user the vital role in the design process we were very much swimming against a stream which tried hard to eliminate the user, to eliminate the designer, and reduce his role to specifying what was required. The properties of the system, the constraints, the objective function, would then be put into a big computer off-line which would churn away and come out with an answer, the principle being here that the designer said what was needed and the computer came back and said if that's what you need this is how you solve your problem. Well, first of all, it's not a good way of solving engineering problems, as all of you no doubt realize. You look at the answer and say, 'That's not what I want to do'. So you then have to go back and try to re-specify the problems and the answer comes out a bit nearer to what you want. The only rational way of approaching that situation is to say that we don't really know at the beginning what is needed and we have to learn by trying to solve the problem. Defining the problem is a large part of the problem itself, and one has a good definition of the problem only at the time that it is solved. There is an essential part for the designer to play in interaction with a system which aids him, clarifies his thinking, works out for him the consequences of the decisions he makes. I think one has to recognize that in saying that, one is going counter to a very deeply held scientific attitude which holds that to rely on human skill at the time the job is done is somehow unscientific and not technologically respectable. From that, my particular interest in the social effects stemmed. Now you have a talk by Mike Cooley, which I think will certainly bear on this kind of problem. All I would like to say to you is this, in designing systems, CAD systems, which will allow a designer to do his job, you are doing many things, and you are contributing to the productivity of the designer. You are possibly creating a situation in which the things he designs will change. He now has different facilities, he can now design a different kind of thing, perhaps more advanced, more complicated. There are several other things that you are doing. You are affecting the number of jobs that there will be. Some kinds of jobs will be lost, some kinds of jobs will be created. That is a problem partly for engineers, but mostly for economists and politicians. But there is one particular thing that you will be affecting, which is very much the province of the engineer technologist, and that is the kind of work people do. The systems that you produce, the systems that you influence by your purchasing policies, the kind of thing that you buy which affects the kind of things that people will develop, will affect the kind of work the designers will do in the future and I would like to sketch two extremes that one could envisage. At one extreme we could set out to eliminate as far as possible all the initiative of the designer and do everything for him in the computer to reduce his decisions to the absolute minimum. Now that may not sound very sensible, but certainly it's what has happened on the shop floor for 150 years or more. All the initiative has been taken away from the shop floor as far as possible and given to the specialized design office. In the same way, one could envisage taking away from the designer as much as possible of his initiative. So that, for example, the geometrical information about a product was entered as quickly as possible. With a little thought, the computer system accepted this and then took responsibility for presenting that tentative design, to specialists in production methods, in corrosion, in stressing who would then make suggested changes. The computer would take the suggested changes, make sure that the final design had gone round the whole route among all these specialists without any further change before it was accepted. One could try to give the computer system great control over the design process. The other way, the other extreme, would be to accept that the computer system has quite different capabilities from the human being. The computer system can ease the designer's work but it can't make new, totally new decisions. It can't see, for example, if a product has some new way of failing which has not been experienced before. A good designer may see this and a computer system never will, and one could aim to give to the designer the maximum possible control over the situation. Giving him the CAD system as a tool for him to use, to ease his work but making him the master and the CAD systems the alternative that I have described, reverses that order. It tries to make the CAD system the master and the designer the servant, and I would suggest to you very strongly that the second way is the way one should try to go. One should accept that human beings have a different kind of ability from machines and one should set out not to eliminate the human being but to use his ability to the best effect. Now that's perhaps somewhat moralizing and the tone of your conference as I notice from the programme is largely technological, but I would like to suggest to you that in your deliberations you will be affecting not only the technology, not only the economics but also the life of people in many ways and in particular the kind of jobs that designers will do. I will stop at that point, and wish you once again a very successful conference.

2. Integration and implementation of computer-aided engineering (CAE) – the strategy for innovative product design in the 1980s

M A Neads

Abstract: Technologies to automate computer-aided drafting and computer-assisted NC tape preparation are available and are beginning to be used widely to help reverse the alarming trends of declining productivity in many industrial economies. However, automating isolated tasks in today's 'build-and-test' product development process, while cost-effective, will not achieve significant time savings, productivity gains and/or strategic benefits, as anticipated by most companies.

The overall mechanical product development process itself must be automated. Products must be developed within the computer. Prototypes should be built to verify and validate computer predictions, instead of being used to find out how a product performs, as is common today.

Extended reaches to improve product performance and quality can be achieved in significantly shorter time through the effective implementation and integration of existing computer-aided engineering and related manufacturing capabilities. Indeed, strategic benefits impacting a company's overall market share, quality image, return on investment and profitability can result from effective CAE implementation.

Benefits do not come easily, however. The CAE process requires change, including change in organizational structure. CAE methods cannot be implemented quickly: three to ten years are required in most companies and industries. CAE implementation and maintenance is expensive: at least 10 per cent of total product development budgets each year is required for CAE software, hardware, maintenance, support and training. An entirely new way of thinking about mechanical product development is necessary in most companies.

Pressures on design engineers

Among these pressures are that buyers are demanding improved efficiency and reliability in mechanical equipment and products; manufacturing executives are demanding design time and cost reductions; government regulations demand concentration on safety, pollution, and noise control – regulations which didn't exist ten years ago. Finally, we have the energy crisis and the demand for the preservation and conservation of our natural resources. Engineers in all industries are facing these pressures.

The other major problem, that everyone is certainly familiar with, is the problem of declining productivity. Not only does declining productivity cause problems within the United Kingdom but the UK has one of the poorest productivity records in terms of the top seven industrialized nations. All of these international competitors face pressures similar to our own, in terms of energy, inflation, etc.

Not only is labour productivity declining, but so is the productivity of our capital equipment in this country.

However, there are hopes for improved productivity, some of which are real. These include the trend toward incentives for increased capital investment in CAE intensified national efforts on research and development and innovation, and finally, the reason that we are here, the *emerging technologies* that are now becoming available to us.

Emerging technology and CAE

Great strides and changes have occurred in mechanical engineering in the last few years. Just read Gene Bylinsky's article in the 5 October issue of *Fortune* magazine entitled, 'A New Industrial Revolution is on the Way'. Most observers feel now that computer-aided manufacturing and computer-aided engineering are 'the manufacturing technologies most likely to raise productivity during the next 10 years'; (New Technologies and Training in Metal Working, National Centre for Productivity, 1978).

We are not talking about the recent advent of turnkey CAD/CAM drafting systems. While these do increase productivity of the drafting task, they do not *dramatically* reduce the time and cost involved in the overall total mechanical development process.

Rather, what we are talking about is a process called computer-aided engineering, which can reduce design time and cost by factors of 30 per cent or more; not the 4 per cent or 5 per cent that CAD/CAM systems provide. The CAE process, then, addresses the entire product development cycle, from concept to finished product.

Emerging technologies – yesterday

To understand how revolutionary technology can be, let's take a look back at only ten or twelve years ago to see what was available to design engineers at that time. You remember 1970? There were no hand-held calculators that are so common today and large resources were not available. Engineers were making only about £1,500 per year. When faced with a new design project – the development of a new car, the design of a bridge, etc – we simply assigned more engineers!

The technology available to engineers in 1970 was archaic by today's standards. We were using time-sharing computers from a 15 character per second noisy teletypewriter terminal. Few were able to get any type of graphical output of results. For mechanical design analysis, very little was available in the terms of computer-assisted design. Mostly, beam finite element models were the best simulation tools available. Very few design analysis computer programs were available; NASTRAN was not yet released to the public. By and large, most design engineers in industry relied on the old textbook approaches and the standard 'build the prototype, test it; modify it, test it: modify it, test it:' methods. Generally, design engineers in industry had very few tools for improving design process productivity.

Emerging technologies – today

However, let's look at what's available to us now. There has been an incredible explosion in the knowledge available to mechanical design engineers. How many of us have been able to keep up with the required reading and continuing education in our speciality?

By 1980, the average engineer's salary had risen to about £8,000 per year. No longer can we simply add more and more engineers to a project. Not only that, there is a shortage of qualified engineers today.

However, technology has come to our rescue, at least partially. For example, there are now minicomputer-based integrated testing systems which allow data from all kinds of vibration, stress, or noise tests to be collected, processed, and displayed in a form that design engineers can understand. No longer are they faced with the tedium of studying strip charts and large print-outs of tabulated numbers.

An excess of computer-assisted terminals, connectable via phone links to large computers, is now available to design engineers. Very complex finite element models can now be generated, analysed, and their results displayed in a matter of minutes.

Today, integrated computer-aided design and drafting equipment accessing software systems for computer-aided engineering allow engineers, through a data-base approach, to be able to go from component design concepts, through component analysis, system performance and evaluation, component evaluation, drafting and documentation, computer-aided manufacturing and advanced prototype performance evaluations.

Total CAE computer facilities are available. The recommended approach involves a hierarchy and distributed data-processing computers, beginning at the lowest level with local user stations. These include 'dumb' graphics terminals and/or graphics terminals with local intelligence (from microprocessors or minicomputers), such as those included in the turnkey systems mentioned earlier. Today's user stations, however, are not standalone, but are connected to higher level computers, either the so-called midrange or product level computers such as the DEC VAX 11/780, or an IBM 4300 series; or may connect directly to mainframe computers such as the CDC Cyber 170 series, or the IBM 3000 series.

Finally, the installation of in-house computers and computer networks does not obviate the worth of world-wide commercial computer networks such as General Electric Information Services. These world-wide networks can provide electronic information for all of the vendor-supplied equipment used by virtually every manufacturer of mechanical systems today. In addition, networks can provide the back-up facilities needed for overflow of corporate mainframes. They also allow the interconnection of a corporation's computers at various locations as well as allow access to data-bases world-wide.

CAE – a designer's revolution

We believe that computer-aided engineering advancements have put mechanical engineers on the threshold of a revolution. It is not surprising that this revolution should finally occur for mechanical engineers in much the same way that the electronics industry underwent the revolution from the 1940s through to the 1970s, transforming large vacuum tube computers that took rooms of space to the VLS integrated circuit familiar to us today.

Design examples

With strong emphasis on computer graphics, computer-aided engineering today allows us to design rapidly this robotic manipulator. Not only can we have a wire frame display, but also a three-dimensional solid geometric model (Figures 1 and 2), which allows us to put a design completely through its paces before anything is committed to hardware (Figure 3). Once we have put the system concept through its initial design

Figure 1 *(see page 9)*

Figure 2 *(see page 9)*

Figure 3 *(see page 9)*

phase, we can concentrate on individual components such as a portion of the robot arm. Using the geometric data-base, it is possible to generate automatically complete finite element meshes, submit this design to a larger computer where the finite element code itself resides and bring the results data back to the graphics display and portray the stress contour output in forms easily understood by the designer (Figure 4). Once we are satisfied with the various components, we can call on the CADCAM drafting systems to produce shop-ready drawings completely dimensioned, with bills of materials.

CAE and computer graphics are now being applied to other mechanical industries, such as the refinery and plant design process business. Computer graphics are being developed to replace the expensive and extensive plastic modelling of new plants (using plastic models of vessels, pipes, etc) to get rid of all the interferences possible in such a complex design. With the addition of colour, added understanding can be provided.

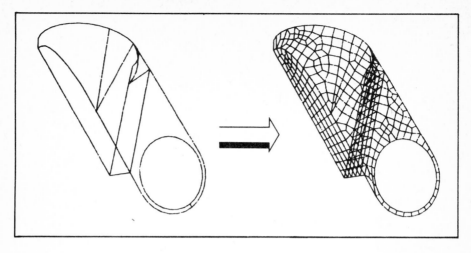

Figure 4 *Boundary edge description of a robot component (left), automatically generated finite element model (right)*

With the solid graphical representation, the equivalent of the hard plastic models, can be generated (Figure 5).

Figure 5 *(see page 9)*

Conclusion

The future is very exciting, and the importance of CAE and its positive impact on engineering design quality and productivity is beyond question. The new CAE approach is computer-based and not only encompasses both CAD and CAM activities, but also goes much further. It automates the entire product development process from conceptual design to release to manufacturing.

If engineering management and top management is willing to measure up to the challenge of integrating and implementing the CAE concepts we will have a promising future. Companies must have a cognizant corporate strategy, based on a five-year plan to achieve this.

Without question, CAE implementation is the 'challenge of the eighties' for most engineering and manufacturing professionals. Implementation of CAE philosophies into product development and manufacturing activities is one process most likely to achieve massive productivity gains and improve the quality of working life.

References

Henke, F R (1981) CAE-Engineering's best bet for productivity improvement. The conference board focus on productivity

Klosterman, A L and Lemon, J R (1969) Building block approach to structural dynamics. ASME publication VIBR 30

Lemon, J R, Tolani, S K and Klosterman, A L (1980) Integration and implementation of computer-aided engineering and related manufacturing capabilities into mechanical product development process. *Saabrucken Jahrestagung*

Neads, M A (October 1981) What is computer-aided engineering? A design case study with strategic benefits. Design engineering conference NEC Paper 1 Session 1a

New technologies and training in metal working, National Centre for Productivity (1978)

M A Neads, Managing Director, SDRC Engineering Services, York House, Stevenage Road, Hitchin, Herts

Discussion

J Butler, Technical Change Centre:

Your presentation seems like an excellent commonsense strategy. What is, I wonder, the fundamental problem? Why is this kind of strategy not being adopted as quickly as many of us would like? What's fundamentally wrong with UK industry? And is it something about UK industry in particular?

M A Neads, General Electric CAE International UK:

I think there is a reticence to try new technology. Companies have to prove to themselves its value (which is quite right). They won't speculate and take risks. As an example, if today you like a particular calculator, would you say to yourself, 'Well, I won't buy that one because in six months' time it will be out of date or half the price. Let's wait and see what's going to happen.' The problem is that if you do wait and get the better equipment – better not in terms of its intrinsic value, but better in terms of its capabilities – you will miss out on all the education, training and productivity which you could have gained had you acted earlier. What puts companies off acting earlier is the large and frightening number of manufacturers trying to sell hardware and software systems today. But until a company actually tries something, they are never going to know whether they are buying the right thing or doing the right thing. The only way to learn is to make mistakes.

We have got around this requirement to make mistakes without it costing us too much and still learning a great deal at the same time. UMIST has a policy to help, with its walk-in type centres, and my company has a productivity centre with this type of equipment, where companies can come in to do pilot studies. There is a move here, and although response isn't great, there is a steady trickle of interested companies coming in. I don't think I have completely answered your question, because I don't know the full answer. I think it's reticence on the part of industry: they have traditionally done their engineering like this and they don't see any immediate need to change.

J Butler:

I didn't expect a complete answer, I simply wanted to introduce that subject. Linked to that I am not exactly sure what you are offering as CAE International. Are you offering new development consultancy for some firms or are you suggesting that you introduce this strategy as management strategy within companies?

M A Neads:

It's really both. We would like to introduce this as a management strategy and although we don't sell hardware, we do offer programs, in consultancy in the area of implementing high technology techniques in the mechanical engineering industries today.

D G Smith, Loughborough University:

I was very interested in your discussion of the product development, from the conceptual phase right through to detail. It answers the question of why there is so much work done in the manufacturing industry, because this is an obvious place for implementation of computing. The question I want to ask is, what is in fact being done at the conceptual end?

M A Neads:

At the conceptual end there are many companies now with three-dimensional modelling packages which can be used in conceptual design all the way from design and styling to modelling components and allowing them to be analysed or simulated together as a total system. It is easy to look at productivity gains in manufacturing because you can calculate how long it takes the NC part programmer to program this part, how long it takes the machine and how long it takes the machinist. At the other end of the scale, it's very much more subjective: what do I get if I spend the money on a system design package to do system design work? How much will I save? The answer is that we don't know. We do know that you can look at many more different design concepts much more quickly and evaluate those in the computer. The engineers know that they are going to end up with better quality designs but the management who looks after the company usually looks for the cheque book and asks, 'What is the productivity gain?' 'How much will I save by buying this equipment?' That's not always the right question to ask, particularly in the area of conceptual design and design innovation.

3. Getting the CADCAM tool working

P F Arthur

Abstract: This paper discusses the approach that needs to be taken by potential purchasers of CADCAM systems. It presents some basic points that need to be considered and also discusses the importance of the education of managers now that CADCAM systems are tools, not toys.

Introduction

It is now a fact that CADCAM systems are working tools in all the sections of the engineering industry. If industry is to get the best out of these systems it must treat them in the same way as any other piece of capital plant in which it invests. Computers have been available and used commercially for over 20 years. In this period the rate of development of computing technology has been very great but the dissemination of knowledge has had some curious aspects. From the mid-sixties universities recognized the important role that computers would play in the future and that therefore undergraduate courses for scientists and engineers at least should contain a course on computing. Unfortunately these courses in general covered little more than the basics of FORTRAN programming. There is an old saying that a little knowledge is a dangerous thing and this has been so in this case.

Whereas a few computer scientists would design and produce, say, a complex mechanical jig to aid production it appears that many engineers felt able or were pressed into designing and developing CAD systems. There have been two important effects of this: (1) the retardation of CADCAM because of the number of unsuccessful systems that resulted but (2) more importantly, it slowed down the training of multi-disciplinary engineers and development teams. It must be recognized that computer science (and the more recent development of CAD as a 'profession') is as specialized a discipline as any branch of engineering but to produce an application package for an engineer it requires the expert knowledge of both the engineer and computer scientists to produce a user-friendly package.

Why use a CADCAM system?

Looking at the computer in simplest terms it can be considered as having two distinct facilities:

1. A processor that can be programmed to carry out a very high-speed arithmetical operation to evaluate expressions or to solve any set of solvable equations, either algebraic or differential.
2. A mass-storage device that can have effectively infinite storage of data that can be searched and retrieved at great speed and input rapidly by a variety of means.

Nearly all applications depend much more on one of these features than the other but an efficient user-friendly system is a subtle blend of the two facilities.

The application of any new technology is usually justified for one of the following reasons:

(a) It allows a new operation to be carried out.
(b) It improves the accuracy or reliability of the operation.
(c) It improves the productivity of its users.
(d) It improves the work place environment.

Thus when we try to apply computers to the engineering design and manufacturing process we must look for techniques that can give one of the above improvements. By analysing the use of time, the areas of lack of knowledge and waste of time and materials we can identify areas that can benefit from the use of a CADCAM system.

Now it is not the object to evaluate areas for the application of CADCAM but to discuss the next stage to look at the decision process on the viability of an application and the choice and implementation of the package.

Buying a CADCAM tool

For many years computers have been the tools of research and management services divisions. The hardware in the past required special environment controls and teams of operators to run it. From the software point of view in the early days all applications software was developed in-house and not bought in. Because of the growth of large teams of development programmers in the sixties and as a justification to maintain these empires, managers have rejected the purchase of externally developed software marketed on a commercial basis. Computer technology is no longer a new technology and computer technology is now a common tool in many areas without the users being conscious of the fact that they are using computers. A good example of this is word-processing. The CADCAM systems have been developed to a high degree of user-friendliness and ease of use. The evaluation of a CADCAM system can be carried out in the same way as any a piece of capital equipment. There are a number of important consequences of that statement. Since the CADCAM system is a tool for one of the engineering departments, the evaluation and selection should be the responsibility of the engineering departments, not the computer data processing department.

Engineering departments have been buying capital tools for many years and the acceptance of the basic principle that those concerned do not become involved in the selection of a CADCAM tool. There is no reason why those concerned in the inner working of the tool need become involved in the computing hardware and software details that are irrelevant to the decision.

The points in the evaluation

Let us forget that we are evaluating a CADCAM system for purchase but instead imagine a piece of capital equipment of the type that we are accustomed to purchasing and therefore evaluating. The following points must come high on the list to be looked at:

1. Define the task that is to be performed.
2. Define the features of a system required to perform a task.
3. Establish performance figures for the tool.
4. Carry out financial justification for a number of alternatives.
5. Establish short list of suppliers.
6. Look at short list under the following heading:
   (a) reliability of the tool
   (b) maintenance offered by the supplier
   (c) training and support offered by the supplier
   (d) product development plans
   (e) flexibility of tool to cope with possible changes in product.

Having set down this evaluation program let us give some thought as to application to a particular aspect of CADCAM, namely design and drafting.

If we consider design and drafting, then in a lot of design departments the majority of items designed will fall into the same category. There is a distinct difference in the purpose of using a CADCAM system between the first two and last two categories. As I said earlier, a computer can enable us either to do tasks much more efficiently or to do tasks that cannot be done, or cannot easily be done, by hand. The first two categories are of the former type, the second two of the latter type and the CADCAM systems fundamentally different. It is therefore vitally important to establish the task to be performed by the CADCAM systems and define the objectives.

Features of the system

Not only is there a vast difference in the capabilities of the four categories of the system, but there is also a difference in the abilities of the user of the systems between the extreme types. This is particularly so in the USA. For convenience let us reduce the number of categories to two two-dimensional and three-dimensional systems. The two-dimensional systems will be used by draftsmen rather than designers and their use is to produce drawings in a highly efficient manner. An important feature is therefore the user interface, ie how easy is it to use.

All systems are driven by commands whether input of characters is through a keyboard or from a menu tablet. The commands should be structured in such a way that they are both easy to learn and require a minimum number of characters to be input and do not depend upon a complex syntax. In only a few offices are the users using the system 100 per cent of their working day. For part-time users the simpler the command structure the higher a level of competence they will achieve. From the computer point of view, the simpler the command structure the less processor power it will absorb. A complex powerful command structure may be able to allow complex operations in a single command but it obviously requires much greater power to interpret the command and the use of simpler commands may use smaller amounts of the processor.

No two offices are identical and hence an important feature of the system is the flexibility to allow it to be fitted to each site without modification of the program. Let us consider a couple of points that can be helpful. Whilst there are national standards for dimensioning and annotating drawings these allow a fair deal of flexibility and each office will define a subset of the options to which it will adhere. Thus a dimensioning facility written in a user language that allows the user to set up strings of system commands with calculation, decisions and storage commands will allow the system manager to tune dimensioning to his requirements without the overhead of a totally general system.

There are many applications where drawings or sections of drawings are achieved by identical strings of commands but with different numerical values; again the user language can be used to advantage in this situation.

Many drawings contain large numbers of systems or standard parts. The use of libraries of symbols or parts can add greatly to the productivity achieved by use of the system. The easy creation of, addition to and change of libraries can be an important factor.

Looking at three-dimensional systems, we have a different set of points in mind. Here, with designers being the users, we can concentrate much more on the technical facilities provided. Three-dimensional systems absorb a large amount of computer-processing power and to reduce this some systems make subtle changes to a model, not always apparent to the user, in which accuracy can be lost. Since this type of system is often used to provide precise definitions of solids and surfaces, such a feature of a system is a major disadvantage if not making the system useless. In these areas of what is still a new and developing technology it is important to ensure that the computing science behind the system is sufficient (within the mathematical meaning).

Performance of the system

The emphasis in this section must be on the complete working system. It is important that it is a design system designed to optimize the components rather than just a collection of components. For a drawing system the end product is a drawing and hence a system is incomplete without a workable facility to produce the final drawings on paper or film. For a drafting system the architecture and power of the hardware must be such as to ensure the quick response of the system to interactive inputs. It is necessary to have a significant number of drawings available on the system at any time. The size of the disc will, together with storage space for drawings, determine the number of drawings that the system can hold. The size of drawings is usually dependent on the detail in the drawing but will vary substantially from system to system.

Financial justification

The simplest case for a financial justification is when the additional costs and savings made can be easily calculated. For a drafting system using dedicated hardware the former can be easily calculated, it is the sum of

the hardware cost (or a proportion of it where mixed applications can occur);
the software cost;
the hardware maintenance over the planned life of the system (this is significant over five years up to 60 per cent of initial cost);
the power used by the system.

The savings are more difficult to quantify; they obviously include the savings in direct cost of drawings but larger savings may accrue from higher physical drawing quality as well as better design quality. These can have substantial effects on the shop floor. This higher-quality drawing is produced more quickly and the items manufactured more quickly, thereby reducing lead time and work in progress. CAD systems are unlikely to produce vast reductions in manpower. They can, however, help absorb efficiently peaks in drawing loads.

For the surface and solid modelling system their benefits come almost entirely from improved quality of design and often from down-stream packages for automatic or semi-automatic NC tape production. It is difficult to put a price against quality and uniformity of information. All that can be said is that there is a rapidly growing awareness of the need for a communication highway within a manufacturing organization, going from concept to delivery.

The short list

At this point we should have a detailed document specifying the features we are looking for in the system and those which may be commonly offered by the available system. We have carried out a financial justification on this basis and have established a budget cost for the system through its planned life. We can now establish a short list of potential suppliers. It is now that I would like to bring you back firmly to my initial statement for at this point we are about to be dazzled by all the buzz words, glitter and 'razzmatazz' of the American sales bandwagon and the British salesman underselling his product. The point I raise again is that we are buying a tool and buying as if it were a hacksaw, hammer or machine tool, not a flashy computer. So now we look in odd terms at what each supplier has behind him to back up his products playing devil's advocate and assuming that everything that can go wrong will go wrong.

What has the supplier to offer?

Reliability

This falls into two categories, hardware and software. Aspects that are worth enquiring about are the source of the hardware – is it his own manufacture, if so, how many has he built, where is it built and can you talk to other users? If it is from a recognized hardware supplier and unaltered then one can have perhaps more confidence in it. At least if you want to use it for other purposes you can do so. Look also at the options on peripherals and numbers that can be added. You may have your favourites already accepted by users and unions and you would like more of these. The final question is, can it run on more than one manufacturer's hardware? You may have spare capacity on an existing machine or you may be able to upgrade cheaply or you might quite simply wish to stick with your current supplier of hardware.

From the software point of view there is nothing like having a go and pressing the keys yourself. If you can drive it and it keeps going, anyone can. If necessary pay for a trial, but do this properly, ie consider it preferable to take on a licence for, say, limited usage on-site, rather than 'play' at benchmarks.

Maintenance

Regardless of the source of the hardware you need to establish where your nearest engineer lives, how many cover your area and what spares are held. An engineer based four hours drive away is likely to leave you out for 24 hours – it is always a part he does not have with him that has failed. It is also helpful to find out how they get the part shipped to you in these circumstances. Do not forget to find out the detailed costs and what you actually get in your maintenance contract.

On the software side ask about the maintenance team, where is it, what are the procedures for reporting faults and what action is taken?

Training and support

Where and when does the supplier run training courses? It is necessary to establish what is provided in the price you are paying and what it costs for additional training. It may be worth looking at the language the manuals are written in. Watch out for American masquerading as English! Enquire what customer support is available. Does the supplier give a particular contact man whom you get to know and who knows your organization and how you use the system?

Product development plans

Computer systems, like all other products, need to be continually developed to stay abreast of the competition. You might find out how you, the user, can influence developments for the product. There are also the questions of upwards compatibility or releases. Will the system in three years' time be able to read in the drawing I am doing today?

Flexibility

No organization has a stable function over five years which is the period of time that we must consider to be the life of a CADCAM system. In some areas one may be able to predict reasonably accurately the changes that will take place, in others not so. It is important, therefore, to consider the breadth of the work area with which a proposed system can cope. It is important in this context to see the fundamental difference between a turnkey system in which the hardware and software cannot be separated and the system that is standard hardware controlled by an often independently purchased application package. In the latter situation the software may be changed or the hardware used for a quite different purpose. In the former this is not possible. Thus flexibility must be considered both from the spectrum of the work that the system can service and ultimately its salvage value. Finally it is also worth considering the optimistic view of how easy it is to expand the system in terms of number of terminals. Often the latter can offer more cost-effective expansion than the former.

Conclusion

In concluding I would like to draw together a number of important points regarding CADCAM systems:

1. CADCAM tools are today professional-engineered products and should be treated as such.
2. The evaluation of a CADCAM system and the justification for its purchase can be carried out in the same way as the purchase of any other item of capital expenditure.
3. The questions of maintenance, development and training are as necessary for the CADCAM tool as the machine shop tool and require the same management responsibility.
4. Do not buy on promises; find a system that can do what you want it to do today, and has the future plans to grow with you.
5. There are a lot of people who need to be educated not about how CADCAM systems work but in how to use the tool. How many purchasers of machine tools know anything about machine tools? There need to be no more users of CADCAM systems who know how they are put together.

Acknowledgement

The author would like to acknowledge the help and information given by Compeda Limited in the preparation of this paper.

P F Arthur, Compeda Ltd, Walkern Road, Stevenage, Herts

Discussion

F Weeks, Newcastle Polytechnic:

We have expensive pieces of hardware containing very sophisticated software capable of doing all sorts of tasks for us and yet very often the person who is starting out to use this equipment is so unfamiliar with the keyboard that he cannot even find an individual key. What is the answer to this? Do you, for example, provide separate tuition in the use of the keyboard or are companies like Howard's looking at alternatives to the keyboard as an input device?

P Arthur, Compeda Ltd:

It's a very serious problem. People have tried a number of options, and it's still an area where we have a lot of research and development still to do. I think progress has been made recently in the structure of input languages, but between the different systems there is still a vast difference, for instance in the language structure for say large turnkey systems such as Computervision, Applicon, and the smaller software packages such as Dragon. One of our customers has a very large turnkey system and a large Dragon installation. I asked him how he decided which system to use for any particular task. He said that if he can possibly do it on Dragon he does because he gets higher efficiency in the use of the hardware resource and because of the ease of learning the Dragon language I have many more people who can drive it. If I use my turnkey system there are only four people who can drive it. An operator has to work full time to remain effective on the turnkey system and I do not find it worthwhile having spare operators to cover for sickness and holidays as they cannot maintain an adequate level of proficiency on less than full time. In some ways I think we have more difficulty in getting users to accept CAD techniques in what they are doing than trying to teach them to type. I think the difficulty we have is that we offer both tablet and keyboard input.

J S Gero, University of Sydney:

I was very glad to hear you raise the point that choosing software is really no different from choosing other things. However, I think you are a little unfair. Choosing software is very different. We all know how to buy cars; we know a lot about cars; even if you are totally non-technical, you know a lot. There are a lot of computers around, but if you want to buy yourself a particular sort of package, it's very hard to find people who have exactly that package so that you can learn from their experience. The trouble is the computer salesman is a bit like a second-hand car salesman. Have you ever seen the logo on top of the riding school that says, 'We have a horse for every rider. If you are a quiet rider, we have a quiet horse, if you are a rough rider, we have rough horses and if you don't like riding, we have horses that don't like to be ridden'. You make the point in your paper that you need to play devil's advocate, but to play devil's advocate you have to know what questions to ask. Most people getting involved in CAD don't known what questions to ask.

P Arthur:

That is why the CADCAM Association has been formed to educate people in CADCAM techniques and facilities.

A R Guy, Hatfield Polytechnic:

I think we all know that the cost of hardware has reduced significantly and that the cost of software is equally low, even though we use the modern software techniques. Do you think this trend will continue, or will the proposed techniques of automated software actually increase the cost?

P Arthur:

The use of new high-level languages and software development and quality assurance procedures will undoubtedly reduce the cost of software per unit of functionality. However, this may be counteracted by a demand for more facilities.

4. Encouraging new technology links between industrial and teaching institutions: Aston Science Park

K Foster

Abstract: The future of the UK manufacturing industries is the concern of many people today and there is a popular belief that the growth in jobs in the UK will come from small firms operating in areas of high technology. In the USA there are many successful science parks but in the UK the science park concept has been much less successful. This paper reviews the US situation and lays down the basis upon which the Aston Science Park has been established.

The job generation process

A report in 1979 by Fothergill and Gudgin, entitled the 'Job Generation Process in Britain' (Fothergill and Gudgin, 1979), which examines the changes in employment in the East Midlands, and in particular Leicestershire, gives a background to the current interest in the encouragement of the formation of new small companies and in the creation of schemes that will help towards this end. The first point the authors make is that out of a total of 567,000 employees in manufacturing in 1968, 150,000 jobs were lost by 1975 and slightly fewer created, giving a net loss over that period of 8,800 jobs. Thus, the net change in employment is the difference between two large numbers. The broad conclusions on changes in employment were as follows:

1. Net changes in employment are modest among all sizes and types of firms.
2. Small plants, particularly the independent ones, provide a small net increase in employment, but large plants show a net loss.
3. Employment gains from new firms over this seven-year period equalled only a little over 4 per cent of base year employment.

However, records were available for Leicestershire from 1947 to the present, and show that if one takes the long view of employment, new firms are considerably more important, in that companies created in 1947 had grown to provide 23 per cent of manufacturing employment by 1979.

The two reasons for the importance of new firms is that they experience substantial net growth even after their very early years, and also as new firms grow older, the loss of jobs due to closures diminishes, more than offsetting the fact that the growth among surviving firms declines.

The authors also turn their attention to the growth of firms and the results lead to the conclusion that the net growth of employment is likely to come from relatively small firms, i.e. those employing between one and 25 people. However, the total number employed in companies of this size is a relatively small proportion of the total employed in manufacturing. Thus, if the formation of new companies and the encouragement of existing small companies is to have any impact on employment, it will have to be carried out on a large scale. This conclusion forms the basis of the present government policies, and points to the need for a large number of activities aimed at removing obstacles in the way of any entrepreneurs who are likely to take up the challenge.

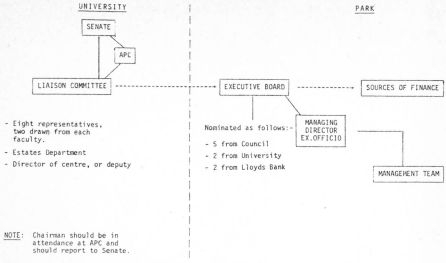

Figure 1 *Structure of the board of directors and liaison committee*

Helping small firms

In a recent joint report from the Department of Industry and Shell UK Limited (Department of Industry and Shell UK, 1982), Segal discusses a number of local initiatives of recent years that have been aimed at promoting the small firms sector by the provision of premises plus supporting services available to tenant firms on a shared basis. He discusses the provision of conventional common services, computer-based common services and the role of science parks.

The provision of conventional common services in seven schemes was studied in detail, covering some 250 tenant firms in a variety of manufacturing and service activities, with no firm having more than 14 employees. The conventional services were:

(a) Office support including physical facilities, (eg conference room) and administrative support (eg telephone answering).
(b) Business advice, generally of a first-stop and signposting kind.
(c) Machinery and equipment, generally for prototype development purposes.

The study highlighted the fact that many of the schemes were successful because of the continuing shortage of premises with unit sizes below 1,500 sq. ft. and having 'easy IN, easy OUT' tenancies that are generally favoured by small firms starting up. Such firms prefer older converted premises for psychological rather than necessarily cost reasons.

The common services apparently most valued by tenants were basic and minimal, including telephone answering, mail handling, reception and conference facilities and security provision. Tenants by and large accepted that services should be priced on a commercial basis, and strongly preferred their payments for shared services to be kept quite separate from those for rental, and that the former should be in accordance with actual use.

The Report goes on to say that the single most important determinant of the overall performance of the scheme is the quality of its management, which should go far beyond being an on-site landlord, but involves selecting suitable tenants, giving business advice and support without interfering in tenants' affairs, ensuring proper use of the scheme's facilities, and of cultivating a special spirit among the tenants. It appears that there are few schemes in operation in which the concept of shared services is properly defined in relation to the particular services offered, the markets aimed at, and the manner of promoting and delivering the services.

As far as computer-based common services are concerned, the Report points that technological advances rapidly reduce the scale of the organization for which the use of computers is potentially a relevant and economic consideration, but that there has been little penetration to date of the small firms market in the UK. First, small firms do not appreciate the full benefits that computers can bring them and do not have the technical abilities to realize these benefits, and second, hardware and software suppliers, bureaux, service companies and consultants have as yet made little effort to cater to small firms.

The common services approach offers an attractive way of filling this market gap, but has yet to be fully developed. Such a scheme would be an enhanced conventional common services scheme, with a special emphasis on raising the level of awareness of the benefits of computerization and also the provision of training, advice, programming services and other support. The expertise could lie in three distinct areas, ie financial and management accounting, technical application areas such as computer-aided design for engineering and architectural purposes, and also the use of computers for information retrieval from external data banks.

In order to achieve the necessary economies of scale and to offer profitable opportunities to the operator, it is pointed out that a scheme would normally need to cater to off-site companies as well as those on-site. This point will be taken up later.

The Report also considers the impact of science parks, and since this is the main theme of the present paper, the conclusions will be considered in more detail in the next section.

Science parks

A useful article on science parks in *Building Design* (Building Design, 1981) pointed out that in America, science parks are not limited to suburban locations but also help create employment in the inner city areas. In America the typical science park is closely linked with a university specializing in advanced technological research, the university often developing the land and sometimes the buildings.

The first science park was established in 1951 by Fred Turman, Stanford University's Dean of Engineering, who organized the development of some 3,200 ha. of farm land owned by the university as a place where growing electronics firms could become established. Hewlett Packard was one of the original companies, but in 1956 further expansion followed the invention of the transistor by William Shockley. A year later eight of his men broke away to found Fairchilds Semiconductor, and this in turn has spawned 38 new companies of which Apple Computer is one. Critical to the process has been the willingness of entrepreneurs who have already made their own capital gains to reinvest in new business.

A number of science parks can also be found along the east coast of North America. Some are in inner city areas of cities like Boston and Philadelphia, but all science parks offer high-quality working environments, recreational facilities and prestige surroundings. Whilst the linking feature of the science parks in the US is their proximity to centres of technological innovation, the location is widespread and diverse, and some have an ethnic character.

The Boston Community Development Corporation is developing a science park for blacks in an area of high unemployment, and the Wang Corporation headquarters in a declining Massachusetts town is mainly for Chinese workers.

The Philadelphia Science Centre is a mixture of converted old factories, new multi-story offices and light industrial spaces, pedestrianized streets and intensive landscaping, whilst the Wang has its marketing centre in an old mill with mellowed brickwork, ivy-clad walls and a stone-faced clock.

Along the west coast, the architecture is rather different with the Apple Computer's building being reminiscent of an early Spanish mission. The point made in the article is whilst the architectural styles of the States' science parks may be unrecognizable to the designers of high-technology offices and factories here, their varied locations are an object lesson in development. With more imagination and insight, inner cities could be used here, as in the States, as seed beds for innovative new firms.

Segal (Department of Industry and Shell UK, 1982) distinguishes two main types of science park:

1. High-quality, low-density development in a park-like environment, possibly but not necessarily undertaken near to or in association with a higher educational or research institute. Such schemes cater for mobile R & D or high-technology projects of major companies, small but fast-growing advanced technology companies seeking to enhance their image and working environment, and national or other research institutions. There are several science parks of this kind in Britain, of which only a few are successful in real estate investment terms, and the article (Building Design, 1981) does give a useful description of the Birchwood Park at Warrington and the Astec West Park in Bristol.

2. Schemes offering 'incubator' space for high-technology projects, with the emphasis on contiguity with and quality and range of support from a university or other institution. The prospective tenants could be
   - entrepreneurs directly from the university.
   - small firms from elsewhere, engaged in 'leading edge' technologies that need to be situated near to the university.
   - small specialized firms which the university constitutes an important market.
   - other small firms that want to be on or near a university campus, even if this is not strictly necessary for the business purposes.
   with the successful 'graduates' of the incubator schemes being expected to move on to the first category of science parks as they grow and have more specific requirements of building and location.

In 'Helping small firms start up and grow' (Department of Industry and Shell UK Ltd, 1982), the point is made that the market for incubator schemes is relatively small and difficult, and there is as yet no operational experience in Britain. Such schemes offer significant potential for stimulation and technological upgrading of the small firms sector, but this potential will not be realized unless:

1. There is a strong emphasis on marketing the schemes and making them attractive to prospective tenants. New and very young high-technology ventures do not require premises and other facilities that are more sophisticated than those required by conventional firms.

2. There is a strong capability for strengthening the commercial aspects of new ventures which are likely to be product – and process – rather than market-orientated. Entrepreneurs in high technology need straightforward business and management advice and support.

3. There is a much more effective mechanism for identifying and mobilizing the universities' intellectual and physical resources relevant to a particular problem.

4. There is an imaginative management capable of fostering a fertile environment for technological ventures and exploiting all the many different opportunities for university-industry exchange.

The Aston Science Park

Bearing in mind the need to regenerate industry in the West Midlands, and the need to create jobs, the City of Birmingham bought a site close to the University and has set aside £2.5m for the refurbishing of the buildings. The City and Lloyds Bank have each contributed £1m to set up a Venture Capital Fund, the interest from which will initially be used to support a company running the Park. The objectives of the Park are to encourage the formation and growth of a wide range of new, small and technologically oriented companies, and through the links that these companies will develop with the City and the University, to improve employment prospects in Birmingham and improve the economic strength in the West Midlands. In order to achieve a high growth rate, the intention is to select 'leading-edge' companies that can make the best use of the University technical facilities and management expertise, some of examples of which are as follows:

- Microprocessors and interface devices
- Instrumentation exploiting recent innovations
- Communications, man-machine interfaces
- Computer systems, computer-aided design and manufacture
- Artificial intelligence, robotics
- Bio-technology and bio-instrumentation
- Energy studies in processes and transportation
- New materials and processes.

To date, of a large number of enquiries received, negotiations have taken place with some 22 companies and discussions are continuing with ten of these.

During the last year of continuous discussion between the City, the University and Lloyds Bank, the nature of the operation of the Park has become more clearly identified, as follows:

1. To provide small units of flexible office and laboratory space with flexible leasing arrangements, and thus act as an incubator scheme for small companies.
   The City will provide other space in the locality for any companies that outgrow the initial premises.
2. To provide common services of a conventional kind, ie
   - building management
   - secretarial facilities, with word-processing services and an adequate telephone system
   - conference room(s) with teleconferencing capability
   - lecture room with visual aids
3. Technical and management advice as required.
4. More advanced common services based on computers, ie CAD.

The majority of companies applying for space in the Park will already have a source of Venture Capital, but, in addition, the Park itself will provide a link to financial services, including insurance, accounting and sources of Venture Capital and also legal and patent services.

Flexible office and laboratory space in small units could be provided by local government or private sources in any part of the country, but the contiguity to the University does offer unique features. The building management will be provided by the University at cost, but the value for money is likely to be high because of the considerable expertise within the Estates Division of the University in the provision of facilities for the type of high technology that one is looking for. Appendix A lists these services in more detail, and it is interesting to see the extent of the demands that are likely to be placed.

The former UBM factory occupies two sites in Dartmouth Street, Aston, next to the Middle Ring Road and the Aston Expressway.

The northern site, A, of 1¼ acres, contains buildings suitable for refurbishment and conversion, an underground car park and direct bridge link to the University campus. There is 4,000 sq ft of offices on the 4-storey frontage facing Dartmouth Street. These are to be converted for laboratory/office-based uses (for example: information technology, electronic workshops). The main 40,000 sq ft factory building is to be

dressmakers and the high technology companies come in. The turnover is normally pretty high and will allow the space to be returned to the pool whenever a new high technology company comes along.

K Foster:

A very good point. Over the months our ideas have been changing and in my personal view it's very much more along those lines. I gave a similar talk to a small international group in Sweden recently. One member had studied the start up of companies in Sweden over a two-year period, looking at the high growth rate companies and the lesser growth rate companies, and came to precisely that conclusion.

I had to judge an innovation competition in Shropshire for the *Shropshire Star*, and we gave the first prize to an engineering company. The second prize went to a company with real commercial success and exactly the kind of growth rate we wanted. They had moved into three premises in 18 months and they were doubling their turnover each year – they made teddy bears. (An innovative teddy bear!)

A Llewelyn, CADCAM Association:

I was interested in your observation of the relationship between University expertise and the Science Park and the point you made about there being some divide. Have you decided what the nature of that divide and of the employment either side is? Do you know of the situation in Toulouse where there is an aerospace complex and high technology and where the university people have two jobs: one part of the week is in a job the other side of the divide, in so to speak your Science Park? I wondered if you had considered that.

K Foster:

I spent four days in Toulouse about 18 months ago, and ten days ago I was in Trondheim looking at a very similar situation there. In both you have high technology; the aerospace industry in Toulouse, and Trondheim, a research institute which is now as big as the university which started it, some 30 years ago, and which is mainly financed by industrial grants. The interesting thing is that as far as one can see they have had very little impact on the local employment. In Trondheim, they say it has had none at all and they are interested in seeing how you can actually set up to do what we are trying to do, which in their case is to draw out the high technology into the small companies starting up and thereby increase employment in the future. We didn't set off with that aim in mind. We set off with simply drawing companies into the Park and then establishing the conditions where they would be able to grow fast. So the Park is absolutely separate from the University. There are at the moment no common posts between them, and this would be left entirely up to negotiation with companies in the Park and the University. The aim of the game is to provide the facilities, make the people in the companies feel they are part of the club and then provide the expertise that stops too many failures in the earlier years.

Appendix A: Birmingham Technology Limited

Estate management and services

In accordance with the concept of maximum interaction between the Park and University, those areas where the day-to-day servicing of Aston Science Park centre can be best undertaken using the existing structure within the University are outlined below:

For the part of the financial year remaining after the opening of the site, an estimated budget for services will be provided under the headings listed below. The estimated costs provided in this paper are based on a notional annual cost for the centre assuming normal operation, at April 1982 out-turn prices. The first year's budget will allow for the phased completion of the centre, the probable low level of occupancy, and the contract retention period for building works.

A discussion with the University staff officer has resulted in agreement in principle that additional staff requirements to meet the anticipated load will be added to current establishment figures within the University. Such increase in establishment would be conditional upon the availability of appropriate finance and should such finance not be forthcoming the establishment will be reduced accordingly at the first opportunity.

General Maintenance

This function which will be the responsibility of the works manager within the Estates and Buildings Department and covers all maintenance matters relative to buildings, services and site and specifically includes the gardening function internally and externally.

General Servicing

This function will be the responsibility of the services manager within the Estates and Buildings Department and covers the following matters:

(a) Cleaning/Portering
Responsibilities would relate to the general cleaning and servicing to agreed standards and schedules of communal areas and grounds, including toilets and refuse collection. The cleaning of 'lettable areas' would be undertaken as part of the agreement negotiated with individual tenants with the Science Park recouping a specialist service charge for the amount billed by the University. The impact of special functions, eg conferences, which could be contained by reinforcing the level of activity with other University staff would be considered independently outside the basic work pattern and costed accordingly.

(b) Fire
The provision and maintenance of adequate fire fighting equipment and procedures, including statutory requirements for fire drills, would be undertaken as an extension of the University security/fire officer's responsibilities. Emergency procedures would be worked into the general University system (see Security).
Costs to be charged to the Park would include an agreed proportion of staff time costs plus material costs.

(c) Security
As for Fire, general security procedures will be interlinked with the centralized University system with direct access by telephone and radio to the control centre in the main building – South entrance.
The necessary staffing to provide a security presence on site would be financed by the Science Park, and the scale of this provision would be governed by the specific needs of the company and its tenants. It is possible that with the Park in full operation, the provision may be two barrier attendants to cover a sixteen-hour period, 5 days a week, and costs shown are on this basis. Additional arrangements to cover for 'out-of-hours' opening are under investigation.

(d) Telephones
All units will be provided with new equipment similar to that now being installed in the University which will give direct access within the Science park and into the University system. Users will be able to install private wires, but the advantages of the link with the University system include free dialling to departments, use of the exchange lines to London, and the range of push-button

facilities available on each hand-set. Monitoring and costing will be provided by the University for individual extensions within the Park. The Park will be billed by the University for telephone costs and the services provided, including costs incurred in arranging the movement of extensions required to the system. Users will then receive an account from the Science Park. Much of the capital equipment required will be installed under the buildings contract and a reduction in rental is to be determined.

(e) Post

There will be an external Post Office connection and delivery, but in addition it is proposed to link the Science Park into the University's internal post system, with its own collection and delivery service. The cost would be minimal and is not shown as a separate item below. This service will facilitate communication between users and University departments, and provide easy access to library services, including the University 'telex'.

Appendix B: Aston Science Park – (services available from the University)

Computers and microprocessors

The University has a microprocessor unit which already provides an assistance to numerous companies, drawing on expertise where necessary from individual University departments. The central computer facility is excellent, having also access to the regional network and to data-bases elsewhere. A link between the Science Park and the University computer system will be provided.

Laboratory and test facilities

The annual equipment grant to the University is £1,222,000 and this is supplemented by equipment from research grants and contracts. University personnel are highly skilled in the use of the particular, specialized equipment. The utilization is not continuous, and it has always been the practice to lease either time on the equipment or services in response to requests from companies. Small companies must not carry the burden of high-cost equipment or specialist staff for intermittent work until the business growth is sufficient to justify investment. The cost of the use of the right piece of University equipment at the right time is small and the benefits can be dramatic.

The charge for hiring the equipment would be based upon the depreciation at current values, together with a charge for the time of any technicians required plus overheads.

Library

Obtaining key information quickly is vital to a young vigorous company and the access to the data must be immediate.

The University library is modern, specializing in commercial and technical books and carries the important national and international periodicals and learned society papers. It provides an inter-library loan facility and fast copying service and has extended opening hours.

Education

The University provides full-time and sandwich undergraduate courses, post-graduate courses and also short, specialists courses.

Undergraduate

During their vacation periods, full-time undergraduate students and, during their six months industrial training periods, university-based sandwich students are a source of trained, mobile manpower that would normally not be available to a small company. They can solve the problems that are important to the company, but which are beyond the capacity of the small core of permanent staff.

Postgraduate

At the postgraduate level, MSc courses are given in most of the technical subjects, and the MBA course is provided in the management centre. With both types of course, an essential ingredient is a long project carried out in depth by the students, and the majority of these projects are carried out on industrial or commercial topics. The students have a greater degree of competence and the projects are longer than at the undergraduate level, so that the results of the projects are often dramatic, providing, for example, working computer software, complete design studies of a new product, and so on.

Short Courses

The short, specialist courses that are mounted by the various departments of the University each year are designed to provide a service to industry and the local community. The courses range from highly technical ones in specialist areas to the broader course aimed at giving an awareness of the subject for managers in industry.

Extension Education

A new initiative is the launching of TVI, extension education based on video cassettes of lectures and demonstrations. A number of advanced courses are also available on a modular or part-time basis, and companies in the Science Park would be expected to take advantage of these facilities.

Consulting

The function of an academic member of the University is to teach and carry out research in his specific subject area. This may involve codifying present practice, developing new theoretical and experimental techniques, designing new equipment or looking towards future trends. When a very specific problem has to be solved, the use of University personnel is highly cost-effective, because it very significantly shortens the customary extended time-scale for development work. At the other extreme of company activity, it is essential to use specialist expertise to formulate a technical and commercial strategy. A company would normally make use of more than one consultant in order to have a balanced advisory team available on a regular basis. Experience has shown that this method of operation significantly improves the chances of commercial success.

Consultancy contracts may be placed through the University or directly with individuals by negotiation. Agreements may be for payments for work at a daily rate, a retainer for a period of time, or work carried out in return for stock options, with the latter providing the incentive to create a successful company in the shortest possible time.

Use of sports, recreation and arts facilities

The University has excellent gymnasiums, squash and badminton courts and a swimming pool. There is also an Arts Centre with cinema, theatre and craft workshops. Several restaurants on the campus also provide convenient meeting points. Personal contacts are important for the rapid exchange of information, and a community spirit is essential to the success of the enterprise. Employees of the companies on the Science Park may be able to use the facilities on a basis similar to that of the staff of the University, depending on the demand.

Part 2:
Ergonomics

5. Education for human-centred systems

M J E Cooley

Abstract: Technological change, in both the fields of skilled manual and intellectual work, displays a strong tendency toward de-skilling. This is already evident in the case of CADCAM. It will be counterproductive in the long term to accept CAD education as merely teaching operators to use such equipment.

It is argued that possibilities exist to design systems which are human-centred and which enhance skill. Examples are provided in respect of NC machine tools and CAD systems. It will be necessary to 'educate the educators' in the nature and potential of these alternative systems.

CAD should, as the name implies, be used to aid designers in better and more creatively carrying out their design assignments. Indeed Arthur Llewelyn, former director of the CAD Centre at Cambridge, has repeatedly and so correctly asserted that computers should not be used as a means of diminishing skills or eliminating designers but rather as tools for improving their ability to carry out creative tasks (Llewelyn, 1973).

Unfortunately the history of 'tools' in general suggests that such a laudable outcome may by no means be automatic. There are grounds for believing that the form and nature of 'tools' and the inbuilt assumptions of their design frequently bring about dramatic transformations in the labour process they were initially supposed to assist. When Crompton invented his hand-mule to ease his own labour and render it more productive, he could hardly have anticipated that within 50 years it would have been so directly turned against the spinner to eliminate his skill and destroy the control which that skill gave him over his work.

An analysis of the development of machine tools would also support the view that it brings in its wake dramatic de-skilling. Likewise there are good grounds for believing that if we perceive the design of CAD as 'tools for designers' on the same basis as that which applied to skilled manual workers in the past, we may begin to repeat in the field of intellectual work many of the mistakes made at earlier historical stages when skilled manual work was subjected to technological change.

Industrial sociologists (Blauner, 1964), (Faunce, 1965), (Shepard, 1971) assured us that the introduction of computers would mean that the previously fragmented parts of the labour process would be integrated by the computer, thereby providing the workers with a panoramic view of the labour process and elevating them to the level of 'systems managers'. Those, such as the author, who worked in the aerospace industry during the evolution of numerically controlled machine tools could observe that some of the most highly skilled work on the shop floor such as turning, universal milling, jig boring and inspection was gradually being eliminated by these NC systems. Indeed the work was being transformed so that the worker was gradually being reduced to a 'machine appendage'. So extensive and dramatic has this de-skilling become that a recent report suggested that the ideal operators for some of these NC systems are mentally retarded workers and they specifically advocate a mental age of 12 (Cooley, 1980). Had the objective been to provide work for the mentally retarded, clearly this would have been laudable. What, however, we witnessed was the destruction of some of the most highly skilled work on the shopfloor – the very educational seed-bed from which future generations of skill could be expected to emerge.

Given that the inbuilt criteria for 'good design' is the elimination of uncertainty and the introduction of predictability, repeatability and mathematical quantifiability, it is not surprising that human beings (who in systems terms are perceived to represent uncertainty) are marginalized by the systems designers. Since skill is closely related to the ability to handle uncertainty it will be seen that a systems approach is inherently de-skilling.

Indeed Professor Noble of the MIT has recently pointed out that given this form of design criteria 'designing for idiots is the highest expression of the engineering art' (Noble, 1982).

The integration of CAD and CAM adds to this an entirely new significance. The main historical tendency is to eliminate the drawing as a means of communication between design and manufacturing. Thus workers on the shop floor will be denied the dignity of receiving a drawing and then conceptualizing how to turn the two-dimensional data into a three-dimensional artefact. In future the data from the design base will be transmitted directly to the manufacturing system. This extends the tradition of part-programming, the process by which the NC tool motions are converted to finished tapes. Conventional (symbolic) part-programming languages require that a part-programmer, upon deciding how a part is to be machined, describes the desired tooled motions by a series of symbolic commands. These commands are used to define geometric entities, that is points, lines and surfaces which may be given symbolic names. In practice the part-programming languages require the operator to synthesize the desired tool motions from a restricted available vocabulary of symbolic commands. However, all this is doing is attempting to build into the machine the intelligence that would have been exercised by the skilled worker in going through the labour process. It is now possible, by using computerized equipment in a symbiotic or human-enhancing fashion, to link it to the skills of the human being and define the tool motions without symbolic description. Such a method is called analogic part-programming (Gossard and von Turkovich, 1978). In this type of part-programming the tool motion information is conveyed in analogue form by turning a crank or moving a joystick or some other hand/eye co-ordination task, using readout with precision adequate for the machining process. Using a dynamic visual display of the entire working area of the machine tool including the workpiece, the fixtures, the cutting tool and its position, the skilled craftsmen can directly input the required tool motion to 'machine' the workpiece in the display. Such a system which may be described as 'programming by doing' would represent a sharp contrast to the main historical tendency towards symbolic programming. It would require no knowledge of conventional programming languages because the necessity to describe symbolically the desired tool motions would be eliminated. This is achieved by designing a system whereby the information regarding a cut is conveyed in a manner closely resembling the *conceptual process* of the skilled machinist. Thus it would be necessary to maintain and enhance the ability of a range of craftsmen and women who would work in parallel with the system. Significant research has been carried out in this field (Gossard, 1975) yet in spite of its obvious advantages it is not received with any enthusiasm by large corporations or indeed funding bodies. That this is so would appear to be an entirely 'political' judgement rather than a technological one.

Other forms of operator programmable systems are now being developed (Boon, 1980), (Martin, 1982), and part of the writer's research at the Open University is to examine the extent to which systems of this kind can be used to provide for human-centred flexible manufacturing systems which would include computer-aided design. Such systems would require that two traditions operate in parallel, namely the 'craft' approach and the 'scientific' one. Thus it would be possible to draw on the conceptual knowledge of the skilled worker, that precious tacit knowledge which Polyani spoke about (Polyani, 1962), and link this in a creative fashion with advanced technology. It will be seen that it would be quite mistaken to organize educational programmes simply to acquaint workers with the means by which they would operate the 'tool'. This would eliminate the skilled environment in which workers develop their tacit knowledge. Thus it would be unacceptable to think of education as merely acquainting the worker with the use of the 'tools' and naïvely assuming that this would provide them with the entire range of skills required in machining. Such a view would be as ludicrous as suggesting that since a typewriter may be a tool for a novelist, learning to type will cause people to be novelists! Precisely the same applies in respect of education for CAD. To learn to drive the system can be a comparatively trivial task and in some instances can be learned in a very short length of time. In Figure 1 the shaded area represents engineering design with no 7 the simple drafting device and no 13 an interactive design system. Because somebody had been trained to operate one of these systems it can hardly be assumed that they had also been trained to *design* any more than the operator of the NC machine with a mental age of 12 can be said to understand the

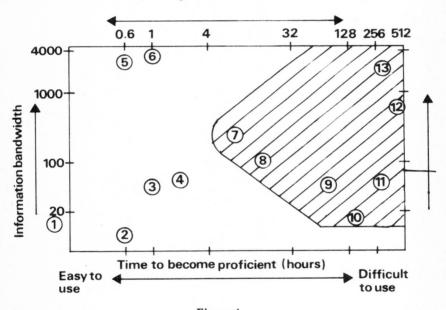

Figure 1

nature of the metal cutting processes. Indeed there are already dramatic examples from industry which would support this contention. The first generation of CAD designers who have not had the opportunity of developing conventional design skills are now operating in industry. As they design an artefact it is for them a series of XYZ co-ordinates and is dramatically abstracted from the real world. A case was cited recently on one of these 'CAD designers' who was designing an igniter for an after-burner. In the course of the design activity he succeeded in getting the decimal place one place to the right. Given the abstraction from the real world and the lack of any *feel for design* he did not notice this and had NC tapes generated. These in turn were taken to the shop floor where de-skilled workers frantically looked for material ten times bigger in every direction. They actually succeeded in manufacturing an igniter which was ten times bigger in every direction! (New technology: Society employment and skill, 1981). When confronted with the physical manifestation of this extraordinary mistake the so-called designer didn't even suspect that anything was amiss. *The designer had lost the sense of feel for design*. There is nothing mysterious about the sense of feel for design. It is based on years and years of experience and contributes to what may be called the art of designing. Rosenbrock has pointed out

> My own conclusion is that engineering is an art rather than a science and by saying this I imply a higher not a lower status. Scientific knowledge and mathematical analysis enter into engineering in an indispensible way and their role will continually increase. But engineering contains also elements of experience and judgement and regard for social considerations and a most effective way of using human labour. These partly embody knowledge which has not yet been reduced to exact mathematical form. They also embody value judgements which are not amenable to the scientific method (Rosenbrock, 1977).

Rosenbrock has questioned the underlying assumptions of the manner in which we are developing CAD systems. He charges firstly that the present techniques fail to exploit the opportunity which interactive computing can offer. The computer and human mind have quite different but complimentary abilities. The computer excels in analysis and numerical computation. The human mind excels in pattern recognition, the assessment of complicated situations and the intuitive leap to new solutions. If these different abilities can be combined they amount to something much more powerful and effective than anything we have had before. Rosenbrock objects to the 'automated manual' type of CAD system since it represents as he says 'a loss of nerve, a loss of belief in human abilities and a further unthinking application of the division of labour' (Rosenbrock, 1977).

As in the case of turning cited above, Rosenbrock sees two paths open in respect of CAD. The first is to accept the skill and knowledge of the designer and to attempt to give designers improved techniques and facilities for exercising their knowledge and skill. Such a system would demand a truly interactive use of computers in a way that allows the very different capabilities of the computer and the human mind to be used to the full. The alternative he suggests is 'to sub-divide and codify the design process incorporating the knowledge of existing designers so that it is reduced to a sequence of simple choices' (Rosenbrock, 1977). This he points out would lead to a de-skilling such that the job could be done by a person with less training and experience. Rosenbrock has demonstrated human-enhancing, human-centred alternatives by developing a CAD system with graphic output to develop displays from which the designer can assess stability, speed of response, sensitivity to disturbance and other properties of a controlled system. If having looked at the displays the performance of the system is not satisfactory, displays will suggest how it can be improved (Figure 2). In this respect the

Figure 2 *(see page 9)*

display carries on a long tradition of early pencil and paper methods but of course they bring with them much greater computing power. Thus as with the lathe and skilled turner, so also with the CAD system and the designer: possibilities do exist for a symbiotic relationship between the designer and the equipment. In both cases tacit knowledge and experience is accepted as valid and is enhanced and developed. In Rosenbrock's case it was necessary to examine the underlying mathematical techniques involved in control systems design (Rosenbrock, 1974). This work does demonstrate in embryo that there are possibilities for human-enhancing, human-centred systems, if we are prepared to explore them. We are at a unique historical turning-point in which there is a danger that we will close off the options that are still open to us. Some of the present educational programmes for CAD are quite laudible at one level in that they likewise accept that design skill is a prerequisite and seek to build upon it (Experimental trainees enter job market, 1982), (Course in CAD, 1982). At a more fundamental level it could be held that they are intellectually parasitic. They are merely an educational superstructure upon a base of skill which they exploit but fail to replenish as a store for judgements based on tacit knowledge. What we tend to witness is a process in which the designers are rendered more passive and the systems become more active. Education for human-centred systems would ensure that the designer is active and the system is passive (Thimbley, 1980).

It would also mean that the 'students' were exposed to those situations and environments which provides them with the ability on the one hand to handle uncertainties and on the other hand to take creative intuitive steps. This is in contrast to the present design of CAD systems and the educational programmes to support them which are widely at variance with the circumstances that appear historically to have provided an environment for creativity (Beveridge, 1961), Eiseley, 1962), (Fabun, D, 1962). Such a form of education would provide the engineer with practical experience of the design activity in a real world situation. It would also accept that human knowledge in the widest sense of information processing is dramatically greater than the artificial intelligence systems which are likely to be increasingly built into CAD systems (Cooley, 1980). (Figure 3). Thus CAD education for these human-centred systems would ensure

Figure 3 *(see page 9)*

that CAD was seen only as a supplement to human knowledge, not a replacement for it. It would not be an end in itself but merely one element in that totality of theory and practice which constitutes the 'education' of creative designers as distinct from the 'training' of hackers (Weizenbaum, 1976). It would ensure that two traditions were developed in parallel such that symbiotic systems (Licklider, 1960) become possible and that polytechnics, universities and industry at large recognizes that the most precious asset it has is the skill, ingenuity and creativity of design staff. A prerequisite to all this will be to 'educate the educators' to understand that there are alternatives to the rather Tayloristic CAD systems handed down to them by the vendors.

References

Beveridge, W I B (1961) *The art of scientific investigation*. Mercury Books: London
Blauner, R (1964) *Alienation and freedom*. University of Chicago Press: Chicago
Boon, J *et al.* (1980) Development of operator programmable NC lathes. UMIST
Cooley, M J E (1980) *Architect or bee?* Langley Technical Services: Slough
Course in CAD (1982) Teesside Polytechnic
Eiseley, L (1962) *The mind as nature*. Harper & Row: New York
Experimental trainees enter job market (1982) CADCAM **8**: 5
Fabun, D (1962) You and creativity. *Kaiser Aluminium News* **25**: 3
Faunce, W (1965) Automation and the division of labour. *Social Problems* **13**: 153
Gossard, D and von Turkovich, B (1978) *Analogic part programming with interactive graphics*. CIRP 27
Gossard, D (1975) PhD thesis: MIT
Licklider, J D R (1960) Man-machine symbiosis. IRE Trans. Electron 2: 4-11
Llewelyn, A (1973) Computer aided draughting systems. IPC: Guildford
Martin, T (September 1982) Human software requirements engineering for computer-controlled manufacturing systems. Baden-Baden forthcoming in IFAC Conference
New technology: Society employment and skill (1981) Council for Science and Society: London, working party report: 41
Noble, D (1982) Introduction to US edition of Cooley, M, *Architect or bee?* Southend Press: Boston
Polyani, M (1962) Tacit know: its bearing on some problems of philosophy. *Review of Modern Physics*: 601-5
Rosenbrock, H H (1974) *Computer aided control system design*. Academic Press: London
Rosenbrock, H H (1977) The future of control, *Automatica*, Vol 13 (1977)
Shepard, J (1971) *Automation and alienation: a study of office and factory workers*. MIT Press: Massachusetts
Thimbley, H cited in Rader M Wingert (1980) *CAD in Great Britain and West Germany: Current trends and impacts*. KFZ: Karlsruhe
Weizenbaum, J (1976) *Computer power and human reason*. W H Freeman: San Francisco

M Cooley, 95 Sussex Place, Slough, Berks SL1 1NN

Discussion

M K Gupta, Central Machine Tool Institute, India:

If the system absorbs the skill of a skilled or experienced designer or operator then an unskilled operator or a fresh designer can work with the system and manufacture the parts with the same efficiency as the experienced and skilled designer or operator. This means that the importance of the skilled operator and designer is lost. Can you comment upon this?

M Cooley, Open University:

I agree that that is the manner the systems tend to be designed. That is because we apply the criteria of natural sciences to all our so-called scientific problems. There are three main elements in our scientific methodology and if they are not present we don't regard it as valid science; the process must be predictable, repeatable and mathematically quantifiable. This eliminates by its nature skill and human judgement, and I think that this is a major defect in our methodology. If we regard the tacit knowledge and skill of people as an asset rather than a liability we would design it the other way round. I would question a part of your statement. Will unskilled operators be able to deal with new problems, new uncertainties outside the range at which they were trained? Increasingly, we are training people to drive systems rather than educating them to think and act as designers. I think we should be addressing that problem rather than critically denouncing the existing systems; most of them seem to me to do the unfortunate things I have outlined. As engineers and scientists, it's important that we think creatively about how we regard human knowledge as an asset, with all the interest and motivation that comes with it, and link this together with a system so that you get a true symbiosis. We are at a unique historical turning-point, where there are real dangers that we are going to repeat in the field of intellectual work many of the mistakes we made in Europe at earlier historical stages when we de-skilled vast areas of manual work. I don't see the alternative as simply retaining the old skills in their own form, but rather enhancing those skills and building upon them, still retaining a training educational process through which people gain this very important tacit knowledge. I would urge everybody to read the papers by Polyani on this subject.

A Llewelyn, CADCAM Association:

You made the point that systems are not at all designed in the best manner. The first generation of systems of the type you describe were in fact designed in the early 70s, for example GNC, PDMS. These systems just cannot be used unless you have a skilled engineer, and many of the criticisms levelled against these systems are precisely because they are used by people who haven't the necessary experience and judgement. Such systems are not easy to design, almost by their very nature they have to be designed in a partnership of contributory expertise. We are now at a stage of going through one generation and starting on the next, so I am merely making the comment, hoping you will agree that all is not gloom. We have made a start of ten years or so.

M Cooley:

Yes, I absolutely agree with that. I opened my paper by referring to a statement made in 1972 that these systems should be used to aid human skills and ability rather than replace them, but tragically the systems approach tends to eliminate human knowledge and ability which it perceives as an uncertainty. That in my view is a defect in our present design methodology. In case any of you feel I am over-emphasizing the point, I would like to quote from a leading systems designer in the United States. He says we should look on human beings as human materials so that we can think of them as we think of metal parts, electrical power or chemical reactions. Then we have succeeded in placing human materials on the same footing as any other material and can begin to proceed with our problems of systems design. He goes on to say that there are, however, many disadvantages in the use of these human operating numers; they are somewhat fragile, they are subject to fatigue, obsolescence, disease and even death. They are frequently stupid, unreliable and have a limited memory capacity. Beyond all this they sometimes seek to design their own circuitry. That in a material is unforgivable and any system utilizing them must devise appropriate safeguards. When I first read that, I thought it simply must be tongue-in-cheek. I wrote to Albert Boguslaw and he pointed out that this is precisely the way in which the systems designers at a major corporation in the United States regard human beings and that from my viewpoint is quite unacceptable. It's unacceptable in humanitarian terms, it's unacceptable in cultural terms and it's very counter-productive in the long term. I would assert that the greater asset any society has is the skill, the ingenuity, the creativity, the sheer enthusiasm of its ordinary people. If we lose that we create a very strange, unstable and dangerous form of society. At this unique historical turning-point where the number of CAD systems is as yet quite small, I am urging you to consider designing them so that we enhance human skill. We will end up with systems that are more productive, more creative and in the long term would be more acceptable to human beings.

J S Gero, University of Sydney:

Dr Cooley has raised what is probably the most significant of the issues in this conference; the question of long-term gain as opposed to short-term gain. Certainly externalizing knowledge which increases productivity immediately, de-skilling people and, more importantly, limiting possibilities for human beings is a very dangerous game to be playing. No one has yet even dreamed of the fact that human beings do things in ways that computers will never do, certainly not in the foreseeable future. This applies at every level. There is no way we can expect a computer to do some of the most trivial novel tasks that every human being does every day of his or her life.

6. CAD and the human operator

E C Kingsley

Abstract: Equipment designers are increasingly turning to CAD tools to speed up their design processes. However, there is often a tendency under traditional design methods to leave consideration of the human operator to the end of the design process. It is likely, therefore, that the introduction of rapid CAD methods will increase rather than decrease this tendency.

To counteract this tendency the CAD industry needs first to provide designers with CAD tools that can take account of human operator's requirements; and second, to apply human factors in the design of its own CAD work stations. By promoting human factors within the CAD industry in this manner it is quite feasible to expect the work places and products of the future to be more rather than less 'humanized'.

Introduction

All of us have experienced at some time or another the frustration, annoyance and discomfort caused by a lack of consideration of human factors (or ergonomics) in the environment in which we find ourselves.

For instance, as any car driver will tell you, it is only very rarely that one finds a car in which the seat can be adjusted to a position that is actually comfortable for long periods; in which it is also possible to reach and operate all the controls without stretching, without strain, and without fumbling; in which one can actually see the corner of the car from the driving position, and in which the displays are clear and unambiguous. Having found this 'ideal' car and got used to it, one then finds that in transferring to another car, all the controls are in different places, and some inevitably work in opposite ways.

This is not intended to be a condemnation of the motor industry, which on the whole is now addressing itself to human factors in a very earnest manner, but merely points out how human factors considerations affect us all daily in our use of equipment and environments. Just as in the design of home environments one needs to consider reaching to objects, the lighting of rooms, the suppression of noise, and the maintenance of a pleasant climate, so, too, in the office, the factory, and the shop all these 'human factors' are important. In addition, the design of the working environment must include consideration of the way in which people are expected to carry out their jobs, and how the organization of the work reflects the human operator's mental and physical capabilities.

In all aspects of design, computer aids are playing an ever more important part, indeed in many industries companies without computer-aided design systems are finding it increasingly difficult to compete with those companies that have already adopted CAD. The purpose of this paper is to point out that in modernizing the way in which we design our products we must not lose sight of the needs of both operators of the products we design and of operators of the computer-aided design systems that are being used to derive those products.

Human factors considerations

Human factors started to be given serious consideration during the Second World War when it became apparent that the efficient operation of military equipment (eg tanks and radar work stations) could be significantly improved by taking greater account of the operators' needs in the overall environment. Since World War II, human factors have become firmly established as a scientific field and have been applied in transportation, materials handling, manufacturing, architectural and many other design areas – all of which are now open to more recent developments in the computerization of the design (and indeed, manufacturing) process.

In the design of equipment and work places, human factors has implications in all the following areas: work space layout, design and layout of controls, design and layout of displays, illumination of the work place, visibility within and around it, access to and egress from the work place, seat design, access for maintenance, noise factors, heat factors, vibration, and finally the design of the tasks that the operator will have to carry out.

If human factors aspects are to be considered in the new computerized design processes then CAD tools must be developed that can evaluate these aspects. To date, the CAD industry has only skimmed the surface in producing these evaluative tools which are outlined in the following chapter. In the application of human factors to the CAD work station itself, the industry has done rather better and there is now beginning to dawn a widespread awareness of the human factors aspects of, for instance, VDU work station design and the design of the interaction style between the CAD program and the user of that program. These developments are briefly described in chapter 4.

Computer-aided human factors evaluation

Under traditional design methods human factors are often given a relatively low priority and hence are left until a fairly late stage in the design process. Thus human factors specialists are often presented with a nearly completed prototype and told 'Now tell us what's wrong with it'. The disadvantage of that approach is that it is often prohibitively expensive to make the modifications that are actually required to improve the human factors aspects. This means that only one or two trivial alterations actually get made and overall the design is little better than it was before.

What is really required is to get human factors evaluations carried out at a very early stage in the design process so that any major problems can be sorted out before they become fixed features of the overall design. The problem for the human factors specialist is that the aspects he is trying to examine are primarily three-dimensional issues and yet it is likely that if he is invited to carry out assessments at an early stage he will have only two-dimensional drawings to work from.

This is where the CAD approach can help by allowing the designer to examine the human factors problems using three-dimensional computerized models of the equipment or work place, at a preliminary rather than prototype stage in the design process.

To date, the programs that have been developed to fulfil this requirement have almost exclusively been derived for the transportation industries, primarily automotive and aerospace. The main requirement for a program that can analyse the human factors aspects of a design is that it contains some form of data that describes a man. In most cases the requirement is for a full three-dimensional representation of the human form in combination with specific analytical routines or facilities.

Current programs that fall into this category include the following: SAMMIE, a man-machine evaluation system developed by Nottingham University and commercially available through Compeda Ltd, (Kingsley, 1982) (Figures 1-3); CYBERMAN/TIM, developed by the Chrysler Corporation as a system to aid automotive interior design (Waterman and Washburn, 1978); COMBIMAN, developed at the Wright-Patterson air force base in Ohio, USA, for aiding the design of military aircraft cockpit interiors

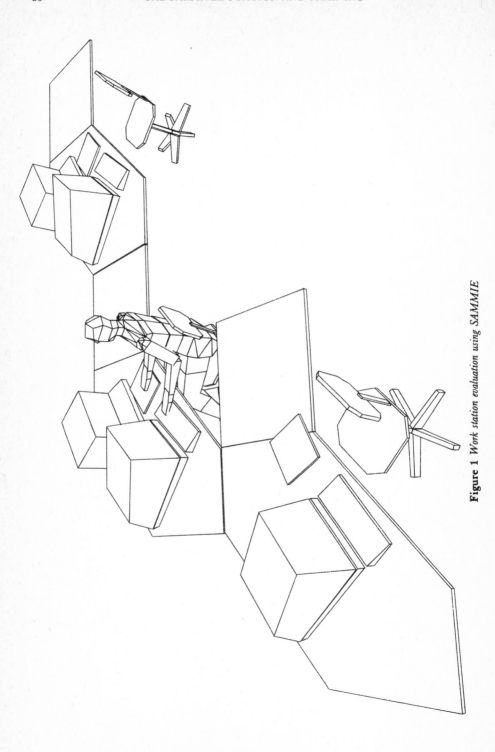

Figure 1 *Work station evaluation using SAMMIE*

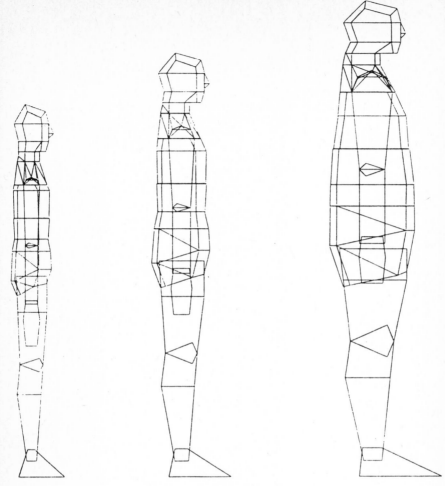

Figure 2 *A range of man-models*

Figure 3 *(see page 9)*

(Bapu, Evans, Kikta and Korna, 1980); BOEMAN, developed as an in-house system by the Boeing Corporation again for cockpit design purposes (Ryan, 1971); BUFORD, developed by Rockwell International as an in-house system for the cockpit design studies; CAR developed at the US Naval Air Development Centre for reach analysis only (Harris, Bennett and Dow, 1980); BUBBLEMAN, developed purely for man visualization at Pennsylvania University (Badler and Smoliar, 1979); and Willmert's models for crash simulation programs SIMULA and PROMETHEUS.

These programs mentioned above vary greatly in the extent to which (a) they are both graphical and interactive, (b) they are analytical, (c) they are capable of providing realistic-looking models of both the man and the work place, and (d) they are capable of being applied in more than just one particular design environment.

For instance, CAR is a non-graphical batch program, whereas the others are all graphical but varying in their interactive capability. Similarly, BUBBLEMAN and

BUFORD contain no inherent analytical routines whereas SAMMIE can analyse reach, vision, and fit problems. CYBERMAN and SAMMIE, for instance, provide very realistic-looking man models whereas Willmert's models are highly schematic. Finally, SAMMIE and BUFORD are the only programs that can be used to model almost any environment or equipment whereas all the others are limited either in the type of work place they can model or in their capability for modelling a work place at all.

It is obvious that there is a long way to go yet in providing programs that can satisfy the following criteria: (a) that they are general purpose interactive systems, (b) that they are expandable to provide the capability for addressing more than just a very limited number of human factors aspects, and (c) that they be capable of interfacing to existing three-dimensional computer-aided engineering modelling systems. Recent developments at Compeda are aimed at upgrading SAMMIE to the point where it fully satisfies these criteria but there is ample scope for further developments amongst all the programs cited to extend the range of human factors problems that they can refer to.

Human factors in the design of CAD systems

In applying human factors to the design of products and work places using programs like those mentioned in the previous chapter, what could be more logical than devoting the same effort to analysing the human factors aspects of the CAD work station itself?

As visitors to any of the recent international computer graphics exhibitions will know, there are an increasing number of computer terminal manufacturers who will advertise their products as being 'ergonomically designed'. This is occasionally an overstatement of the case but on the whole the proliferation of articles, books and research reports on the human factors of VDU work stations (Cakir, Hart and Stewart, 1980; Alder, Daniel and Kanarick; Stewart, Ostberg and Mackay, 1974) has had a positive and beneficial influence on the manufacturers concerned.

Evaluation of the special problems posed by people working at computer terminals for long periods of time started to be taken seriously in the mid 1970s with articles from the occupational health as well as the human factors practitioners (Stewart, Ostberg and Mackay, 1974; Ostberg, 1975). Of particular concern was the notion, now largely discredited, that there might be a radiation hazard from working for long periods in front of cathode ray tubes (Bruton, 1972).

Another aspect, but one that has not diminished in importance, is the problem that many operators experience, namely eyestrain or visual fatigue. Eyestrain can be caused by a number of factors, not only the quality of the image on the screen but also the ambient lighting in the room where the operator is working. One has to take account of screen reflections, the contrast between the screen brightness and the background brightness, the lighting of adjacent surfaces for laying out drawings and text, and the effects of adjacent work stations on each other.

As far as the work station 'furniture' is concerned, the avoidance of muscular strain and fatigue can only be avoided through careful design of the screen itself, the keyboard console, the operator's chair, and the furniture used as a 'table' for the terminal. For instance, screens should be capable of both vertical adjustment and tilting, the keyboard should be independent from the screen and separately adjustable for height, the operator's chair should be fully adjustable providing firm support for the lumber spine at variable overall height and tilt positions, and finally the furniture should provide space for laying out or holding documents at angles which allow them to be easily examined without excessive movement away from the keyboard operating position (Business Equipment Trade Association, 1980).

Of equal importance to the human factors of the hardware involved in the CAD work station is the human factors of the software. In other words we must examine how the software has to be used and how these methods of operation can be improved to reduce frustration, confusion, loss of patience and other negative states of mind.

Typically, this will involve analysis of the types of commands required by the program; the consistency between similar commands for different aspects, the type, frequency, and detail of the messages the program sends the user; and the screen. Documentation for programs can never be underestimated in their value to the program user in reducing frustration and confusion in operating programs. There is a very great need for human factors to be applied here, too, in designing manuals so that they are easy to read, digest, and refer to again.

Conclusion

One of the most frequent lay criticisms of computers in general is that they are 'dehumanizing' in their effects on society, by leading to automation and the loss of traditional work roles, as well as causing the people who operate them to behave like automatons themselves. Obviously negative images like that do not do the CAD cause any good and it is therefore up to those who believe in the beneficial effects of CAD to counteract these images with ones which demonstrate that at least the CAD industry *does* care about human factors and *is* actively promoting it.

The purpose of this paper has been to point out two of the ways in which the CAD industry can consider the human operator, namely (a) to provide computer-aided design programs that permit the evaluation of human factors aspects of a design at an early stage in the design process, and (b) to give full consideration to the human factors aspects of its own CAD work stations and the environments in which those work stations are used.

If both of these developments take place to the extent that is really required then there is every reason to expect that the work places and products of the future designed using CAD methods will be more humanized, rather than less humanized than their traditionally designed predecessors.

References

Alder, D G, Daniel, R W and Kanarick, A F (1972) Keyboard design and operation – a review of the major issues. *Human Factors* **14**: 275-93
Badler, N I and Smoliar, S W (March 1979) Digital representation of human movement. *Computing Surveys* **2**: 1
Bapu, P, Evans, S, Kikta, P and Korna, M (1980) COMBIMAN. University of Dayton Research Institute and J McDaniel Air Force Aerospace Medical Research Laboratory
Bruton, D M (1972) Medical aspects of cathode ray tube displays. Trans. Soc. Occ. Med. **22**: 56-7
Business Equipment Trade Association (1980) A guide to users of business equipment incorporating
    visual display units. BETA: London
Cakir, A, Hart, D J and Stewart, T F M (1980) *Visual display terminals.* John Wiley. Chichester
Harris, R M, Bennett, J and Dow, L A (June 1980) CAR-II-a revised model for crew assessment of reach. US Naval Air Development Centre
Kingsley, E C (1982) SAMMIE – 3D modelling for human factors evaluations. Butterworth Scientific Inc. Proceedings CAD 82 (Brighton, UK: 30 March–1 April 1982)
Ostberg, O (November 1975) CRT's pose health problems for operators. *International Journal of Occupational Health and Safety*
Ryan, P W (November 1971) Concept Geometry Evaluation. Boeing Co.
Stewart, T F M, Ostberg, O and Mackay, C J (1974) *Computer terminal ergonomics: a review of recent human factors literature.* Statskontoret, Stockholm
Waterman, D and Washburn, C T (1978) CYBERMAN – a human factors design tool. Society of Automotive Engineers Technical Paper Series no 780283

E C Kingsley, Prime CADCAM Ltd, Walkern Road, Stevenage, Herts SG1 3QP

Discussion

J Butler, Technical Change Centre:

If an industrial company now recruits a recent ergonomics graduate is he already likely to have had experience of SAMMIE?

E Kingsley, Compeda Ltd:

He is likely to have heard of SAMMIE. But it's unlikely that he will have used SAMMIE unless he has been to Nottingham University or Loughborough University which both have SAMMIE at the moment. Speaking as a purist ergonomist, one might say that ergonomics encompasses such a wide spectrum of subjects and unless you can address the whole spectrum it's not worth using an aid like SAMMIE. What may not be appreciated is exactly where CAD fits into the industrial environment, how important it is and how using CAD aid in human factors can have a great impact simply by making people who design the equipment, notice human factors and take account of them. In this way you may be able to influence other areas of human factors that SAMMIE doesn't refer to simply by using something which is appealing, has an immediate impact and is a very powerful communication tool for putting over human factors aspects, albeit in a limited range of problems, to engineers and to designers. It has a great potential but whether this will be fully realized is another matter.

S R Bennett, Ford Motor Company:

Have you noticed any difference in ergonomic requirements in different countries, particularly Germany? Do you have a feel for the amount of time that people can normally sit in front of terminals before they ought to be taken away? Do you have a view of the optimum colour for a terminal?

E Kingsley:

Taking the first point about Germany, I work with German companies, and the German auto industry primarily, and they do have different requirements. These mainly concern the different European legislation requirements and perhaps also the way in which they think of the interior space of the car. They do have slightly different priorities perhaps, but on the whole we are talking about the differences between different industries rather than different requirements between different countries. I am thinking mainly of the UK, Germany and America, which are the three areas in which I have had most experience.

How long can somebody stay in front of a terminal before they have to be dragged away? It depends very largely on the language that is used to define the interface between the man sitting at the terminal and the programme. I know people who will walk away from the terminal after a minute if the terminal gives them no indication that it is doing something for them but can't actually show them at the moment. There are a lot of people who find this very frustrating and there are many ways in which you can let people know that the terminal is working for them although it can't show anything. There is one manufacturer that has a little bee that flies around the screen. It's a silly idea but it does keep you sitting there and it does avoid the feeling, 'Well I have just been left here on my own, and for all I know the program crashed.' There are many little tricks like that which a true engineer might say are daft, but it does reduce the frustration of working at a CAD terminal for too long, particularly for people who are not very experienced users of computer systems and are just getting used to it. And you need things like a 'help' facility and possibly a tutor facility where you can get a second level of detail about the commands that are available to you, giving more information on what each command does.

What colour for a terminal? Not red. A grey or dark green terminal is very pleasant. It's a matter of personal choice but I would say subtle colours rather than garish ones. A lot depends on the type of work that is being done, whether you have drawings and you are continually having to look from the screen to the drawing, whether they are blue drawings, what colour the text is, etc. There is a whole host of factors that you have to consider – such as the colour of the walls in the room, the plants around – so I can't give you a straight answer on that.

D G Smith, Loughborough University:

Could you tell us the advantages of using a program such as SAMMIE for the evaluation of a work station over actually getting some terminals and moving them around a desk and putting out the areas for drawings and so on?

E Kingsley:

Do you mean the advantages of doing it on the computer terminal *vis-à-vis* the real situation? The main advantage is that you can do it before you have the terminals, the room, or even before you have built the room, so that you can plan that space, not only to answer the question 'Can I set up my CAD room in the available space'? but also 'Can I set it up so that it's going to be a workable CAD environment where people are prepared to spend the whole day working and not get reflections from the window on their screen, doors opening and closing behind them, light glaring on their screen, another operator causing them irritation' and so on. So it's essentially a question of advanced planning, making sure you have got it right before you put the installation in.

D G Smith:

You are relying a lot on the experience you have gained simply because terminals are now available, not because you have used SAMMIE. In that case, what is the point of using SAMMIE?

E Kingsley:

Well, you can move things around if you already have the terminals. The way in which you would use SAMMIE is before that stage, when designing the work station environment so that you build it right. If you have already built it you may find SAMMIE useful in trying out a number of different possibilities and weighing up their relative merits without having to move everything around, rebuilding the walls, making an extension, raising the ceiling height or changing the position of the lights. Some of these things are very difficult to do in practice but very easy to do on a computer.

A L Johnson, Cambridge University:

You showed SAMMIE welding the front of an aeroplane from a gantry of some sort. When an analyst sets up SAMMIE in a position like that, is it simply his guess as to the posture the workman is going to adopt or does SAMMIE have built in algorithms to adopt the most comfortable posture that workmen naturally would adopt to reach a particular nut?

E Kingsley:

SAMMIE is an aid to the designer. It doesn't replace the designer. It doesn't replace the human factors expert. It doesn't replace the engineer. So the engineer will have to decide what posture is a good posture for doing that particular operation and then see, using SAMMIE, whether it's possible to do it in that environment. He may try that, find that the environment is wrong for that posture and have to think of another posture to put the man in. In that situation, the program will tell him whether the joints or any particular joint is within a comfortable range of movement or whether it's between comfortable and a maximum range of movement. If it's outside the maximum range of movement, for instance turning his head 180 degrees backwards, it won't let him do that because it's not possible with a real human being. It has constraints built in to stop you doing silly things like that.

What was the last thing you said?

A L Johnson:

Is there any feedback as to the discomfort that SAMMIE experiences in trying to get into that posture?

E Kingsley:

No, there isn't, other than having a table of what all the joints positions are, their angles and where they are in 3D space. If you wanted to, and this is where I am talking about an expandable analytical capability, it would be perfectly feasible if you have the data available to add on another module to the program to decide whether that posture is sustainable for x period, y period, or z period (ie broad period bands). Each posture that the program comes up with could be categorized into one of those time periods. Anything's possible but at the moment the program doesn't do that, although there is no reason why it shouldn't.

R Astley, Pilkington Brothers PLC:

What population does SAMMIE represent? Are alternatives available and can the user input his own anthropometric data?

E Kingsley:

Yes, the data that SAMMIE contains in the latest version of the program is based both on the Drafus data-base and on the RAF data concerning 2,000 airmen. There are perfectly easy facilities within the program for substituting any of that data with any other data, so if you are interested in British housewives aged 17-23, you can, if you have that data available, put that in a SAMMIE and a man-model or woman-model will then be derived using that data.

C Jakeman, Engineering Sciences Data Unit:

Can Mr Kingsley give us a standard way of calling up a 'help' menu? Is there an accepted industry standard?

E Kingsley:

I don't know if there is an accepted industry standard but in the SAMMIE program at any stage in its operation you can type in 'help', and it will give you then a list of the available commands to you at any one stage. If you go in for the tutor mode it will go one stage further than that and tell you what those commands do and where that command might lead on to another group of commands. The 'help' facility is a memory jogger when you know what you want to do but can't think of the command. The tutor prompt will give you an answer to 'What do I do next?'

Part 3:
Training – general

7. The training needs of CADCAM

N Spoonley

Abstract: The training needs of CADCAM are described with reference to the widest effect on organization. Solution may be found by considering all training media including computer-based education using similar technology to that used for CADCAM.

There is no clear definition of CADCAM, or for training. To some extent they are both concepts and this is well illustrated when one looks at the literature on either topic. There seems to be little limitation to what can appear in the journals, as most describe the application of CADCAM or techniques of training. The agenda for this conference equally reflects the broad nature of the topic.

As an early presentation it seems appropriate to retain the broad canvas and not to narrow down to a tightly defined aspect. Both topics have one major objective and that is the improvement of productivity. The objective of productivity improvement should be to increase the profits that an enterprise can make if it applies the correct techniques. No one should be under the illusion that all CADCAM, or indeed all training will necessarily increase the profitability. It has inevitably to be done selectively and carefully in both areas.

But first, let us look at CADCAM. The impact of CADCAM ranges probably much further than many of you will wish to recognize. The technology of CADCAM can severely change the lives of many individuals, and, it is not always clearly for the better. Every individual is entitled to their own views as to what is right for them and they will inevitably perceive that some of the changes caused by the introduction of CADCAM techniques have negative implications. Indeed, it would be stupid of anyone to believe that only the advantages should be recognized, as most of us know that there is always a reluctance to accept change. Clearly, we must consider the sort of changes that will be created and attempt to ensure the highest probability that they are generally for a good reason. We cannot limit our vision of these changes solely to the engineers and technicians but we must include a whole range of other occupations that are often forgotten. It is necessary that we do recognize the extent of the change that CADCAM creates and those people who are driving this new technology into our society have a large educational responsibility far beyond the engineering profession.

The people most directly affected by CADCAM are those whose jobs may be replaced or radically altered. The majority of these are probably technical people and possibly managers. But then there is a range of people who will have to be provided with new skills in order to use the techniques and devices that CADCAM causes. There are obvious opportunities for equipment maintenance, operators etc. The engineer who should now have new tools at his elbow or the plant manager with new information, will have to be retrained and re-educated. All this seems very obvious, but then, there are the people yet to come. There is no use in having an education industry of colleges, universities and schools which produce only people who fit into the old order of activity. Thus, there is a need to change curricula and develop new courses and modify old ones, within the education system in an overall sense. This is an exceedingly long process, and we must clearly recognize that any change of this nature is a difficult one to introduce.

New products will probably emerge onto the marketplace. Initially, they are the CADCAM products themselves which hitherto have not been seen. But they, by being used, should create new and effective products simply from the versatility of the new tools. We therefore have the educational aspects of the media and the marketing men. One could possibly cite the original advertisement for Fiat in advertising the Strada car as being the end result of educating the media.

One major impact of CADCAM which is often totally ignored is that it creates an operation which is capital-intensive with lower operating costs. There is a significant build-up of assets on company balance sheets and the financial structure of an enterprise begins to look very different. This change is an exceedingly hard one to swallow for many companies. However, it is people who look at assets and profit and loss statements and the numbers associated with companies who often make the decisions with regard to all forms of investment. These people have to recognize that CADCAM or automation or whatever it is called, can affect the profitability of a company, and this can be seen in the numbers, and effective implementable plans can be created. It is worth questioning precisely who will create such plans. The input required is far beyond the general knowledge of most of the financial people and they will not understand the benefits of CADCAM and recognize how to incorporate them in their plannings. For this reason, it is likely that they will resist the change and continue to create plans that make it difficult to invest and allow effective introduction of CADCAM.

I have mentioned a number of areas where CADCAM will inevitably cause changes. I believe the introduction of modern technology into organizations is an inevitability that generally the timescale and method of introduction can be planned effectively. It must be planned so as to maximize the profitability of the organization and we must continue to recognize that is the intended objective. I will not however subscribe to the enveloping view that has recently been expressed as 'Automate or Liquidate'. We must all recognize that one of the major impacts of CADCAM is to change the way of life of people. They are also part of the assets of organizations and the use of them costs money. It is all too simple to take the easy road and draw up plans which take people out of the equations. There are quite likely to be many situations whereby productivity can be improved without CADCAM, or with a partial CADCAM solution and equally, there will be plenty of ways by which introducing CADCAM can destroy a company. The blind assertion that all automation engenders new profitability should not be accepted and adequate planning must be done to justify the change.

I maintained that the introduction of new technology is inevitable and that therefore implies that change is inevitable. If we accept that, then we have to recognize how to manage the change and therein we come to training. We cannot create change without simultaneously creating the tools to allow that change to be effectively implemented for all concerned. I should like to suggest that the very techniques that we call CADCAM have within themselves a capability for education and training which we must use to the utmost.

There are close similarities between CADCAM and computer-based education. They are both dealing with people to a very large extent and they are both applying computer techniques together with all the interfacing aspects. We must recognize that as these tools are available to us, we ought to be able to plan the appropriate training around them. I believe that training packages on CADCAM systems should become a natural part of the system itself.

These should not be solely for the people who are immediately affected, but should range across all the various categories that I have already mentioned. We should use the technology that we have, whether it be video tapes, books, films etc and these should all be enhanced with computer-based education material. Of all the new techniques that we see around us, computer-based education probably offers the greatest promise for assisting the introduction of CADCAM than any other. It brings the same technology to all the groups affected and thereby enhances their familiarity.

It is worthwhile looking at some of the properties of modern computer-based education systems. In this context, I am considering systems such as PLATO, MENTOR, TICCIT etc. These are sophisticated, like CADCAM and have modern terminals, and specialist attributes solely geared for the motivation of the student. All of us at some stage have no doubt played Space Invaders or some other form of video game. The amount of creative design that has gone into the entertainment side of the use of modern technology is quite high. Slowly, we are seeing some of this creative design being turned to educational purposes and the challenge for the future is really to see how the addiction and attraction generated in the entertainment field can be married with the use of training materials. As is well known, I have been associated with PLATO for some years, and we have seen within that system how sparks of creative genius have created educational packages that verge on being addictive. What is also interesting, is, that these particular pieces of material last a considerable period of time before they are felt to be outmoded. However, the principal advantage for trainee purposes in using computer-based education, is the degree of interaction between the material and the student. This is very similar to CADCAM. I could argue that the principal benefits of computer-aided design stem from the foreshortening of traditional timescales in the processing of information. This is what computers do when they operate in an interactive mode. Thus, a student can be guided, directed, assessed and redirected according to his capabilities for learning. This powerful technique will change according to his capacity and possibly his mood. With this capability we have the ability to ensure that the training is effective and that the student can demonstrate that it has been well received.

With the ever-developing range of communication facilities and their increasing speed and capability, the widespread availability of good educational material will be enhanced. The remote factory will be capable of receiving just as good educational material as if it was in the middle of a well-populated city. There are many interesting aspects of the use of computers in training, as possibly there are in CADCAM. There are many experts in the audience here and a number of them are delivering papers.

CADCAM as part of new technology is an inevitability in the manufacturing and engineering scene in Britain. It must be planned and evaluated effectively, so that the principal objective of achieving higher productivity and profitability is maintained. This requires changes in a wide-ranging set of people and attitudes and this should be accomplished by effective education and training. We should use similar technology alongside CADCAM to deliver this training and we should make it a habit to extend its availability to all who may be affected as a matter of principle.

N Spoonley, Control Data Ltd, 179-199 Shaftesbury Avenue, London WC2H 8AR

8. New directions in training

P G Raymont

Abstract: The rapid rate of technological change both poses a challenge to trainers and offers technological tools to meet the challenge. Educational technology is to be understood broadly as a set of tools for the design of effective education. Essential steps in establishing a training programme are the identification of training objectives, design of the learning situation and evaluation/improvement of the training. Training methods (large group, small group, individual) should be distinguished from the media used (human, real, video, audio, computer, print). A shift to individual learning is taking place and appropriate media with computer support, hinging around the programmed instruction concept is a recommended approach. Mixed media packages are considered the best. A future technology with much potential for training is the computer-controlled videodisc. Open and distance learning employing individualized learning methods should be used by CADCAM suppliers since this form of training is most convenient for the user.

Introduction

The rapid development of technology in the past few decades has provided at once a challenge and a possibility of meeting that challenge to those concerned with training people for tasks in industry.

The challenge arises from the ever-increasing rate of technological change, requiring people to acquire new skills every few years, rather than, as in the past, learning one set of skills to last a lifetime. The sheer volume of training and retraining needed has increased very substantially, to the level of straining the resources available for training to breaking-point. At the same time technology has made available a whole range of new equipment which can be exploited to improve the efficiency and effectiveness of training. Relevant developments include audio, video and computer systems. Whilst it would be deluding oneself to argue that these technologies have been applied to training problems with consistent success, nevertheless there are numerous examples of excellent uses, and it seems we are on the threshold of significant advances. The aim of this paper is to discuss the methodology needed to capitalize on the present potential, in both conceptual and technological terms, relating the generalities where appropriate to the particular problems of CADCAM training.

Educational technology

The term 'Educational Technology' is frequently misunderstood. It is often taken as referring to the use of technological devices for educational purposes. But this is to mistake the means for the end. Educational technology may be more properly understood as a set of tools for the design of effective education. The overall sequence encompassed by an educational technology approach to a training problem may be explained as follows (Rowntree, 1974):

Identify the training objectives

This should encompass an analysis of the aims of the training, a description of the trainees and their assumed existing skills and knowledge, a specification of the objectives of the training in behavioural terms with a design for tests to determine if the objectives have been met.

Design of the learning situation

The learning sequence needs to be determined. Then the appropriate media and methods for each element of learning selected and the necessary materials and learning situations designed.

Evaluation/improvement

The learning as designed should then be evaluated and the results monitored. This may result in revision of the objectives, methods, media etc, to improve the effectiveness of the learning.

In the above brief description two things are to be noted. The first is a shift of emphasis from 'teaching' to 'learning'. The traditional concept of classroom presentation gave prime importance to the teacher or lecturer; attempts to improve training quality and throughput were based on improving his performance through various facilities such as visual aids, notes for students, exercises with solutions, case study materials, etc and relied ultimately on his wide teaching skills for maximum results.

The new concept places the trainee at the centre of the learning situation and supports him with aids such as programmed texts, audio and video tapes, and workbooks, which he uses on his own (or in small groups) to work at his own pace and at times convenient to himself and his management. One very important aid which the trainee has in this environment is the tutor who meets him (or the group) on occasion to guide and motivate him (them), and provides that human interaction which is essential to sort out problems and misunderstandings quickly. It also gives the tutor the opportunity to put the particular aspect being studied into the broader perspective of the subject as a whole. As mentioned above, the tutor will also require some aids of a more conventional type.

The second point to be noted is the emphasis on training objectives. A clear analysis of what the training is trying to achieve is essential if we are to know whether or not the training has succeeded. The objectives should be linked to objective testing procedures. Thus in defining objectives one should avoid words such as 'appreciate', 'understand', etc (how does one know if the trainee has achieved the desired result?) and instead use words implying the existence of objective tests, eg 'list', 'distinguish', etc. The tests should be 'criterion based', by which we mean that, whatever the skill to be taught is, the criteria for assessing that skill needs to be stated. The assessment is then simply: does the trainee meet the criteria or not? This is very different from conventional examination approaches which tend to assess some sort of overall performance, so that an examination can be passed even when some quite basic and fundamental skills are not mastered. If a person is being trained to operate a system, for example a CAD pipework design system, then it is important that he satisfies criteria set down for using each of the main facilities of the system which will be used in the application on which he is working. It is not enough that he has some overall understanding of the system.

Learning modes and media

In the introduction we mentioned the uses of various technologies in training, and in the last section we spoke of the design of learning situations. In designing a learning situation one needs to distinguish the method of learning from the *media* which will be used. Methods of learning may be conveniently divided into

– large group, eg lectures or conventional classroom methods

   – small group, which may include discussion groups, seminars, group exercises, etc
   – individual, ie where the learner is working on his own.

The media for presenting training include

   – the human voice/gestures
   – real or model objects or systems
   – films/TV/video
   – audio material
   – computers (micros or terminals)
   – written or printed text/graphics.

It is clear that large group methods are increasingly falling into disfavour for training in industry, in spite of the apparent economic advantages of training a large group together. Reasons for this include the well-known ineffectiveness of lectures, the problems of different pre-knowledge, rate of comprehension, etc of the individuals making up the group and the difficulty of assessing the individuals progress when in a large group.

In computer training in particular there has been a marked increase in the use of individual methods of learning over the past few years and this is a trend which is expected to continue. A recent National Computing Centre survey of data processing training showed that some 15 per cent of training budgets were being spent on materials for individual learning with the following breakdown:

video	43%
programmed texts	29%
audio	17%
other	10%

Media for individual learning

The trend towards individual learning has been made possible by technology. The only human form individual learning can take is the one-to-one tutor to learner situation and this is clearly impossible economically to implement on more than a very limited basis.

The earliest work on individual instruction used the idea of 'programmed instruction'. This was implemented either through printed text/graphics or through the now defunct 'teaching machines' which automated the page-turning mechanics of using printed materials. The concepts and advantages are well known (Hartley, 1972): Briefly, the material to be learned is itemized into a series of 'frames', and the sequence of frames controlled according to trainee responses to questions posed at appropriate points.

The method depends on being able to formulate questions which have unambiguous and easily evaluated answers. Programmed instruction has been implemented on computer systems, some of which seem to do little more than use the computer as a page-turning teaching machine: however the use of a computer can enable sophisticated structures of branching between frames to be implemented and can also enable quite complex evaluations to be carried out. Furthermore, data on trainee performance can be collected and analysed. Used together with computer graphics the facilities which can be made available through a computer system are very powerful indeed. The PLATO system of Control Data is perhaps the most powerful and best known of such systems. For a general review of such systems see Lewis and Tagg, 1980. CADCAM training ought to lend itself to the use of such methods. In point of fact programmed instruction texts are still widely used, particularly by IBM for computer training. However it is now recognized that the use of any one of the media to the exclusion of others is demotivating to trainees. Hence more variety is now incorporated in multi-media training packages. For example, many commercially available PLATO courses also incorporate reading assignments, audio cassette and video cassette elements.

Programmed instruction is often more flexible than the strict programmed text style described above. It is sometimes incorporated in a 'workbook' which contains text and graphics for comprehension, tests, examples, problems with solutions etc, as well as sections of conventionally programmed material. Audio cassettes can be used for such things as talking trainees through procedures, recorded interviews and discussions, descriptions of equipment (so that the trainee can look at the equipment or photograph of it, whilst hearing the description, thus saving a lot of the irritating switching of visual attention which would arise from using printed text), and for many other purposes. Audio cassettes have the advantage that cassette players are portable and very common, even domestically, so that there are opportunities to listen to the recorded material in a variety of locations including the trainee's own home. Video cassettes are currently the most common form of visual material where moving pictures or animation are required. The old idea of video recording a lecturer (the so-called 'talking head' syndrome) has been largely superseded by the more imaginative use of visual devices including animations, a lot having been learnt from the techniques of television documentaries.

The following two tables from Raymont, 1981 summarize cost-effectiveness considerations on individualized instruction generally and some factors concerning the effectiveness of particular methods.

Table 1

Conventional courses	Individualized instruction
Cost	
Labour-intensive therefore tuition costs are rising sharply.	Falling costs of equipment for audio-visual presentation.
Associated travel and accommodation costs are also rising rapidly.	No associated travel costs.
The services of the trainee are totally lost to the installation for the period of the course.	The training can be interspersed with normal work activities. Cost of 'courseware' production are substantial therefore the purchase or hire of materials likely to remain quite expensive.
Effectiveness	
Tending to cater for lowest common denominator in speed and level of learning.	Enables trainee to enter at the right level and work at his own pace.
A well established method and therefore fairly static in terms of increased effectiveness due to innovative techniques.	New techniques are continually improving the effectiveness of audio-visual 'courseware'.
Scheduling may be difficult owing to non-availability of a course at the time the training is required.	The training is available immediately when required.

Table 2

Advantages and disadvantages	Some recommended uses
Programmes instruction text	
Portable and easy to use any time, anywhere. Tend to be dull, so best use in short time slots. Does not easily cope with topics where opinions, interpretations and diversity of methods dominate.	To provide prerequisite knowledge for some other form of training when the trainee lacks this. To give machine specific follow-up to general conceptual knowledge.
Audio-cassette	
Relatively portable. Can be used to get over points by dramatization, discussion or interview, as well as by straightforward instruction. Lacks visual impact but can be accompanied by visual material or even linked to slides in a tape/slide presentation.	For talking trainees through procedures. To provide orientation and motivation elements in training.
Video-cassette	
Expensive and only usable where the equipment is installed (but this is getting cheaper and more portable). Gives full visual impact with movement and, normally colour where necessary. Permits a full range of media techniques including animation.	To show techniques and equipment in action. To show dynamic procedures (eg the building up of a program structure). To demonstrate human interaction.
Computer-assisted learning	
Little effective off-the-shelf material is available as yet and doing your own can be very expensive. Allows for a considerable degree of interaction and flexibility in learning paths.	For training in data entry procedures or similar operational procedures.

Videodisc developments

There is one particular development which seems to offer much for the future which is worth brief discussion. This is the advent of the videodisc. Several companies are planning mass marketing of videodiscs shortly, which will inevitably bring costs down into line with current video cassette costs. The advantage of the videodisc for training is the immediate access to any frame on the disc. Each side of a videodisc can accommodate about 50,000 frames (equivalent to nearly half an hour continuous playing time). Audio material can also be put on the disc, even computer programs can be stored on it. The most exciting potential arises from having a videodisc controlled by a microprocessor linked to a computer-based training system. Thus, the trainee's answer to a question can route him to either a particular frame of information to study or to an animation to watch, or to a straight piece of video or to a computer program (eg a simulation) downloaded from the videodisc. This can provide a much more sophisticated system of computer-based training than has hitherto been possible. To date only pilot facilities have existed but some very promising work has been done (David, 1981).

An interesting, but relatively unsophisticated, commercial exploitation of videodisc has been made by IBM in their Guided Learning Centres (1981), where videodisc and System 34 work station sessions are integrated into a programmed instruction environment. The ability to store computer displays on videodisc should make this a powerful tool in CADCAM training.

Open and distance learning

The experience of several institutions, but most noteworthy the Open University in the UK, has led to an increasing interest in open and distance learning.

Open learning may be defined as a system to free learning from the constraints of locations and times, availability of course places, restrictive entry criteria, inappropriate learning methods, and logistic difficulties associated with conventional training. Distance learning is a set of techniques to facilitate learning at a site physically remote from the trainer, for example by correspondence, television, radio or telephone methods. Thus, whilst it is clear that distance learning may contribute substantially to the achievement of open learning objectives, open learning is a broader concept and does not necessarily imply the use of distance learning. An in-company programme of computer-based training on a flexibly scheduled basis in a company training centre would implement an open learning system, but would not use distance learning at all.

The potential for open and distance learning in technical training has not been systematically explored though some proposals have been made (Manpower Services Commission, 1981). It is clear that the methods for individualized learning described above would figure largely in implementing any such programme. The appeal of such a programme is the convenience to the learner of the open approach. The suppliers of CADCAM equipment and systems should have a considerable interest in an open approach to training in the use of their products as perhaps the most 'user-friendly' system of training.

Conclusion

It is not possible to generalize about CADCAM training since the requirement is so diverse (David, 1981). It encompasses the appreciation training of managers, the training of CADCAM experts (who will advise on the systems to be used by the users) and the training of CADCAM systems builders. However, it is clear that in this whole range there will be many applications for the techniques discussed in this paper.

References

David, B T (Lausanne, July 1981) *Computer aided design education – an overview*. In proceedings of world conference on computers in education. North-Holland: Amsterdam: 1982
Guided learning centres – an environment for learning. (1981) Basingstoke, IBM UK (Form 41-8 111)
Hartley, J (ed) (1972) *Strategies for programmed instruction: An educational technology*. Butterworth: London
Lewis, R and Tagg, E D (eds) (1980) *Computer assisted learning*. North-Holland: Amsterdam
Manpower Services Commission: London (1981) An open-tech programme
Panel discussion 3B in proceedings of world conference on computers in education. (Lausanne, July 1981) North-Holland: Amsterdam: 1982

Raymont, P G (7 July-August 1981) In favour of individualism. *Data Processing* **23**: 29-30
Rowntree, D (1974) *Educational technology in curriculum development*. Harper & Row: London

P G Raymont, The National Computing Centre Ltd, Oxford Road, Manchester M1 7ED

Discussion (refers to chapters 7 and 8)

A I Llewelyn, CADCAM Association:

Mr Raymont, you made the point that education and training is now a diffuse area. There has always been a question in education as to whether you are training people for skills or training them conceptually. What do you see as the implications of this for government and industry? Who is in charge, who is funding what? You mentioned a lot of agencies, but what are they doing, who is running with the ball?

P Raymont, The National Computing Centre:

I am often asked the question about the distinction between education and training. I will repeat a piece of wisdom which was passed on to me by one of Her Majesty's Inspectors who, when I asked him about this, enquired whether I had any children and whether they were at school? I said 'yes'. 'If they came home from school and said that they had had some sex education today you would probably be quite interested, but if they came home and said they had had some sex training you would probably be a very worried man.' That's the best explanation of the distinction I have ever heard. I must apologize to all visitors from overseas for talking about problems in this country, but there are really two agencies here, the Manpower Services Commission which largely concerns itself with shorter courses which are, generally speaking, training orientated, and the Department of Education and Science who concern themselves with much longer rather more academic courses which are perhaps more conceptual in nature. My belief is that in actual fact the Manpower Services Commission is very much alive to some of the current problems created by information technology and computing in general and is very keen to have programmes of training which will help to overcome the skill shortages but I think it is open to persuasion and I think if a case can be made and proof can be given that a certain kind of specialist needs a training course and that training course is comparatively short, under a year, then there is every possibility of getting support from the Manpower Services Commission. When it comes to the *education* of people in CADCAM then I think there are rather more difficulties, in that the whole cycle of development of new courses in universities and colleges takes a long time. What you first proposed may well be out of date by the time the cycle is completed.

N Spoonley, Control Data Ltd:

I used to run the Control Data Institutes which train programmers, engineers and technicians. One of the objectives of the Manpower Services Commission is to place people in jobs. If you train people in a school which is sponsored by the MSC and fail to find those people jobs, that school will disappear and the Manpower Services Commission will get very dispirited about encouraging such funding again. Part of the problem really does go right back to the industry itself. It's no use me training people in Leeds if all the jobs are in Bristol. A lot of people are not prepared to travel, particularly in the junior age groups, to get their training, so it is very delicate, going to the Manpower Services Commission and saying that there is a need for 30 programmers in the North-West and asking them to fund the training of those unemployed or semi-employed people and then place them in jobs. What it requires is for the industry to be really receptive. It's no use saying we need more skilled people because you are not going to get them unless their training objectives are very carefully defined by the industry so that they do get jobs. Otherwise the Manpower Services Commission will just cut the budget.

Unidentified contributor:

I suspect that many people will wonder how to approach the MSC. I would guess that almost all CADCAM is by definition in the engineering industry and it would seem to me that to approach the Engineering Industry Training Board which is still in existence is perhaps the right route to take.

S Bennet, Ford Motor Company:

Has there been any analysis done on the cost-effectiveness of systems like PLATO?

N Spoonley:

It really is an exceedingly difficult job to establish a comparative cost-effectiveness, but if you sit down and do a very careful analysis of the existing costs of training and the productive loss as a consequence of performing that training and compare it with an alternative method, you will often find that the new methods are more beneficial. If you take a salesman into a company who needs sales training, product training on your products and some field skills, and there isn't a product course for a month, you should really take out his salary as part of his loss to the company because he is doing nothing. If he is being put out into the field as a novice alongside an experienced person, you will recognize that the latter will have a lower performance over that period and that ought to be counted in. If you could ensure that the day a person arrives in the company he is given a ticket to a learning centre and the training starts the day he arrives, it's not difficult to calculate that induction-type training is exceedingly cost-effective.

A Llewelyn:

You haven't followed up qualitative assessment of people undergoing different training with a view to assessing their effectiveness. Personnel managers do this all the time and a number of firms are now interested in seeing whether some of these new schemes do prove effective in use.

N Spoonley:

The situation is very similar to CADCAM technology itself. There is a large group of people who believe that if you use new technology it has to be evaluated, checked, tried and perfected before anyone will use it, and this also applies to computer-based education, whereas the existing techniques of overhead slides, stand-up lectures, etc, are not evaluated at all. I would challenge whether industry personnel departments are really doing a serious evaluation. I think they want proof of new techniques but they do not analyse old techniques to determine how they can be improved by putting in the new technologies. I think this is a major delaying factor on the introduction of new technology.

P Raymont:

People are voting with their feet in industry. All the evidence is that in computer training the big companies are moving more and more to individualized instruction, not just computer-based training but audio-visual packages, etc. That is the big growth area, so people are obviously finding it effective or they would not continue to use it.

R W Howard, Construction Industry Computing Association:

On the subject of getting the right balance between teacher-centred instruction and learner-centred instruction, we are involved in mid-career training courses for people in the construction industry and we have found that we were going too far towards computer-centred learning. We find that the problem arises of getting their attention back again once they are involved with computer-centred learning. It becomes impossible for the lecturer to get their attention back to give them the next lesson or tell them what the next stage is. We have had to produce a small device which enables us to switch off the screens in a primitive network fashion, and superimpose on their screens either lecture notes or something which the lecturer has typed into a console. A balance is important to get the best of both worlds.

D Greatrex, Engineering Industry Training Board:

I will confirm Patrick Raymont's comment that the EITB are taking significant initiatives in promoting new technology. We are also active in providing grant aid and advisers in project services in the field of CAD, for instance. I would like to ask the two speakers what they see as being essential training needs of trainers in industry faced with undertaking and providing appropriate training?

P Raymont:

The first thing that occurs to me is that the training should not necessarily be technology specific. The point I was making about defining objectives is very important and although one might expect all trainers to be familiar with that method of working when you investigate the real situation, it is far from true. Therefore emphasis would have to be given to defining the training objectives, the training media and so on. There is probably a need for an expansion of the training available in those techniques. This would have to be backed up with the necessary technological training in how to use computer-based training systems, video tapes, video discs and so on. There are already a number of agencies already working in these various fields – all that is needed is an expansion of the amount of training available. Finally, I would say that it is wrong when developing training of this type to give too much emphasis to the technology in the sense of the hardware and one needs to give more attention to the concepts that need to be used.

N Spoonley:

There isn't a simple answer to that question. I think one has to recognize that the title 'trainer' itself is probably wrong. There are obviously going to be many trainers around, but a large number of people are going to be training managers or training co-ordinators, but what they are doing is simply managing a process. Control Data have taken a set of schools which were instructor dominated, run with just 'chalk and talk' sessions, and they are now run completely as computer-based training schools. The students move in and out of booths and in and out of small group tutor sessions. We have had to train the instructors in a computer-based environment. It's possible that there are a variety of ways to counsel such people because their attitude to their job is going to change. You have to put in the people who will help them understand why it's changing. It's no good taking a trainer and just giving him a whole set of new techniques, because if he is totally infused with the process of one on one delivery or class delivery, which a lot of teachers are, then he may as well find another job if you are going to totally change that training system. If, however, he is really enthused by seeing the students leave with their certificates or whatever, as a lot of other teachers are, then that person would probably be enchanted by some of the new techniques around. That has to be identified in the individual trainers so they really recognize whether they should be in for a career change or whether they want to embrace the new technology.

9. Skills and knowledge requirements for CADCAM

S A Abbas and A Coultas

Abstract: This paper concentrates upon the mechanical engineering industry, to examine skills and knowledge requirements for implementation and usage of CADCAM and then to suggest ways in which these might be met.

The authors' experience in their own organizations is drawn upon to substantiate the points made and the ideas put forward, so that the practical situation is juxtaposed with the formal teaching content. Some engineering-user experience is given and two approaches to training and re-training are outlined.

Introduction

In a previous paper presented in 1978 (Abbas, Coultas and Lee, 1978) we analysed emerging CAD job categories and examined some educational and training schemes being developed at that time. Our thinking was based largely on the proceedings of the first CAD ED conference held in 1977. Now, five years on, it is useful to re-appraise the situation in the light of recent technological advances and wider awareness of CADCAM.

CADCAM has a very broad sweep. In our paper we shall concentrate mainly on our own direct experience, and develop views deriving from that. The industrial/academic interchange and collaboration that has existed between WTCS (Whessoe Technical and Computing Systems) and Teesside Polytechnic in CADCAM activity has been of significant advantage in forming our views. Aspects of such co-operation include R and D projects, undergraduate training and projects, course development and revision, and professional activities as illustrated by the joint work in the formation of the CADCAM Association. In order to see clearly the perspective from which this paper is written, it is important to briefly summarize the work of the two bodies, WTCS and the CAD Group of Teesside Polytechnic.

WTCS is part of the Whessoe Group of companies involved in the design and manufacture of plant for the process and energy industries. WTCS provides a technical computing service to other companies of the group and also to external firms. Its major software products cover the areas of pipework analysis, plant modelling, pressure vessel design, heat exchanger design, cylindrical and spherical storage vessel design and chemical data-base and enquiry. These products are available to clients through the WTCS bureau service, other commercial bureaux, or mounted on the clients' computers. WTCS also have systems acquired from elsewhere, such as finite element packages, to be used as design aids. The computer systems are based on the DEC family, consisting of the Series 10, Series 20, and the VAX. They are linked and access is available via national networks.

CADCAM activity at Teesside Polytechnic is centred around the CAD Group, which is at present based in the Department of Computer Science. Staffing and facilities for the Group are funded both internally by the Polytechnic and externally by sources such as the EEC, local industry and SERC. In addition, staff based in other departments in the Polytechnic – at present primarily design and mechanical engineering – work with the core group in specific activities, eg new teaching schemes and research and development projects. The group encourages and fosters the incorporation of CADCAM in existing courses, and develops new educational and re-training courses. Its major research interests are computer graphics and 3D modelling and visualization, microcomputer-based design and drafting and data-bases for CAD. Hardware is currently based on a Univac 1110 and a Prime computer, with a large number of graphics devices of various sorts attached to them and to microcomputers.

In the next section we shall look more closely at the activities and functions current at Whessoe Technical and Computing Systems. We shall discuss briefly their discerned observations of engineering users in their take-up of CADCAM. In the succeeding section we shall describe two schemes for education and training in CADCAM at Teesside Polytechnic to illustrate some current approaches. One example is drawn from undergraduate teaching on formal courses; the other is part of a pilot project funded by the EEC for re-training unemployed mechanical engineers. Finally, we shall derive some conclusions and pointers for the future.

Industrial user experience

Before looking at the engineering user's experience of CADCAM systems it is helpful to outline the application areas at WTCS with which we are concerned. Systems are provided and supported for the activities listed below.

- 3D modelling of process plant (Figure 1)
- Stress analysis of products such as pressure vessels, nuclear vessels special purpose structures and pipework, using finite element and other analysis packages with graphics aids (Figure 2)
- Detailed design of equipment based on international Codes of Practice (Figures 3 and 4)
- Cost estimating for materials and labour
- Overall project planning
- Material control
- Workshop scheduling and direct shop floor feedback
- Manufacturing with purpose-built and conventional NC and CNC equipment.

The movement is towards integration, with systems generating and passing information to other systems through the mechanism of design and manufacturing data-bases. Hardware implementations are across the spectrum of micros, minis and mainframes.

It can be observed that the engineering community has in general embraced CADCAM enthusiastically and used it effectively. Recognizing that there is still much progress to be made in the development of powerful adaptable systems, in man/machine communication and in work station design, our engineers have coped remarkably well with the challenge and inherent difficulties of the technology. This says much for the traditional education, training and experience acquisitions process which generates the flexible skills and the ability to engineer an approach to applying new hardware/software engineering tools. This adaptability and resilience is not readily apparent in a number of other professions.

It is useful to make brief observations in three categories, namely the young engineer, the mature engineer and engineering management.

Young engineers of today, recently graduated, are approaching the use of the technology with a confidence born of their formal training in general computing and increasingly in CADCAM. With this background they are accommodating existing in-company systems and applying them quite naturally in their work. Interestingly, they are becoming catalytic and acting as pressure points on senior management for the extension and further deployment of the technology. The old lesson of the 60s generation of graduate training schemes is emerging however, typified by the comment '. . . the best method of learning was by modelling a "real" job – more problems were

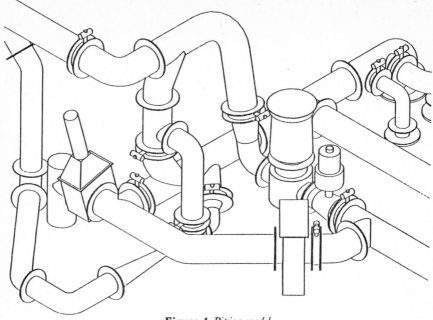

Figure 1 *Piping model*

Figure 2 *(see page 9)*

Figure 3 *(see page 9)*

Figure 4 *(see page 9)*

created for me by practical engineering cases than "are dreamt of in your
philosophy," . . .'

The new generation of CADCAM systems, where they have been carefully matched
to the problem situation and where training has been planned and thorough, have been
adopted enthusiastically by many mature design engineers and draftsmen. It is still
noticeable that there is resistance to using some elements of CADCAM systems because
of the non-user-friendly terminal keyboard, particularly amongst the older generation of
engineers. We should be sensitive to this resistance and understand that many of today's
terminals, graphics devices and work stations are ergonomically unsatisfactory. The skill
requirement to use CADCAM will be impacted markedly when systems allowing more
natural forms of man/machine communication are implemented at the hardware level
on a single screen surface resembling traditional tools like the drawing board.

At the engineering management level there is still much work to be done and
knowledge to be imparted. Many managers are not equipped to specify, select and
manage the implementation of CADCAM systems solutions to their problems, and this
applies in large as well as small companies. A recent example of a senior engineering
manager's involvement in the selection and implementation of an advanced 3D
modelling system meant that he had to work with system developers in specifying:

 – training to give maximum acceptance and utilization;
 – input and output requirements with the freedom to apply his own engineering
   data;
 – response requirements and screen commands;

– user protocols to minimize the intricacies of data management requirements for large computers;
– processing cost bounds;
– functionality and reliability characteristics of work station equipment;
– the degree of system tailoring to meet specific requirements;
– fault reporting mechanisms.

Working with a responsive CADCAM group this involvement has resulted in a highly effective production capability. Surveying the situation in industry in general, it is apparent that the inability of senior and middle management to specify systems adequately for their needs has led to many disappointments.

Two illustrative approaches to CADCAM education and training

We can identify two broad areas which have distinctive education and training needs: formal undergraduate courses and re-training courses for mature engineers. The former implies re-thinking traditional curricula in the light of current developments so that young engineers graduating from universities and polytechnics in the 1980s and 1990s are capable of using up-to-date technology with confidence. The latter is concerned with utilization of existing experienced engineering manpower, otherwise doomed to wastage, by grafting on CADCAM know-how. In this section we shall describe current approaches in these two directions as taken at Teesside Polytechnic, which should help to illustrate some of the general problems.

Undergraduate courses

We shall concentrate here mainly on the mechanical engineering sector, and consider the incorporation of CADCAM in existing courses. Our discussion here is based upon a detailed paper presented earlier this year at a conference in Newcastle (Abbas, 1982).

In the previous section we looked at some of the functions being performed by CADCAM-related personnel in industry today. It appears that mechanical engineers graduating from universities and polytechnics should be ready to assume roles in industry in which they should be able to work effectively with computers in design, development, manufacture, control, maintenance or other activities. There are now large and increasing numbers of computer scientists being trained. We can expect them, when they enter industry, to work on aspects such as data-base development, languages, systems design and artificial intelligence. They would work closely in teams with engineers in devising applications systems and the need for adequate two-way communication between the computer scientist and engineer would be profound.

With a broad acceptance now that student engineers should be exposed to computing and applications throughout and across the whole curriculum, where does CADCAM fit into the picture? Indeed, should one be specifically teaching CADCAM? These questions are still being debated, but there are some important issues that can be identified. These hinge upon the conventional breakdown into subjects in traditional curricula.

The key point is that CADCAM is concerned with the integration of different subject areas, and that any teaching schemes and projects should emphasize aspects of data creation, analysis and information flow. The analogy with design teaching itself is fairly close, in that we are not very concerned with the detailed techniques of analysis of the components of the design but rather with how the components relate to each other in the make-up of the whole system. The use of CADCAM techniques makes it possible to tackle substantially larger-scale projects than would be possible conventionally in equivalent time slots in courses. In Abbas, 1982, this theme is expanded upon, with an example, in more detail.

At Teesside, at the present time, CADCAM is not taught as a separate subject in the undergraduate course in mechanical engineering. A good deal of time is given to it in final year design, and there has to be close co-operation with the teaching of production technology, in which numerical control and manufacturing systems with the use of packages such as GNC (Graphical Numerical Control) are covered. As integration of the design and manufacture processes progresses, there is a growing need to re-think the teaching approach in these areas.

It is recognized that CADCAM teaching has a high resource requirement, both in terms of trained staff (scarce) and facilities (expensive) and this is the major factor holding back progress in educational institutions. It is of interest to outline briefly the hardware and software now available at Teesside Polytechnic specifically for CADCAM and their utilization in two particular undergraduate courses very different in character from each other, mechanical engineering and interior design.

A crucial step in advancing CADCAM in the Polytechnic was to acquire facilities specifically for CADCAM usage, run by a CAD Group separately from the general purpose Polytechnic computer systems. This CADCAM hardware consists of a Prime 250-II computer, with five CADCAM work stations, each consisting of a graphics display, a VDU and a tablet, plus a drum plotter, printer and paper tape punch. The peripherals were supplied by Counting House, and together with the major software packages, GDS for drafting and GNC for machining, constitute their Integrated Technical System. Other software mounted on the Prime 250-II includes the PADL-1 geometric modeller and the VAMP 3D colour visualization package. Finite element packages, eg PAFEC, are mounted on a Prime 750 and are used from the work stations via Primenet. These systems are used by mechanical engineering students in their final year.

Interior design undergraduates base their CAD work on Apple computers, of which there are a number situated in their department, with digitizers and plotters, using a 3D wire frame visualization package called CAPITAL. Increasingly, as these students gain familiarity with the microcomputer-based systems, they begin to hit limitations, and move on to use the packages on the Prime.

These resources are the base level necessary for the teaching of CADCAM. By putting together the hardware/software components ourselves, rather than buying a turnkey system, we can retain considerable flexibility, and expand, enhance or modify with some freedom, which is necessary in the teaching environment.

A retraining course

In the UK at the present time, under depressed economic conditions, there are large numbers of people with good qualifications and valuable experience in design and manufacture, but who are threatened with redundancy because their skills are no longer required in a society which is undergoing structural changes. It is attractive to believe that the solid industrial experience of many such people could be harnessed in a very significant way, by grafting on to it an exposure to modern computing developments. However, this is new ground, with very little related experience to draw upon, and there appeared to be a clear need for pilot studies to establish some guide-lines. Since a pilot project in the field of CADCAM was going to be costly, some methods of funding were sought. It was found that the EEC Social Fund would provide a suitable mechanism, since it contains a Pilot Project element and deals specifically with problems associated with the introduction of new technology. At this pilot stage, it was decided to concentrate on the mechanical engineering sector of industry. It was expected that with careful design of the project the experience and results obtained would provide very useful pointers towards future schemes in other fields, eg civil and structural engineering, building and architecture.

With backing from the CADCAM Association, the Polytechnic applied to the EEC Social Fund, and in mid-1981 was awarded a grant of £50,000 to assist the pilot project. Further assistance was provided by the Manpower Services Commission, who agreed to give TOPS grants to some of the unemployed candidates who could not find sponsorship from other sources. The project, as proposed, was to last for one year. The core is a full time course lasting a total of 19 weeks, including 12 weeks of intensive teaching and seven weeks of unsupervised project work and exercises. After the course, the students are to be assisted in finding suitable employment and the initial stages of their work will be followed up and monitored in an attempt to evaluate the relevance of the course and the students' response to it.

The course content and scope was fairly wide. After an introduction to computing systems and programming, the following were the major topics covered:

- computer graphics
- drafting systems
- numerical control
- robotics
- microprocessor applications
- finite element methods

The topics were treated in a practical fashion and the industrial application aspects were emphasized. Although theoretical and mathematical background were not dealt with in depth, it was considered important that the material should not be covered too superficially and 'blindly'. A compromise had to be found between these approaches, and clearly one of the essential lessons for the course organizers was in establishing a suitable level of compromise.

At the course planning stage, indications were that there might be some student sponsorship on a release basis by firms. However, it was found finally that only three firms put forward employees for attendance on the course. The reason given in every case, was that the pressure of the present depressed economic circumstances had caused most firms to trim down personnel to such an extent that there was no spare manpower to release at this time. The three students who were sponsored by their firms were all on 'severance terms', ie they were to be made redundant in the future. The remainder, nine in number, were all unemployed, or under short-term notice of redundancy. The Manpower Services Commission had agreed to provide eight grants under their TOPS scheme to unemployed candidates and so one student was self-financed. The 12 students ranged from 23 to 55 years. Their qualifications were HNC, HND, or degree, all in mechanical engineering. Their backgrounds were varied: design engineering, drafting, maintenance, general management and production. Thus the nature of jobs envisaged for the students after leaving the course would be varied. It was believed that the spectrum of career paths might include: CADCAM systems operations and supervision; CADCAM systems sales, marketing or support; general management in a technical computing environment; manufacturing systems supervision. However, it is recognized that particular individuals might discover certain interests and aptitudes which would direct them along certain paths of a deeper technical nature, examples being finite element techniques or applications programming work.

At the time of writing the course is nearing its end, and the trainees are looking for jobs. A full report evaluating the project will be produced for the EEC towards the end of 1982.

Largely as a consequence of the experience gained in running this project, another course has been devised, to start in September 1982. This new course will last for one full year, and is wholly sponsored by the Manpower Services Commission through TOPS grants for unemployed candidates.

Conclusion

The five years that have passed since the first CAD ED conference have seen CADCAM becoming commonplace and maturing into everyday use in engineering. The manner in which the accommodation by practising engineers has taken place has been interesting to observe. Some engineers have embraced the technology successfully; others from the same generation have struggled. Many examples of improved engineering and increased productivity are evident: equally there have been many disappointments and indeed failures to meet objectives and requirements of taking up the technology.

As illustrated by the particular industrial environment covered in this paper, new kinds of roles are opening up for engineers. The situation is currently highly dynamic, and it is clear that we must be prepared to be flexible and adaptable in re-defining job responsibilities within our organizations at short intervals, if necessary. With increasing size and complexity of systems, the importance of team work and co-operation across disciplines becomes aparent; the interface between the engineer and the computer scientist must allow efficient and sympathetic communication.

To keep pace with the changing industrial scene, it is necessary for teaching institutions continously to re-examine their courses and change curricula where appropriate. At the present time, more so than at any other, the need for a close dialogue between industry and academic institutions appears to be crucial for success to be achieved.

References

Abbas, S A (April 1982) CAD curriculum requirements in mechanical engineering. Teaching of CAD, conference paper, Newcastle Polytechnic
Abbas, S A, Coultas, A and Lee, B S (September 1978) CAD education – the way ahead. Interactive techniques in computer-aided design, conference paper, Bologna

A Coultas, Whessoe Technical and Computing Systems Ltd, Brinkburn Road, Darlington DL3 6DS
S A Abbas, Reader in CAD, Teesside Polytechnic, Borough Road, Middlesbrough, Cleveland

10. The teaching of CAD – a review of the proceedings of a conference held at Newcastle upon Tyne Polytechnic, April 1982

R Schofield, D F Sheldon and F Weeks

Abstract: The increasing application of CAD systems in UK industry is considered and the implications of this for higher education are reviewed. Current and projected ideas for the teaching of CAD in polytechnics have been drawn from the papers and discussion at the Newcastle conference. These are presented within the themes of expertise required, curriculum implications, teaching experiences, the use of packages and resource implications.

Introduction

The heads of engineering departments in polytechnics meeting in 1981 considered the increasing application of computer-aided design systems in industry and discussed the significance of this in relation to the courses for which they were responsible. As a result of these discussions it was decided to hold a conference on the teaching of computer-aided design at Newcastle upon Tyne Polytechnic in April 1982.

It was hoped that the conference would provide a forum for the exchange of experiences and for the discussion of common problems. Since it was felt that the two sectors of higher education faced different methods of course validation and resource provision, it was decided to invite papers and attendance from the maintained sector only to ensure a commonality of outlook. The organizing panel decided that the scope of the conference should include mathematical models, design analysis, graphics and the interface to manufacture, but that the technology of CAM should be excluded.

Nine papers in all were presented in three sessions on the themes of curriculum developments, teaching experiences and the use of packages. The conference opened and closed with keynote addresses by invited speakers. The conference arrangements allowed for extensive discussion, which achieved successfully the organizers' aim of encouraging an effective exchange of opinion and experiences. The aim of the present paper is to review the conference papers and discussion so that the ideas may be considered by a wider audience.

Full details of the conference papers and authors are included in Appendix I. It should be noted that where the section headings in this paper correspond to the conference themes, the ideas reported are not necessarily all drawn from the papers in that particular session.

The need

A recent study (Centre of Engineering Design, 1981) predicts a considerable growth in the application of computer aided engineering in UK industry. The number of computer-aided design and drafting (CADD) installations in service is envisaged to rise from 100 in 1980 to 1,000 in 1990 and 3,000 by the end of the century. On the basis of this forecast, the number of trained people required for the engineering functions of design and manufacture is assessed to be 92,000 in 1985 rising to 208,000 by 1990. Of these, the greater proportion (58,700 and 153,500 respectively) are described as system operators who might be seen as deriving from system suppliers training courses, in-house training 'on the job' or from updating short courses. A significant number described as system tailors, system buyers and training staff might be seen to require a more fundamental and extended study of CADD, to be provided as part of higher education courses.

The relevant figures for predicted national needs, extracted from the study, are shown below:

	1985	1990	2000
System tailors	5,555	17,700	46,713
System buyers	10,000	13,000	13,000
Training staff	1,000	3,000	3,000

These figures are compared with the total output (in relevant subjects) of current higher education schemes validated by CNAA (3,318 people per annum) and the Technician Education Council (3,487 per annum). The paper observes that this higher education output is almost entirely earmarked for other purposes and at the time of writing (April 1981) has little CADCAM content. The lead time in education implies that there will be no new output on new CAD subjects before 1986 at the earliest.

The authors of the study acknowledge that there is a possibility of a considerable margin for error in this type of prediction, but emphasize that an underestimate is just as likely as an overestimate. In any event, it can hardly be doubted that there is an urgent need for an educational response to this new technology.

Expertise required

The need for education in computer-aided design seems well established. As Riley observed in his paper, engineering departments are in danger of being caught in a pincer movement between industry and schools, with the former increasingly using CADCAM techniques and the latter providing an intake to higher education, already skilled in interactive computing techniques and programming.

In considering educational provision for the subject, it is desirable to clarify what is meant by CAD. Abbas refers to a lack of understanding of the nature of computer-aided design in engineering circles as evidenced by the use of definitions such as, 'the use of computers in engineering'. In educational programmes, computer methods (including graphics as well as analysis) may be used in individual subjects and for the solution of design problems. The question is raised as to whether it is necessary to think of CAD as a topic, or whether the desired knowledge could be transferred by teaching computing and allowing computer applications to emerge throughout the curriculum of existing degree and non-degree courses.

Abbas identifies the key point (and this is echoed in other papers) that, 'CAD as a topic should really be about integration of the various subjects in the curriculum, using computer technology as the vehicle and design as the pivot'. Use of the term 'computer-aided design' is taken to mean that the results of the computer analysis or graphics should be used within the design process and this, in turn, implies information transmission or linkages between different activities of the design process and manufacture. On the same theme, McLeod and Weeks see the use of a common object description in design, manufacture and planning, together with the increasing speed with which information can be fed back into design, reducing the traditional separation between design and manufacture. It is suggested that this should be reflected in improved integration within the curriculum.

In assessing the educational requirements for CAD, it is necessary to distinguish between training and education. The former may be regarded as involving the acquisition of 'application specific' knowledge and skills, while the latter implies the transmission of an identifiable body of fundamental principles of general application, together with illustrations of their application to contemporary problems. It follows that the former is more appropriate for short intensive courses, primarily for those already practising in industry, while the latter approach is essential in full time degree and higher diploma courses. The products of these courses emerge from the educational process after three or four years, with their skills only becoming fully operational up to ten years after the formulation of the course content.

All the conference papers were concerned with education rather than training. From them and the subsequent discussion, it is clear that many people are moving towards the identification of a clear set of fundamentals, which might form the basis of a separate subject of CAD or CADCAM, but there is still some way to go. Some authors concentrated on computer-aided design, including graphics, while others such as Riley and Cockerham describe a link to CAM and Hargreaves saw the wider view implied by the term 'computer-aided engineering' as being more relevant to future needs.

Llewellyn in his keynote address set the scene for the educational requirement by showing a view of industry with CADCAM systems (involved in analysis, data-handling and problem solving), linked to Robots (implying artificial intelligence and learning systems) and Flexible Manufacturing Systems. On the theme of integration, he visualized the technologies converging, with important implications for education. The trend in manufacturing industry was towards flexible manufacture in which it would be possible to change the product by changing the software rather than the plant. This was expected to give a better return on capital invested and manpower. There were three key elements in this development: computers, numerically controlled machines and robots.

Curriculum implications

All were agreed that a fundamental requirement is a knowledge of computer programming, a fairly general view being that it is most satisfactory for this to be taught by engineers. Most started with BASIC, taking advantage of the availability of cheap interactive facilities, and introduced FORTRAN at a later stage. At Huddersfield, however, FORTRAN is taught first to avoid the problem of students preferring to rely on the easier language where BASIC is taught first. It is desirable to provide illustrations of applying the programming knowledge to engineering problems. The general practice is for this to occur naturally in other subjects, but at Teesside a special subject is provided.

There was general agreement too, on the need for students to experience the use of pre-written programs (either developed in house or commercial packages), but a marked variation of view existed on the proper balance between giving the students extensive experience of writing their own programs and of using those produced by others.

On the one hand Danks suggests that students should not only have to write programs and submit assignments to demonstrate their ability, but should be required to produce a critical appraisal and constructive criticism of existing programs into which particular problems have been deliberately incorporated. At the other end of the spectrum, there is a call for a greater emphasis on providing experience of the preparation of problems for numerical solution and on the use of pre-written packages.

A variation of view on the role of engineering drawing tuition also emerged. While Danks referred to 'the need for concepts to be translated into the basic means of engineering communication, the engineering drawing'; Riley and others implied that CADCAM developments might make the drawing in its present form irrelevant in certain cases. Nevertheless, it seems likely that a knowledge of orthographic projection will continue to be required. McLeod and Weeks suggest, however, on the basis of their experiences with 3D representation of engineering components and geometric modelling in particular, that CAD practitioners will require an enhanced appreciation of three-dimensional space. Their belief that this may be inhibited by too early an emphasis on orthographic projection received some support in discussion.

There was general agreement that computer graphics should be included in the curriculum, with Baxter arguing that all engineering courses should include some element of this. He believes that the principles are best learned by applying graphic techniques to the solution of real problems involving the creation of graphics software. Illustrations are included of software developed by students to produce and manipulate a logo, a cam profile and tool path plots. Several papers refer to the use of linkage design and the determination of loci as suitable illustrations. At the other end of the spectrum reference is made to the desirability of including experience of computer aided drafting.

All of these developments have resource implications which will be referred to later.

Drawing on his industrial experience, Warman suggests that many graduates have difficulty relating to their studies to the practice of real engineering and makes a plea for academics to get involved with industry on real tasks.

Teaching experiences

Most papers refer to CAD topics being introduced as part of the engineering design element of the course, although two argue for the introduction of a separate CAD subject. At Sheffield a 40-hour CAD module is proposed for the final year of the degree course. Two alternative arrangements are to be available; one as part of the mechanical engineering option emphasizes engineering analysis, whilst object definition and automatic manufacture are to be the dominant themes of that for the production engineering option. The objective of the modules is quoted as, 'to provide a broad appreciation of technical and economic benefits, range of available hardware and software, understanding of underlying principles'. Case studies are used to provide students with the opportunity to appreciate the problems faced by industry in deciding if, and what, CADCAM facilities can be used to good effect in the company.

Experience of using a commercial package such as APT is found to provide sufficient insight into the calculation and logic contained within it to enable the students to appreciate the possibility of producing a purpose built facility with limited facilities on a small computer. Reference is made to the development of a CAM system for a small company using a microcomputer.

Hargreaves refers to a degree course which is highly design-oriented and includes computer-aided design as a subject in its own right introduced in the third year and fourth year. Student project work results in the development of analysis packages and experience is provided of the problems of preparing interactive programs. Good programming and documentation practices are encouraged by providing students with a marking scheme. A linkage design program is used to introduce the techniques of computer graphics. Finite element applications are introduced in the final year and the development of a FE program is a project requirement.

Several papers refer to CAD being introduced into final year projects.

Two of the papers refer to the use of a computer-aided drafting package. Metcalf describes the experience at Nottingham, where a CAD studio contains several graphic terminals, a digitizer and some conventional drawing boards so that terminals are used as intensively as possible. A commercial software package is available.

Reference is made to the fact that naïve users will inevitably make keyboard mistakes. The system should enable such errors to be quickly redeemed. Exercises should emphasize the routines that are available for completing the drawing once the basic shape has been defined, since it is in this area that significant gains can be made. The requirements to produce a small structured drawing, which includes a range of line types and uses a few standard facilities, stimulates the students interest and promotes confidence.

Introducing the CAD studio early in the course enables it to be used naturally in later design work ranging from complex assembly and detail drawings to manufacturing drawings required in design and make exercises. Formal class time is minimized with the emphasis on the facility being available on a free or bookable basis. A disciplined method of working is desirable and students are encouraged to prepare effectively in advance of studio sessions.

Several authors referred to the long learning time required with a computer-aided drafting system. Metcalf suggests that this may be minimized by producing a user guide specially for the undergraduate engineer. At Newcastle, a self-written drafting package with simple drawing facilities is used. Students working in groups of 4 or 5 spend about 12 to 15 hours carrying out an exercise on this system. One of the problems is to introduce the students to computer-aided drafting (which is essentially a manual skill) in an educational context. The method used is to ask the students to prepare a report on computerized drafting techniques for an imaginary company which is considering purchase of a system and has one on trial. Simple part drawings are produced together with an extensive report which usually highlights the obvious deficiencies of the system. Most reports conclude that computerized drafting would be beneficial if some improvements were made to the system and certain additions were provided. These often add up to the specifications of a full commercial drafting system. The authors suggest that this exercise should be followed by the opportunity to use a commercial system.

The use of packages

As mentioned earlier, opinions vary on the question of whether students should be educated to use commercial software only or to undertake the development of substantial CADCAM packages themselves. Riley is in no doubt that students should leave higher education able to understand and be actively involved in the writing and developing of CADCAM systems. He believes that students should have the opportunity to develop skills in both interactive graphic design programming and in programming for CNC machine tools. An integrated CADCAM system is described, with software developed in-house to illustrate the input of graphic data, its display and output as a drawing or as a punched tape to control a CNC milling machine. Developments are in progress to provide a direct link between object description and machine tool, as well microcomputer control of a robot and the CNC milling machines to form a flexible manufacturing system.

It is suggested that a CADCAM software package developed for teaching will have different requirements from one which is intended for use in industry. In the latter, the designer is seen as a user only, whereas in education the objective should be to provide engineers with an understanding of, and a capability to develop, CADCAM systems software. Most delegates saw engineering graduates as users of software rather than creators.

Riley's paper describes progress towards a teaching package which includes a 2D sculptured geometry program as well as a 2D standard geometry program. The latter allows artefacts consisting of combinations of straight lines, circles and circular arcs to be defined at the alpha-numeric terminal and passed to manufacture. A drawing option is not included and it is suggested that this reinforces the fact that, for some detail design, the stored computer model can replace the component drawing.

Most courses include some reference to finite element techniques and the conference papers indicate that this is seen as an important segment of CADCAM activity. Taylor argues that the most important skill to be developed is that of creating the finite element model since this has a fundamental influence on the accuracy of the analysis. In spite of FE packages of varying sophistication being available, none eliminates the need for the engineering analyst to thoroughly organize the modelling procedure. Taylor offers the following quotation from a recent paper by Wilson (Wilson, 1980):

> the majority of finite element analyses currently being conducted are in large error and, in many cases, are not better than a good approximate analysis which satisfies statics

The paper describes a consistent procedure for creating finite element models in three dimensions to help overcome this problem. Discussion of Taylor's paper raised the general point that students should be brought to appreciate that computers do not provide universal truth and that there are dangers inherent in applying insufficient knowledge to the use of packages.

Resource implications

The widespread application of CADCAM techniques in industry, which seems likely in the next ten years, presents a revolution for the mechanical engineering industry quite as significant as the development of the transistor in electrical engineering. It is unfortunate that higher education should be required to respond to the challenge of this innovation at a time when resources of manpower and hardware are constrained.

Nevertheless, the general picture to emerge from the conference is that the polytechnics will make maximum use of existing resources even if they fall short of the ideal, while endeavouring to achieve a reasonable allocation to meet educational needs. A major problem is in respect of graphics. While general purpose computing facilities can be used for interactive, analysis, graphic activities require a fast response. Abbas identifies the main factor inhibiting the development of CAD graphics as the computer which is used. He refers to the availability of microcomputer-based drafting systems at about £25,000 which, although limited in power and scope of application, would be satisfactory for teaching purposes. Since they are single-user systems, it is sometimes suggested that their use may be planned as with laboratory equipment, but several authors (including Abbas) believe that this analogy is not appropriate because of the long learning time, even with simple systems. Multi-user systems provide more extensive facilities and a lower cost per user, but are still expensive. Since they are intended for industrial use they may not be entirely suitable for educational purposes. Whichever solution is adopted, it is clear that careful scheduling will be required to ensure maximum effective use of this equipment. In turn this may have implications for other course arrangements.

Many authors refer to the problem of making time available for new subjects in already crowded curricula. At Huddersfield, courses are being restructured to provide more time for computer-aided engineering topics. Warman in his keynote address refers to the problems of adding new information to courses without discarding some. He refers to the pressures to increase the length of courses, but doubts whether this will necessarily produce more and better enginers. Certainly it is essential that course content should be continuously reviewed to ensure that it is relevant to the needs of a rapidly changing technology.

Conclusions

It is clear from the papers reviewed and from the resulting discussions, that polytechnic engineering departments recognize the need for CADCAM education to be included in the courses for which they are responsible. The conference revealed a variety of views on the elements of computer-aided design and the way in which these should be taught, but indicated that a number of people are endeavouring to identify a body of fundamental knowledge, which could justify the inclusion of a separate CADCAM subject within the curriculum.

Discussion of the papers revealed that present developments are generally driven by 'enthusiasts', although it was recognized that any expansion of CAD education would require the expertise to become more widespread. It is likely that a considerable programme of staff development will be required and it may now be appropriate for consideration to be given to the best means of 'educating the educators'.

Availability of resources of study time, manpower, software and hardware will affect the speed and effectiveness with which CAD can be introduced into courses. As Abbas concludes in his paper, 'to obtain the necessary resources it is vital for those with common interests to act together and bring the necessary pressure to bear on the appropriate organization both in industry and government'.

The Newcastle conference demonstrated that there is among polytechnic engineering staff, a nucleus of enthusiasm for the teaching of CAD which needs to be expanded, a desire to share experiences and a willingness to come together to provide the influence to which Abbas refers.

References

Centre of Engineering Design, Cranfield Institute of Technology (April 1981) A study of training needs for Computer-Aided Engineering
Wilson, E L (October 1980) SAP 80 Structural analysis programs for small or large computer systems. Proceedings CEPA Conference

Appendix I

Conference programme

Opening keynote address, A Llewelyn, OBE

Session 1: Curriculum development

1. Curriculum requirements for mechanical engineering, S A Abbas (Teesside Polytechnic)
2. CAD – A numerical and graphical contribution to design teaching in a mechanical engineering course, J Danks (Coventry (Lanchester) Polytechnic)
3. The computer representation of engineering shapes and its implications for the undergraduate curriculum, A J McLeod and F Weeks (Newcastle upon Tyne Polytechnic)

Session 2: Teaching experiences

4. Maximizing the benefits of CAD, R Metcalfe (Trent Polytechnic)
5. The development of CAD teaching for undergraduate mechanical engineers – the Huddersfield experience, T Hargreaves (Huddersfield Polytechnic)
6. Teaching CAD and CAM, G Cockerham (Sheffield City Polytechnic)

Session 3: Use of teaching packages

7. The development of an integrated interactive CADM system, J A Riley (Kingston Polytechnic)
8. A consistent approach to three-dimensional finite element modelling, G T Taylor (Glasgow College of Technology)
9. Computer graphics – its place in the curriculum, R S Baxter (Trent Polytechnic)

Concluding keynote address, Dr E Warman

R Schofield, Leeds Polytechnic, Department of Mechanical Engineering, Calverley Street, Leeds LS1 3HE

D Sheldon, Huddersfield Polytechnic, Department of Mechanical Engineering, Queensgate, Huddersfield HB1 3BD

F Weeks, Newcastle upon Tyne Polytechnic, School of Power Engineering, Ellison Building, Ellison Place, Newcastle NE1 8ST

11. The integration of a commercial CAD package in the teaching of CAE on an undergraduate course

D J Pollard and C Douthwaite

Abstract: The authors present an argument for the combination of commercial software and that written in-house for teaching on an undergraduate course. The manner in which this combination is achieved is described and examples given.

Introduction

Engineers do not need to be convinced that they should use computers. The current position with reference to computers is simply the latest development in a long line of aids that may be thought to have started with Babbage, and engineers have in general always taken advantage of these developments. The advent of microcomputers and multi-access machines has led to the latest advances in interactive use of computers and this type of use makes a significant change to the possibilities.

Traditionally universities and polytechnics have generated software themselves and there has been a certain amount of exchange of programs between 'friends' and the more hard-headed members have launched commercial ventures. With this type of approach software has been relatively cheap and indeed some hardware manufacturers have used software as a loss-leader. However the position has now changed and whilst some contributions may be made in this way it now becomes necessary to take a slightly different view. Development times for software are long, often quoted in tens of man-years therefore once one has some hardware provision it now looks as if in order to make substantial steps in CAE one needs to buy in some software. This should not be viewed as an admission of insufficiency but in the same way as the purchase of a piece of equipment such as a Tensile Testing Machine. The important points are to choose the software carefully and to concentrate on the exploitation of the capabilities of the chosen package. By this method of approach one may establish a sound data-base upon which to build. It is of course imperative that the use of any package should be developed alongside and integrated with other in-house work.

The authors in this paper set out the results of their work to date and their future plans for the use of a commercial package. This package is being used in the department of mechanical engineering at the University of Surrey in the undergraduate teaching on the 2:1:1 thick sandwich course and in connection with the research and industrial-educational activities of the department including the MIDAS Teaching Company Programme.

Objectives of CAD teaching and research

The predicted tenfold growth by 1990 of CAD facilities in industry makes it desirable for all graduate engineers to be familiar with, or at least aware of, the advantages, limitations and costs of CAD systems. Experience of these systems should not be restricted to a few 'lucky' final year students but must be available to all students on a mechanical engineering course.

There is an extensive range of systems available and it would not be possible for our department to demonstrate all of these systems available. In principle a general appreciation must be given by lectures and films where they are available. However, for a general appreciation of the topic to be meaningful some study and experience in depth of a particular area is necessary. In particular this means that some 'hands-on' experience is required, that extends the computing in other areas.

There are three main areas of activity which are present in CAD which can be identified as part of the general picture.

2.1 Preparation of 'in-house' programs for design calculations
2.2 Computer-aided drafting
2.3 Use of specialist software for complex problem analysis

Some experience of these areas may be gained during the industrial year. It is however necessary to prepare students to make the best use of that experience and to cater for those students who will not find it available to them.

Research in CAD needs a sound basis of experience and some closely defined objectives for each individual involved. It seems reasonable to suggest at present that provision of the capability to operate at undergraduate level should take precedence because this will make the CAD facilities available to researchers within the department and provide CAD workers with experience. Research in CAD should evolve from this base with the department's strength and contact of MIDAS, helping to identify the particular objectives for each individual.

(MIDAS, Manufacturing Industries Development Association at Surrey. This is a Teaching Company programme of eight local companies in a consortium called MIDAS funded by SERC/DofI with the aim of introducing computer-aided engineering into small industrial units.)

The aims and objectives for undergraduate teaching may be stated as follows, to provide each student with:

2.4 The ability to understand a high-level language so that he/she can communicate design requirements to systems analysists and to develop/modify existing software.
2.5 A working knowledge of a commercial applications package.
2.6 An understanding of the implications of linked systems which encompass such fields as finite element analysis, numerical machine tools and simulated systems control.
2.7 A general appreciation of the scope and influence of CAD at present and its future possibilities, with respect to the manufacturing process.

The facilities available for this tuition range from use of the main University 4 × Prime 750 multi-user system, through Hewlett Packard 9845 Desk Top Computer, PET/Apple Systems with disc storage to small microprocessor systems based upon the Intel 8080.

Software investigation

The aims of CAD teaching set out above led to the need for staff involved in CAD to become fully aware of the capabilities of current commercial software. About thirty of commercial and university/college packages with varying capabilities have been examined.

The scope of these capabilities is listed below.

3.1 Drafting in terms of 2D detail and assembly
3.2 Three-dimensional modelling
3.3 Finite element analysis in two and three planes
3.4 NC and CNC programs for manufacture
3.5 Circuit diagram and Printed Circuit Board layout

The key to most of these systems is the generation of the basic design data in the form of a layout/detail drawing. Therefore, it is essential that any system implemented in the department is able to interface directly to commercial programs or be linked to an in-house development.

It is clear that for CAD to be developed one must start with a geometric data based drafting package since this is the foundation upon which all the subsequent developments are based.

As a result of the survey of software the authors suggested that there was a package which with compatible interface systems covers most of the requirements specified above and readily forms the foundation of CAD.

The CIS Medusa package has been acquired and was implemented on the University Prime system during 1981, subsequently Prime Computers Inc (USA) has adopted the package as its main data-base for work in CAD/CAE.

The software package

This brings together two approaches to 2D line drawings, a 3D component modeller and a simple mechanism for assembling components into a single system. The system includes also a powerful 3D viewing module which has the dual function of generating perspective views with full hidden line removal, and of assisting in the development of traditional side and end elevation views.

The software comprises a two-dimensional interactive drafting system around which a series of interpretive and analytical modules are arranged. It is a powerful 2D line drawing system produced by combining the data processing capabilities with boolean geometric construction techniques.

It also has the ability to turn two-dimensional drawings into 3D perspective views by using the profile linking mechanism described below for the modeller interface. By choosing viewpoints to align with the major axes, side and end elevations may be drawn in fuller detail than is necessary for defining the object in the first place, in this way, simple applications may be handled by the system. Thus, an installation set up for relatively straightforward or conventional applications (Figure 1), may readily be used to include modelling techniques in subsequent years (Figure 2), without significant changes to the way the system is used. The student is not therefore faced with having to learn to use two systems.

The most fundamental importance of the modules is the 3D modeller, which can interpret 2D drawings, and construct 3D models of the components. This method of driving the modeller provides the student with a powerful and easily controlled 3D tool. The drawing from which 3D models are generated contain four types of information:

4.1   2D views or part views of object components.
4.2   Links defining how these views are related, and hence how to generate the 3D components.
4.3   Information showing how the components are to be assembled into objects and the necessary view points.
4.4   Definitions of which views and sections are to be output. (These are illustrated in Figures 3 and 4.)

The views and sections generated by the modeller may be annotated, cross-hatched and dimensioned to yield fully detailed drawings.

Three further modules are available; a plotter driver, which is used to produce

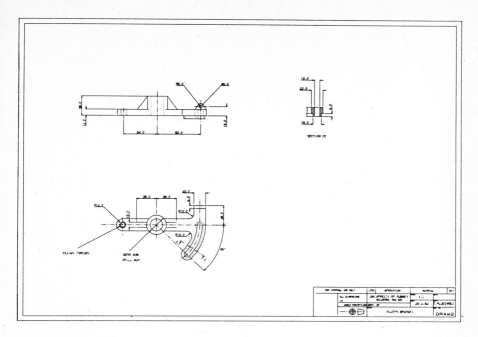

Figure 1 *An illustration of the use of the system as a two-dimensional drawing system*

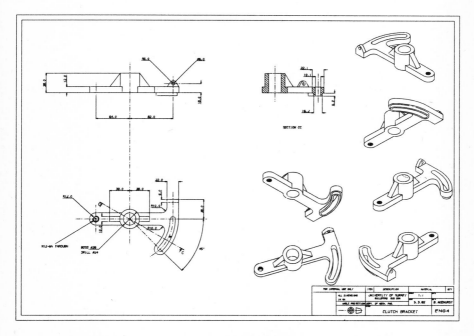

Figure 2 *This diagram illustrates the same problem modelled using the systems three-dimensional capability*

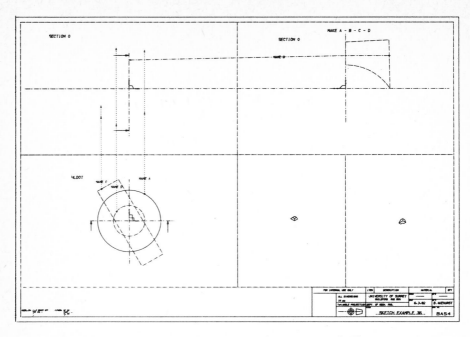

Figure 3 *Layout of information prior to the result shown in Figure 4*

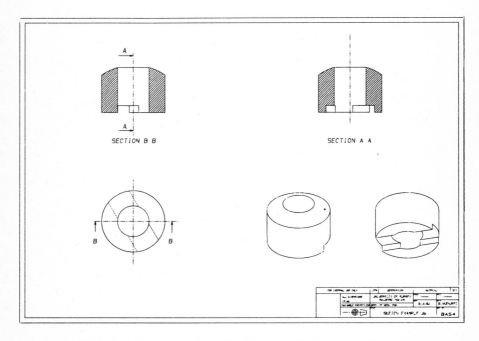

Figure 4 *Output from Figure 3. All views excluding cross-hatching*

finished drawings, and the scheduler, which is used to produce parts lists via the normal computer file output. The final module allows an interactive viewing routine to rotate the model in all directions without altering the drawing set-up.

Work stations

The package is accessed via four interactive graphic work stations, as shown in Figure 5, the sequence of operations required to produce a detailed drawing is given below.

5.1 The student inputs 2D drawings/data via graphics visual display unit, monitor and menu tablet.

5.2 Modeller constructs a 3D model.

5.3 The interpeter generates 2D details, sections, etc from the model. These are then combined with the original drawing.

5.4 The student annotates and dimensions the drawings and adds any details not drawn in initially.

5.5 The plotter driver generates detailed drawings on the drum plotter. (Varying line thickness is achieved by pen selection.)

5.6 The scheduler produces parts lists.

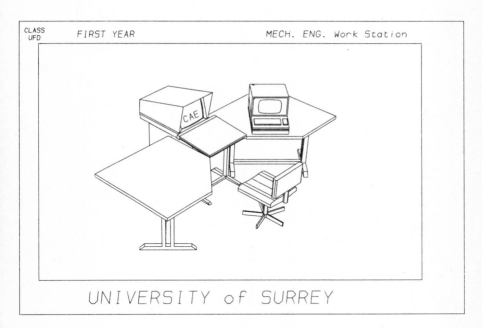

Figure 5 *A typical work station*

Figure 6 shows the interaction of modules, such as NC, which may be interfaced with the installed package. Such software could be purchased or more limited versions designed as final year student or research projects. An example could be in the field of control theory for pneumatic systems to show both graphic and simulated dynamic responses.

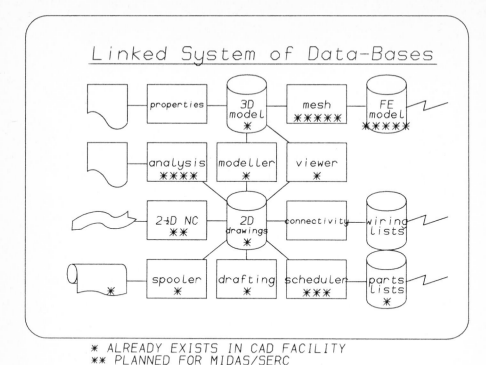

Figure 6 *Interaction of the software with other CAE modules*

Course structure in CAD

Outlined below is the course structure for computing and computer-aided design. This programme is based upon experience gained through 1981-2 using the purchased software and the period 1978-82 using a variety of CAE equipment and software.

First-year computing tuition will be given concentrating on introducing programming and use of the computer for solutions to engineering problems. CAD will be introduced into the design module by a general CAE lecture followed by demonstrations which will be consolidated by at least two individual drawing assignment giving a total 'hands-on' time for each student of not less than 6 hours. Also during the 'Design and Make' project some of the detail or assembly drawings will be produced on the CAD system, this should average about ten hours per group.

It is anticipated that the present system of drawing tuition on the board will not be superseded but will run in conjunction with CAD drafting.

The use of the system continues through the second year as a drafting package alongside further use of the computer for problem solving involving calculation, selection, matching and optimization. Emphasis here is placed upon experience in the specification and preparation of small elementary packages and making use of error traps and messages. In future, as more design packages become available, the use of these will be encouraged within the design assignments.

Final year CAD work should remain initially within the field of project work and the design assignment, although it may be fruitful in due course to consider a CADCAM option. Student projects in this area will continue to be formulated to establish an increasing number of mutually compatible programs.

This development would be likely to exploit the existing PETS and Apples and other mini/micro systems as well as the HP 9845 and the Prime 750 facilities. This would occur through individual research and contracts with industry or via the MIDAS Teaching Company programme. In this way developments in industry can be studied and the department will be able to continue its progress in this area.

After trial runs on a selection of first year and final year students, during 1981-2, the implementation of these plans will commence with all of the first year (September 1982) mechanical engineers using the CAE system. The use of this package will be both an aid to the teaching of the principles of projection and an introduction to computer-aided drafting.

The influence of this early introduction will continue throughout the design courses for all students and particular use of the equipment will be made by second year mechanical students which will prepare them for contact with and use of similar equipment during their industrial year.

Final year mechanical engineering students will utilize the equipment during the design core for CADCAM activities and for other types of analysis which require graphical displays. Projects are offered regularly in this area to final year students and it is anticipated that these will continue to be linked with industry, directly or via the existing MIDAS infrastructure.

The current position

Developments to date within the department of mechanical engineering in CAE have concentrated on CAD. These activities have fallen into three categories, undergraduate-taught courses, undergraduate final year projects and the use of final year projects to develop parts of the taught course.

Developments within the taught course have arisen from the combination of design teaching with computing tuition with examples such as shaft-keyway design for a given duty. Appendix A shows the abridged specification given to students for such an exercise and indicates how selection from stock data, British Standards, etc, forms an important part of the capabilities of CAD.

Final year projects have been used to develop experience of various small systems, such as PETs, in conjunction with local industry interested in low-cost systems. This exercise has been conducted in association with a Teaching Company (Vestec/University of Surrey) and a second more ambitious Teaching Company Programme (MIDAS/University of Surrey) has been commenced to develop this further. The types of project undertaken so far have been set so that the design process is integrated within the work of the company at quotation and manufacturing stages. The company concerned manufactures thin-walled pressure vessels and the projects have developed programs to specify and detail various types of vessel ducting the developed shapes thereof and the assembly of standard components. Another independent project has been undertaken, also using a PET system, to facilitate the drafting of turned components with a non-graphical input, this project has established the maximum size of drawing program that can be loaded on a PET system.

In order to develop the graphical CAD/CAE teaching on the main university computers, various projects have been undertaken using standard packages such as GINOF, GINO-SURF, DESPAC, Prime stress, etc. Whilst these were useful and informative the implementation of the new software has pushed them into the background of the author's thinking. The final year project students were in an ideal position to assist us in the development of our ideas for the use of the new system and have benefited considerably from the experience. With the arrival of the package it was necessary

7.1   to determine how undergraduates could learn to use it most effectively.
7.2   to develop a program of drawing exercises.
7.3   to develop a library of standard parts (see Figures 7, 8 and 9).
7.4   to establish ways in which the modelling capabilities could assist with the perennial problem of the small number of students who take an excessively long time to come to terms with visualization.

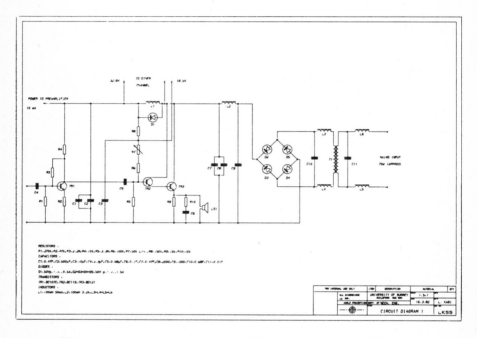

Figure 7 *Drawing using student-generated library of symbols base on BS 33939*

Project students have now developed and tested manuals on the new system using a restricted form of menu, to speed the initial learning process. Drawing exercises have been devised using this restricted menu and a library of standard parts (nuts, bolts, washers, electronic symbols to BSS) is in existence. Typical drawing examples using this library of parts are shown on Figures 7, 8 and 9. Exploring the modelling capabilities of the system has been most useful and has produced a set of computer models of the various preliminary visualization exercises used in drawing tuition which may now be used by a tutor to view the objects from any viewpoint to assist the student with difficulties in this area.

The entire computing/CAD/CAE facilities will be used for a variety of purposes, indeed it is only on this precept can the purchase of the equipment and software be undertaken and future purchases will be dependant upon a demonstration of the viability of this precept. Undergraduate tuition in CAD/CAE will form a substantial proportion of the usage. Use by researchers within the department for research purposes and as a design tool should lead through to CAE developments and the purchase and installation of an CNC machine has been implemented to extent the CAM link. The second teaching company programme should also lead to developments in the use of the

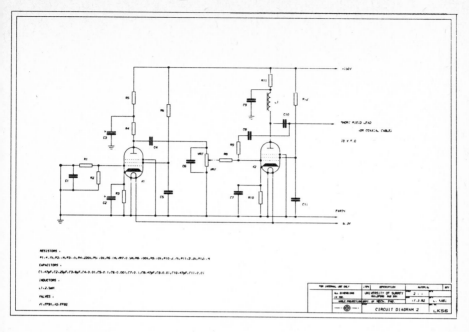

Figure 8 *Symbols are based on a module of 3mm and are 'locked' on a definable grid*

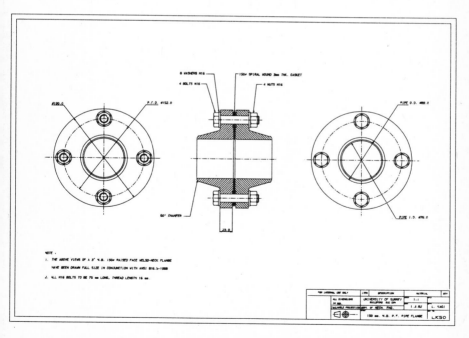

Figure 9 *Example of a typical sub-assembly using the macros (produced by students) of bolts, nuts, washers etc, giving precise sizes as per BS, with the range M3 to M36*

system for CAM and it is hoped that the interface to analysis will also be developed in this way. All of these developments should be of considerable benefit to the undergraduate course both in providing the stimulus of final year projects and building experience for inclusion in the appropriate parts of the taught course.

Conclusion

The authors' experience to date leads them to believe that a commitment to a substantial amount of introductory CAD/CAM/CAE experience is becoming essential to an undergraduate course in mechanical engineering. It is essential that this type of activity is accompanied by research and development work and a Teaching Company Programme seems an ideal way of integrating all of these activities.

Acknowledgements

The authors wish to acknowledge the support given to them by the department of mechanical engineering and its head Professor I M Allison and the computer unit at the University and in particular Mrs V A Harmer who have given their most valuable support to the work.

To the Teaching Company Programme Directorate and especially to Mr S Gent for his encouragement and enthusiasm during the teething period of the MIDAS scheme.

The authors are also indebted to their students in evaluating the proposed tasks and especially to their final year project students.

D Pollard, Department of Mechanical Engineering, University of Surrey, Guildford, Surrey

C Douthwaite, Department of Mechanical Engineering, University of Surrey, Guildford, Surrey

Appendix A: Computer-aided design – solid shafts and keyways

Many design solutions employ the use of electronic calculators to obtain from standard or derived equations a theoretical answer. This has then to be modified to take into account stock and catalogue sizes or indeed material and/or component availability and forms a small part of the process called CAD.

This problem requires the production of computer program reading data (stock sizes, keyway dimensions etc) from external files that could be used in the design office.

The program should be limited to the data given plus the stock lists already available from the assignment *Reading from files*. In a commercial package the full stock list, tolerance zones, preferred lengths would have to be accommodated, likewise different key types.

The program in BASIC should include the following points:

1. The program should be as foolproof as possible.
2. Remarks statements should be included at relevant points.
3. The designer should be able to select the following:
     3.1   Any specific material type.
     3.2   A tabulated output of all four materials.
     3.3   A lowest cost selection, print-out of only *one* material.
     3.4   A print-out of material stock (from previous assignment).
     3.5   A shaft without a keyway.
4. Inputs should be power (kW), speed (rpm) and design factors.
5. The output should include *all* the relevant data necessary for design.
6. Your final report should include:

6.1   Project specification.
6.2   Summary of your program as a CAD tool.
6.3   Computer listing.
6.4   Flowchart of program.
6.5   Como output of a typical RUN use sheet A1 to validate your program.
6.6   Theory and equations used, show with cross-reference to listing line numbers.
6.7   Report/theory/conclusions/on the project.
6.8   Please leave in your own UFD a compiled program called CADSHAFT.

Abridged assignment sheet, full text and example program available on request from authors.

Part 4:
Training – curriculum and training equipment

12. The development of micro-based procedure for the teaching of three-dimensional geometric design

A J Medland and S J Crouch

Abstract: Whilst many people appreciate that CADCAM is an integrated technology covering the whole of design and manufacture since systems development is still in its infancy, many of the industrial installations use only one of the many components of a system. It is however important that at the educational level the concept of CADCAM is taught rather than the detail of industrial application thereby differentiating between education and training. This paper details the background to the development of the microcomputer-based CADCAM education system relevant to mechanical engineering.

Introduction

The implementation of computer-aided design systems within the mechanical manufacturing industry seems to have been directed toward the replacement or enhancement of existing design functions, rather than by the bolder approach of building new design processes which rely upon computer graphics for their success. Whilst there are many interrelated reasons for choosing the former approach, in the main it is seen to contain the minimum of risks for both purchaser and supplier. A replacement exercise provides a simple 'test' of effectiveness as some arbitrary parameter, such as the number of drawings produced with the system, which can be compared to that measured before installation. Here the objective is seen only in terms of the limited activity of a single operation or department. A better test of its effectiveness, in terms of the company profitability, may conversely be to establish how many drawings it in fact saves.

CAD has thus spread within the industry as a patchwork of developing techniques rather than as a single design tool. Large and sophisticated systems are seen to exist where complicated analysis techniques have become necessary. The procedures involved during the design and manufacture of an aircraft have resulted in a need for the solution of complex equations related to both the aerodynamics and strengths of structures. Where conflicting requirements of style, manufacturing technique and technical function need to be explored CAD has been readily accepted. The motor industry was thus quick to see the potential of this analytical approach in design of body shells.

Many other users have approached CAD as a basic drafting aid. Here the system is seen to simply be a graphic generator which can be tailored to produce drawings to any given standard. Whilst this approach may have been due to a conscious need for the company to rationalize its drawing operations, it is often seen to have been stimulated initially by the company's inability to recruit a desired number of draftsmen.

The third method of entry into CAD is seen to have arisen from a need to provide data for the company's numerically controlled production machines. Here computer graphics provides an easier means of generating and visually checking the necessary machine instructions.

None of these approaches (Figure 1) is seen to be sufficiently fundamental to give a good teaching base in the subject of CAD for mechanical engineers. It was thus necessary to develop a computer-based procedure which could provide that central understanding and would naturally lead into all these areas of application.

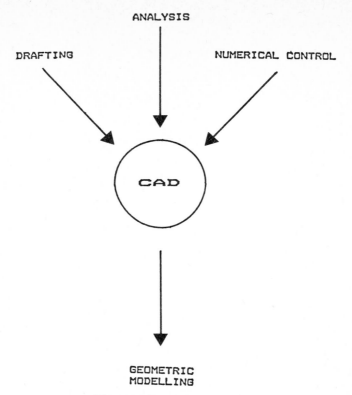

Figure 1 *Approaches to CAD*

Teaching needs

In order to provide a course of instruction which ranged from these fundamental concepts through to specific areas of application it was apparent that a number of CAD systems would be required. Each would be aimed at a different range of skills and be provided in differing numbers to satisfy the projected student usage.

The department (running the Special Engineering Program) had already been provided with a two-station ComputerVision turnkey system through the SERC Teaching Company Scheme. This, through work within the Scheme, provides the focus on real engineering problems and is suitable for demonstrating the application of CAD in areas from electrical layout to NC preparation. The current research programme on this machine also gives an insight into the approaches necessary to turn such systems from simple drafting tools into real design facilities. Here various design programs can be demonstrated including the relationships between components within a machine space and the kinematics analysis of mechanisms during the designing phase (Crouch, 1982); (Medland, 1982); (Medland, 1982).

Owing to the dedicated nature of this machine to the particular needs of the industrial user, it is not configured in the best way for the teaching of computer graphics fundamentals. The need for extensive operator training in the graphics language and operating system, together with the limited work stations, also makes it impractical to employ such a system as a general teaching facility.

It was for these reasons that time and effort was put into developing a simple three-dimensional wire frame modeller for the Apple microcomputer. This suite of programs has been written to allow instructions and geometric constructions to be performed at the tablet in a manner compatible with normal engineering drawing practice. The simple models can be constructed, rotated and operated upon in order to investigate many of the classical geometric constructions. Wireframe models can be constructed (from points, lines and circles) and used to check similar models constructed upon the drawing board. These programs are now an integral part of the year 1 design course in which students are taught to handle three-dimensional space by applying the three techniques of co-ordinate geometry, engineering drawing and computer graphics (Medland).

With the Apples providing the basic teaching role and the ComputerVision the industrial applications, there was seen to be a need to provide a micro-based system which bridged the two (Figure 2).

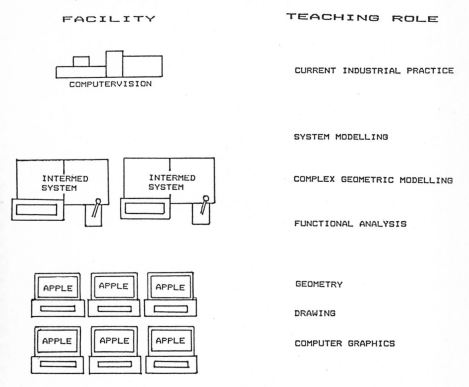

Figure 2 *SEP design facility*

This machine must build upon the geometric model developed in the primary teaching system but contain the ability to be extended to simulate many aspects of the operation of turnkey systems. It must also retain the flexibility to allow new approaches to geometric problems to be investigated and displayed. The following sections describe the development work that is currently underway to test many new ideas for such a system, together with the projected system configuration.

Intermediate graphics system

A number of fundamental decisions were taken which influenced the design of this new graphics system. These were as follows:

Language and program format

The need for ease of teaching and flexible expansion of the system has resulted in the choice of a structured high-level language such as Pascal. Upon successful completion of the various tests on the sub-system the programs will be assembled using an agreed format in Pascal. Here dynamic variables will be employed for the holding and manipulation of data-blocks. Where possible, changes of entities and view axes will be evoked through a standard set of routines.

Structured instruction handling

In many commercial CAD systems no distinctions are made between graphical input mode, viewing instructions and data manipulation and storage. In order to provide the student with the ability to explore the model space, view manipulating instructions can be activated at any time to override any other activity currently in progress. Thus, for example, a line command can be initiated in one view and the start point set. The view can then be changed and the line terminator inserted into that new view. The normal mode of operation is thus in the data entry state, with viewing instructions lying at a higher level and with data handling lying at yet another but lower level (Figure 3).

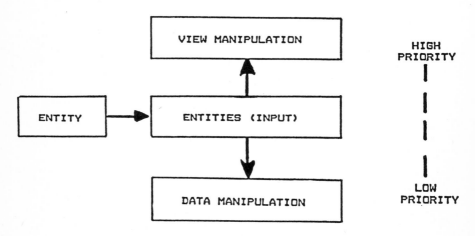

Figure 3 *Structured instruction handling*

During the teaching of data insertion and manipulation it is an advantage to be able to regroup the entities gathered in the data-base. Only entities of a chosen type, associated with a particular feature, part or location, or classified into a definable group will then be displayed or changed. Features can then be selectively displayed or highlighted to provide a clearer interpretation of the activity taking place.

Graphics display and 3-dimensional input

Figure 4 shows the system features.

The graphics display and data input devices must be interrelated via the software to allow the data to be viewed during insertion. The system is thus equipped with a high

Figure 4 *(see page 9)*

resolution screen capable of displaying up to eight grey levels. The shades of grey can be used to provide depth-cueing and the highlighting of features during insertion etc.

Entry of data will be made via a four degree-of-freedom joystick device (the Digiquad). The first two normal freedoms provide the horizontal and vertical positioning of the cursor, whilst the others provide the depth of entry and orientation, if necessary. This joystick control will also be employed in a menu mode to provide a flexible method of data and instruction entry.

A further peripheral that will be used on some computers is the CA3D Design System (Computer Aided 3-Dimensional Design System) (Figure 5). This is a three-

Figure 5 *(see page 9)*

dimensional input device with a programmable force feedback to provide the operator with feel, and with the ability to 'sculpt' shapes held in the associated computer memory.

Graphics entry

Techniques are at present under development to allow many different modes of implicit data entry. Whilst the data entry surface on most CAD systems is set in the plane of the viewing screen, this can in the development system be translated, rotated and scaled as though this surface were a model in its own right. Rays are then projected down from the viewing plane to strike and produce a series of points upon this entry surface.

The entry surface can be constructed as plane perpendicular to the view direction (with the offset of the surface being specified or obtained from an auxiliary view). The dimensions of the surface can be varied during construction to produce both irregular surfaces and non-linear grids (Figure 6). Drawing can thus be performed in two simple

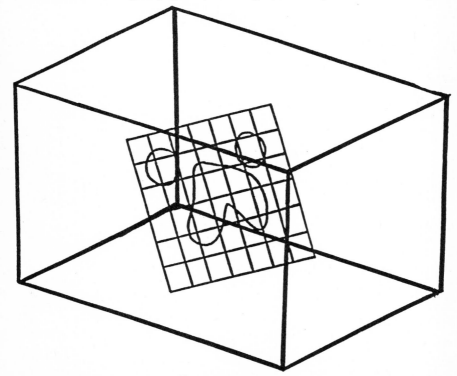

Figure 6 *Construction grid*

modes (as well as the general oblique case). In one the entry surface is parallel with the screen so that lines are inserted by indicating the two end points upon the screen (as the common CAD entry mode). The other occurs when the entry surface is set at right angles to the view screen such that lines inserted as single points produce rays which are projected across the full extent of the entry plane.

The entry surface can be wrapped on existing surfaces (Figure 7), unwrapped for

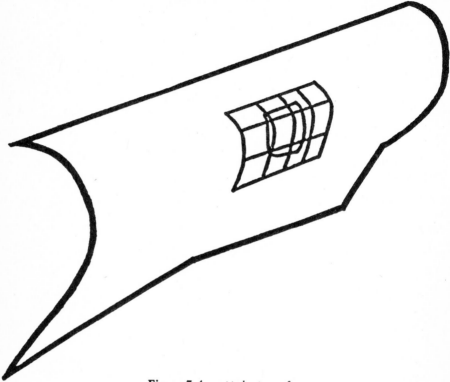

Figure 7 *A wrapped entry surface*

detail construction or surface design, and then wrapped back, complete with the new information.

The hardware

To the user, the intermediate system appears to be a dual screen device with keyboard, and a joystick type of input (Figure 8). At a later date, audio warnings and messages will be provided by a speech generation sub-system.

The first screen provides dynamic menuing, 'help' and warning information, and, in conjunction with the keyboard, facilities for explicit numeric and text input. All system commands for storage, retrieval, plotting, etc are echoed and reported on through this screen; additionally it has a high resolution graphics capability so that it can be used to provide an additional view of the model (to clarify a feature by providing simultaneous viewing from differing view points).

The second screen provides a high resolution graphics display, with subsidiary text capability. This is a monochrome display, but does have eight intensity levels to provide the capabilities of showing shaded shapes and depth cueing.

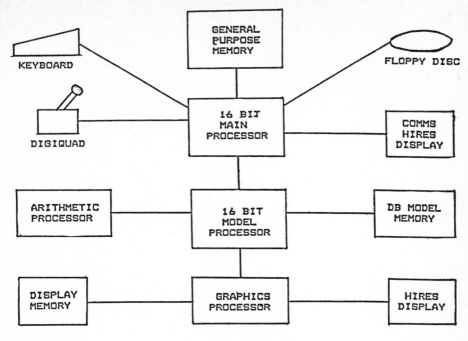

Figure 8 *Hardware*

Associated with this display is a 16-bit processor and memory for view manipulation and entity association.

The software

The software is divided between the processors (Figure 9). Programs in the main processor can be overlayed from the floppy discs.

The file structure which students will be encouraged to use is of particular interest (Figure 10), (Medland, 1982). The structure is divided into four files with specific uses – work space for trial designs and constructions; the spacial file with position and geometric information; the engineering file containing manufacturing information; and the technical file with the designer's references, calculations, equations and graphs (see Figure 9.)

Figure 9 *Software*

Teaching procedures

This intermediate CAD system is designed to provide the following teaching facilities.

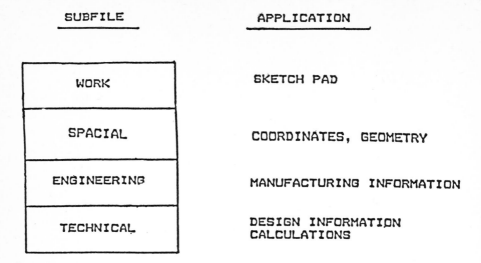

Figure 10 *File structure*

Exploration of three-dimensional space

The students will be instructed in the techniques of inserting data directly into a three-dimensional space. The normal two-dimensional engineering drawing procedure can thus be shown to be a sub-set of this fundamental space. Other sub-sets can be explored (which will include perspective constructions).

Both the Digiquad and the CA3D Design System will allow data to be 'pushed' and 'pulled' through the entry plane. A three-dimensional grid can be employed to force values to 'snap' to adjacent grid points. These features will allow not only wireframe models to be constructed directly into the space but also provides the basic construction techniques necessary for the generation of auxiliary view, sections and intersection of surfaces.

Pictorial representations

As a piece part is designed, the system will calculate and store silhouette information and a reference node for the part giving its orientation and position in the assembly (Medland, 1982). The student can indicate a portion of the assembly he is working on, and the system will search through the stored design information, extracting a list of those piece parts whose silhouettes overlap the portion indicated (Figures 11 and 12). Thus it is relatively straightforward to determine which parts are significant, and which need not be displayed.

By regrouping the entities into a hierarchical format it will be possible to display individual groups of nodes selectively. This will allow the students to compare various approaches to improving the display by including the right visual clues. Remote or hidden detail can be removed or downgraded in intensity in order to generate an impression of the solid. Shading or block infilling can be imposed upon (or between) groups defined as a surface. Here the addition and removal of graphical features can be explored in order to achieve the right visual effect.

Manipulation of engineering details

Here the groups and sub-groups of entities can be used and assembled to describe particular engineering shapes and features. Holes and other metal removing activities

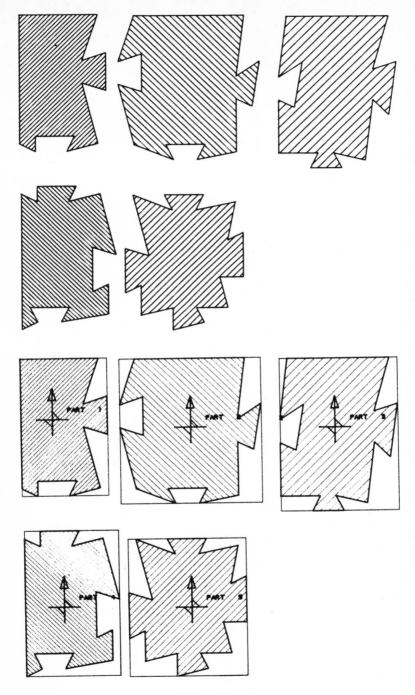

Figure 11

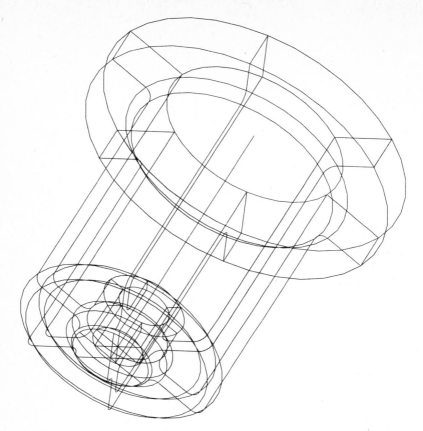

Figure 12

can be expressed as negative solids in the final state. These can then be interpreted as swept solids generated by each metal cutting pass or step. The production engineering operations and, in particular, the numerical control aspects can be fully integrated into the teaching of computer graphics.

These procedures will also develop an understanding of those features which are essential to the operation and manufacture of the component parts. The generated engineering drawings for the part are thus extracted, as sub-sets of the original three-dimensional model, in order to illustrate all of these features. The drawings (including views and details) then become a means of communicating the engineering needs rather than a picture-generating exercise.

Functional representations

Research and development activity currently under way within the department into ways of representing the functional properties of a machine will also be included. Here the kinematic functions representing the movement of a part are also held graphically within the part drawing file (Figure 13) in the specified sub-file. The graphs so produced are generated and handled in the normal way as they have exactly the same features as any other model in the data-base. The functions are thus held simply as their wireframe representations.

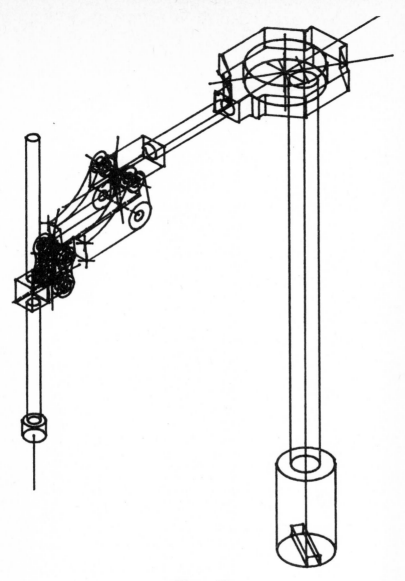

Figure 13

Conclusion

The intermediate CAD system is now being assembled. Many features, as illustrated, have been discussed, tested, developed and prepared for assembly into the main system. Much development work however still remains.

The parallel development of the teaching procedures is also continuing in order that both may be available at the earliest possible time. These teaching facilities will however be built for extension beyond the normal range of a mechanical engineering undergraduate course. Interest has already been expressed in the use of these systems for training and development work within industry.

It is planned to provide courses and systems tailored to the specific needs of an industry. Menus and programs will be developed to simulate many different industrial CAD installations. Full training and the development of specific skills can then take place without the restrictions imposed by the main commercial installation.

References

Crouch, S J (29 April 1982) A university installation – The first year. CV European Users conference: Monaco

Medland, A J (1 April 1982) The development of a CAD system based upon the mechanical engineering design process. CAD 82; Brighton

Medland, A J The development of a suite of programmes for the analysis of mechanisms. *Int. j. of digital systems for ind. automation*

Medland, A J A spatial approach to engineering drawing. *Int. j. of mech eng. education*

A J Medland, Brunel University and Department of Engineering and Management Systems, Uxbridge, Middlesex UB8 3PH

S J Crouch, Department of Engineering and Management Systems, Uxbridge, Middlesex UB8 3PH

Discussion

P Cooley, University of Aston:

Last year at the University of Connecticut, I saw PLATO in operation both in the department of mechanical engineering and the department of chemical engineering. For reasons which I was unable to discern, it was more successful with chemical engineering students. Have you, in your experience with computer-aided learning generally, come across subject areas which are more applicable for this technique?

R Martin, Control Data Corporation:

The tendency with computer-based education and computer systems learning, is to use it for technical subjects. Naturally enough, those subjects which potentially have a high graphical input in the learning process are those which are most appropriate. I would expect certainly the area of mechanical engineering, and maybe electrical engineering, to be highly appropriate. That chemical engineering would seem to be more successful I am most surprised. In the experience of my company, the converse would be true. I would suggest that it reflects personalities more than anything else, because the subject matter and the motivational aspects of the learning material itself depend entirely on the author, so that if you have had a bad night the night before giving a lecture you tend not to give a very good lecture the following day. So it is with computer-assisted learning. If you are not well versed in the skills required to produce motivated interactive material, the students will not succeed as well as someone who is better motivated himself and better skilled in that area.

H C Ward, Teesside Polytechnic:

Mr Allen has pointed out quite emphatically that to use CADCAM you have to have the skills and years of experience in design and drafting. Mr Martin has said that in future these skills are unnecessary to use CADCAM. Would somebody like to comment on this?

R Martin:

I thought I had emphasized the fact that the operation of a CADCAM system is an unskilled operation. However there is an element of de-skilling in it. I would suggest there is a great creativity which can provide a multiplying factor, but the more mundane aspects of drawing office work can be potentially done by a person with a lower grade of skills. I am not saying in any way that it is not a skilled job: the skills are changing, the requirements are changing. You have to be a highly skilled draftsman in order to interact effectively, but it is also true that you need a blend of skills to appreciate what computers can do for you. You need to pull the two together. I am not saying that skills will disappear, far from it.

E A Warman, CADCAM Publications:

I think it would be dangerous to suggest de-skilling, because some people would say that the alterations of drawings now become a trivial task. The very act of altering the design could cause severe consequences to a product and needs the development of a new technique. I think that there are new skills required with the system.

C W Allen, British Aerospace:

There really is not a basic disagreement here. It is the way that we in industry apply our tools of the trade, which include designers and draftsmen. Whenever I talk to my apprentices, one of the things I tell them is that I am not in the business of employing draftsmen for the future. That is without doubt a dying trade. With the advent of computer aids, we do not need the skill of somebody who is limited to just drafting. The benefit will be in using designers for the one thing they have – creativity. Wearing my Institution of Designers hat, we are fighting for designers to be recognized as people who can assist industry today with their creative skills. It is those skills which we need to develop if computer aids are to help us. By building up those data-banks we will get an upgrading of the skills.

A Guy, Hatfield Polytechnic:

Mr Allen, if a draftsman breaks a pencil, he can pick up another one, but if a machine crashes then he has several work stations inactive. Could you comment please on the reliability of your system, and the security of the data?

C W Allen:

The problems that we have with reliability are probably the most important as far as a manager is concerned. Systems today are usually very reliable. Most systems are not on the market unless they have about 98 per cent reliability overall.

It is important of course that you have an extremely good maintenance back-up, and preventive maintenance is as important as being able to pick up the telephone and get a service engineer there when you have a problem. If you have a good company, they will make sure that you don't get into that situation. Obviously, we have our problems. Luckily we now have three systems. I am not advertising anything, but I believe that there is a limit to how many work stations you can hang on one independent senior unit. This is not necessarily saying that I am an advocate of mini-systems or main frames, or anything else, but I would prefer to have a number of CPs so that if something catastrophic happens I will not be completely out of business, because it can get extremely expensive.

There are problems with security. If you are not picking the right sort of person or the right sort of codes, it is obvious that people can get into your data-base. I think it is no more difficult to control this than it is to control the staff that you employ in the normal way. We have to put in security bars to stop people from playing around with data, so that it cannot be changed by anybody. All these security systems must be looked into by the system managers, and a system will be capable of holding things back. I think you have got to worry more about somebody who is making a genuine mistake than about industrial espionage or national security.

D Smith, Davy Computing:

I am very worried about the way computer-based training is spoken about. We hear about teacher-centred instruction and learner-centred instruction and I see computer-based training moving towards machine-centred instruction. We have been hearing about computers being tooled and I would suggest that it is simply a tool for the teacher to use to gain motivation.

R Martin:

I would entirely agree. To be properly interactive, the particular medium of computer-assisted instruction relies on the author, and whilst computer-assisted learning can provide reduced instructor dependence, in no way could I currently foresee it being totally independent of instructors in the long term. Perhaps with the various short-term sequences of learning one can envisage the total independence from an instructor, but for more than a few hours of instruction, an integrated course would demand an instructor as well. I cannot see computer-based learning displacing the instructor.

13. A comprehensive approach for CAD ED curriculum elaboration*

T David

Abstract: Education in CAD becomes very important as interest in computer-aided design increases in more and more varied fields of application. The problem is especially difficult as both professionals and beginning students must be trained. Existing training programs are too limited in scope and number. Therefore new programs, both public and private, are being set up. Their aims and contents are extremely varied.

In our presentation we shall explain a comprehensive approach for the elaboration of a CAD education curriculum. Essentially, we should like to answer the following questions concerning CAD education:

– for whom? (different types and levels of education)
– what teaching methods? (independent or integrated education)
– how to elaborate a particular curriculum?

The ideas expressed here were put into practice during the elaboration of different CAD education curricula.

Introduction

CAD now seems to be achieving maturity. In a growing number of engineering and related fields the possession of, and the ability to use comprehensive CAD systems is a prerequisite for competitiveness and successful operation. Many organizations need to be able to design and develop CAD systems in order to maintain a position of technological leadership. CAD has become a prime element in the most basic of engineering processes, the process of design.

Although there are industries and organizations in which CAD is currently used, there are a vast number of organizations, of all sizes and in many different industries, which have so far remained untouched by CAD. However, the data-processing revolution which is occurring means that technology is now available to virtually all sectors of industry. Hardware costs are decreasing, and options are open for both general purpose or single-application oriented systems. Engineering experience is now being cast into an information-processing system that for the first time provides the potential to codify methodological principles and to apply them in an integrated form to meet the increasingly complex requirements.

The educational implications of such fundamental changes are profound. This systematization must be well conceived and properly implemented, especially as this education must allow the training of both company management and existing technical personnel, as well as the younger generation who will receive training during their pre-professional education. Figure 1 presented by Llewelyn (Llewelyn, 1978) shows the evolution of the demand for training.

*This paper was initially presented at the 3rd World Conference on Computers in Education (Lausanne, July 1981) under the title *Computer-Aided Design Education: An Overview*, and published by North-Holland Publishing Company in the proceedings edited by R Levis and E G Tagg.

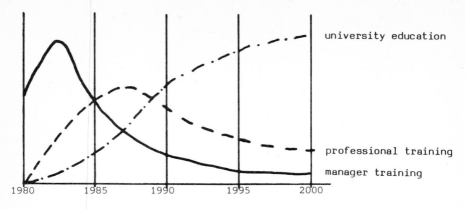

Figure 1 *CAD training and education*

However, very little has been done until recently. Current training programs are few and limited to several privileged fields. Recent interest in CAD has inspired little reaction from most of the classic training organizations (universities, professional schools), but has resulted in intense activity on the part of a number of small companies interested in the potential of CAD. Often, though, this interest tends to be also publicity for the firm's products. The content of this type of training is often too limited and does not fulfil its goals.

It seems therefore important to try to define as objectively as possible what CAD education should be, who should be trained, what should be taught and how. This is what we would like to present in the following paragraphs.

CAD education for whom?

What kind of people, with what skills and basic training, do we need for CAD? This is the first and fundamental question we should examine before drawing up any programs for CAD education.

We have distinguished four types of people concerned with CAD:

- The company *managers* capable of evaluating the impact of CAD and of making the right decisions as to the introduction of CAD in their firm.
- The CAD *users* in a specific field: they use CAD tools in their work to solve their specific problems.
- The CAD *experts* in a given field; they not only use existing tools, but also conceive new CAD tools for use in their field of application.
- The *builders* of CAD systems; their role is to resolve all the underlying computer problems and to furnish a solid framework for the implementation of an application.

Obviously, the educational program is different for each type. In order to define what should be taught, we must first know what CAD is, or more specifically, what is the basis of CAD.

As Figure 2 shows, we feel that CAD is based on three fundamental fields:

- computer science
- design methodology
- the field of application

Before discussing the application of CAD in a given field, we can note two interesting intermediate phases: on the one hand, the use of computer science in design (CD), and on the other hand, design methodology in the given field (DA).

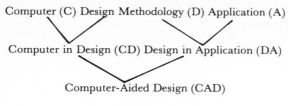

Figure 2 *The basis of CAD*

This approach which can be visualized as a hierarchy allows us to place methods, techniques and tools in this diagram according to their degrees of specificity or generality. This information will be used in the subsequent elaboration of a curriculum.

If we try to schematize the relative importance of these three fields (computer science, design methodology and application), for the different CAD people, we obtain the following results (Figure 3): for managers a working knowledge of all these three fields is necessary. For CAD users knowledge in the application field is more important than computer and design methodology aspects. For CAD experts in-depth knowledge in the application field and in design methodology is required. For CAD system builders computer knowledge is fundamental, design methodology is important and as for application knowledge, experience in different application fields is useful.

	Computer	Design	Application
Managers	working	working	working
CAD Users	limited	limited	working
CAD Experts	working	important	important
CAD Builders	fundamental	important	superficial of different applications

Depth of knowledge

limited	working	important	fundamental

Figure 3 *Depth of types of knowledge for different CAD people*

However, curriculum design must not be limited to the technical aspect of CAD. Considering that computer-aided design is the problem, both human and social aspects are fundamental. At the same time the economic aspects are also important as most often CAD is used in industry. Finally, before choosing a curriculum, all three aspects – technical, economic and social – must be well-defined.

The relative importance of each of these aspects for the four different types of CAD people is shown in the following table (Figure 4).

For the managers: social and economic considerations are certainly dominant with less importance being given to the technical aspects. On the other hand, CAD users must possess operational techniques with a certain comprehension as to the economic consequences of their work in its social context. For the CAD experts high level technical knowledge must be completed by a general view of the social and economic context. The CAD system builders are above all top-level technicians who understand the economic and social consequences of their decisions.

	Technical	Social	Economic
Managers	limited	important	critical
CAD Users	fair	limited	limited
CAD Experts	important	fair	important
CAD Builders	critical	important	important

Relative importance
_____

| limited | fair | important | critical |

Figure 4 *Relative importance of technical, social and economic aspects for different CAD people*

Educational objectives

After this analysis we can define the objectives of the education necessary to train these different CAD people.

- For the first type, the manager, the objective should be to provide the information necessary to permit him to answer his most fundamental question: should CAD be introduced or not into my firm?
- For the second type, the CAD user, the objective is to inform him of the existence of the CAD approach to design in his field and teach him the basic knowledge necessary to enable him to use an existing CAD system.
- For the third type, the CAD expert, the objective is to permit him to define the external specifications of a CAD system and to add new tools to an existing system. He must be able to justify economic and social consequences of his propositions. Knowledge of computer science can be fairly limited as the job requires only external definitions of small improvements within a determined framework.
- For the fourth type, the CAD system builder, the objective is to form a computer specialist oriented toward applications, and more specifically toward the subject CAD. He learns to classify design problems, different design methodologies and the role of the computer in the design process. He must be able to formulate the specifications of CAD systems and the methods for developing these systems, to understand both the general systems approach to CAD and the computer-aided construction technique for CAD systems implementation.

The comprehensive approach to the elaboration of a CAD curriculum

As we have just seen, we need to train in CAD a large number of people who had different backgrounds, who work in varied application fields and who have different functions. Therefore it seems important to organize the design of a CAD curriculum. First of all, an effort should be made to arrive at a common definition of the contents of this curriculum, so that for the same kind of training in different application fields, similar instruction is given and teaching documents can be used profitably.

When necessary more specialized training can be built upon this common foundation.

To achieve this goal, the knowledge necessary for all the application fields as well as for the four types of CAD people must be defined.

But before doing this, it seems interesting to see in what measure this required knowledge is dependent on the field concerned. By superimposing the contents of the CAD education for different fields of application we can structure CAD knowledge in the three classes:

- *Common CAD Knowledge* which must be part of the teaching program in each field. This common part of CAD education is mostly concerned with the methodological part of design and the use of the computer in applications and particularly for design. This instruction constitutes the basis of the CAD education curriculum, which must be progressively adapted, first to the different classes of application and eventually to each application itself.
- *Group CAD Knowledge*, useful for a group of applications. In this case we can create common courses to teach this knowledge in the group of fields concerned.
- *Specific CAD Knowledge* in a given field. This subject matter should be taught in that part of the teaching program which pertains directly to the field and could even be integrated into existing courses.

It seems interesting to explain here the basic common CAD knowledge for each type of CAD person (manager, user, expert, builder).

- Managers:
    - computer systems
    - CAD introduction
    - economic and social consequences
    - elements for choice of CAD systems
    - case studies
- CAD users:
    - computer systems: an overview
    - basic computer structure and operation
    - review of CAD (economic, human and technic aspects)
    - rationalization of the design process
    - man-machine interaction in the use of graphics
    - use of packages and/or CAD turnkey systems
- CAD experts:
  in addition to the CAD users' requirements:
    - development and running of problems in a high level language
    - graphics
    - computer communications
    - data management
    - CAD systems and applications
    - social implications of CAD
- CAD system builders:
  in addition to the knowledge mentioned above
    - systems approach (life cycle: specifications, analysis, synthesis, implementation, appraisal)
    - design methodology (simulation, work organization, modelling . . .)
    - CAD system design methodology (hardware-software engineering, man-machine communication; problem-oriented languages, graphics, data structures and manipulations . . .)

Another fundamental problem in the design of curriculum is the choice of type of instruction. For CAD education, different teaching methods are needed, we can distinguish eight types:

- the *course*, the aim of which is to study in an analytical manner one subject often in a single discipline.
- the *interdisciplinary group*, in which, for a given project, the student puts into practice the knowledge acquired during the courses and he studies the different aspects (economic, sociological, technological, etc . . .) of the studied object, as

well as the design procedure used. The group discusses the impact, the potential and the defects of the methods and tools used, as well as the object obtained.
- *Supervised on-the-job training*, which must be on a sufficiently high professional level.
- the *research group*, in which the student studies a specific problem (theoretical or practical) in depth, to bring to light new solutions and perhaps even new tools.
- the *personal project*. Work similar to that of the research group but of a smaller dimension and more concrete, leading to a formal presentation before a jury.
- *Supervised homework*. Simple exercises based on the contents of the courses.
- *Laboratory experiments*. Practical exercises using the CAD system.
- *Case studies*. Presentation and perhaps practical experiments on a concrete example, often presented by a professional.

Therefore for each unit identified in the curriculum the most appropriate teaching method must be chosen. This choice of method also depend on the type of CAD person:

- for the managers: courses, supervised homework and case studies
- for the CAD users: courses, interdisciplinary group, supervised homework, laboratory experiments and case studies
- for the CAD experts: all the methods listed for the CAD users plus research groups and personal projects
- for the CAD systems builders: all the proposed methods are important for complete training of CAD systems builders

Organization of the instruction

Another fundamental question we must discuss concerns the curriculum organization in time. This may also be seen as question: should CAD education be integrated or autonomous?

We feel that in the case of beginning education, the CAD educational program must be integrated into another curriculum (either in the field of application or in the field of computer science).

For the complementary education of personnel an independent training program organized to provide the most information in the least time seems most appropriate, particularly for the managers and the CAD users. However for the CAD experts and the systems builders the independent training necessary may require a fairly long time.

To build a particular curriculum, we proceed in the following manner: first, we note the knowledge received by the students in the field into which we must integrate our curriculum or in their initial education for professional training (Figure 5.1). Next, we define the knowledge required by the CAD person we wish to train (Figure 5.2). The difference between the knowledge learned in the field of application and the knowledge necessary for the CAD person, constitutes the goal of the CAD education (Figure 5.3). If the difference between these two sets of knowledge is null a specific CAD education is not necessary, as predicted by J. Allan for the year 2,000, (Allan, 1978), because the required knowledge will have already been acquired. In this way we determine the contents of the CAD education and we can, thus, build the specific curriculum for CAD education.

For example in an architecture school, the CAD instruction must include an introduction to computer science as there is none in the general curriculum. On the other hand, in an electronics engineering school, the CAD instruction may be limited to those aspects which are specific to CAD as basic computer science is already a part of the general curriculum.

Specific CAD curriculum elaboration

From the results of this analysis we can in effect build a specific CAD education curriculum, ie the division into teaching units, the definition of their contents, and their position, for instance in a general university curriculum.

1 - appplication field knowledge produced

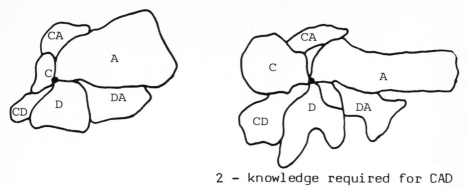

2 - knowledge required for CAD

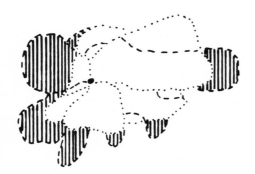

3 - contents of CAD Education

Figure 5 *Elaboration of CAD education curriculum*

Two factors are then important – on the one hand, the choice of teaching methods: a well balanced distribution of courses, practical experience, case studies and research is very important. On the other hand, the question of timing is also important. In the case of an integrated program, when should the CAD education be introduced into the general curriculum? It should be inserted after the basic knowledge in the field of integration has been acquired, but soon enough so that this new CAD knowledge can be used in the elaboration of the end-of-studies project. However, for independent professional training the length of the programs must be determined.

The result of this process is a curriculum which is specific, but is designed in a comprehensive manner and fits into the general context of CAD education.

In this way teaching materials (documents on general CAD knowledge, group CAD knowledge, etc) can be re-used and it is possible to adapt the instruction to the application field concerned for instance by studying examples from this field. In the same manner the specificity of these different types of training can be taken into account:

– for beginning general training (often integrated) a more comprehensive view, concerning varied applications and using a more theoretical approach appears most appropriate.

- for beginning specialized training, which is often independent, a general but more operational approach is recommended.
- for professional training sessions a practical, operational approach is indispensable.

We have applied this approach for the elaboration of CAD curricula in three different contexts:

- the first (David, Zanelli and Z'Graggen, 1979) was the definition of a CAD educational program in an architectural school for the training of both CAD users and CAD experts for CAAD (Computer-aided architectural design). This is an example of integrated CAD training.
- the second (l'ENSIMAG) is a proposal for the creation for a special CAD program at ENSIMAG (National School for Computer Science and Applied Mathematics of Grenoble). In this case engineers from different backgrounds are trained in CAD for a year to become CAD experts or even CAD system builders. This case is an example of an independent training program.
- the third (Micado, 1979) is the definition of the contents of professional training sessions organized by A F Micado for managers, CAD users and CAD experts.

Conclusion

In this paper we discussed the problem of CAD education. We emphasized the fact that different types of CAD people must be trained in different contexts: university or professional education, introductory or advanced training; integrated or independent educational programs.

We also attempted to determine the objectives of these different types of training and then their contents.

Finally we presented a comprehensive approach for elaborating CAD curricula. In this way we think we can organize a well balanced, adaptable program, building from a general view. This should aid the much demanded development of CAD.

At this point in the development of CAD education is it possible to determine who should learn CAD first? We feel the needs of the first three types of CAD people (manager, user and expert) are almost identical and equal importance. At a first glance the CAD systems builders appear to be well trained for their job. However, in reality the situation is less promising. They have had technical and often very specialized training (management, systems, networks), but they are rarely able to apply this knowledge to different fields or to synthesize the different computer specialities into a single system. Because of this fundamental lack in their training, it seems that the CAD systems builders must also be trained specifically for CAD.

References

Abbas, S A, Coultas, A and Lee, B S (ed) (1978) *CAD ED: International Conference on Computer-Aided Design Education*. IPC Science and Technology Press

Allan, J J (1978) *CAD education to the year 2000*. International Conference on Computer-Aided Design Education. IPC Science and Technology Press

David, B, Zanelli, F and Z'Graggen, F-J (1979) Teaching CAD in an architectural school with the SIGMA-ARCHI system. *Computer and Graphics* 4: 193-9

Encarnacao, J (ed) (1980) *Computer-Aided Design: modelling, systems engineering, CAD systems*. CREST advanced course; Springer-Verlag

Llewelyn, A I (1978) CAD study for the commission of the European communities

Micado, A F (12-15 November 1979) Seminaire sur la formation à la CAO; St Pierre de Chartreuse

Proposition pour la création d'une section speciale CAO/FAO à l'ENSIMAG

Purcell, P (September 1980) Computer education in architecture. CAD **12** 5: 239-51

T David, Laboratoire IMAG, B P 53, Grenoble 38042, France

Discussion

K Ramsay, Paisley College:

I was very impressed by the breadth of the application areas that were covered and the teaching of computer-aided techniques. How did you manage to enthuse all the individuals with their various areas of expertise who gave the necessary back-up to provide such a wide spread?

R B Morris, Cambridge University:

The answer to that is not really simple. If you have the computer-aided design terminals next to the standard drawing boards in the open, well-lit drawing office, as people go past, things develop. The whole thing springs from the fact that the course at Cambridge means that everybody covers all these subjects in the first two years. Because of that you get an undergraduate, usually with a bee in his bonnet, who will talk to his own supervisor in his own particular subject and will come back to us and ask, 'Look, can you make it do this?', 'Will you put corners on'; or something like that. We have a suggestions book full of little comments like this. We try to provide what they want in the way of building bricks. The spread has come as much from undergraduates as from ourselves. We are delighted about it as you can imagine; it's beyond anything we had dreamed of five years ago.

A L Johnson, Cambridge University:

At the same time, we have had to become fairly adept jacks of all trades to get involved in anything from thermodynamics structures to mechanics and dynamics, in fact anything that comes along.

K Ramsay:

Can I take it from that that the main load in terms of producing these applications falls on yourselves rather than your colleagues who are the primary experts in the fields?

A L Johnson:

Yes.

F Weeks, Newcastle Polytechnic:

Most of us would feel that the ultimate test of any design is for it to be converted into hardware and to be tested, and many of our courses include design, make and test activities of the kind that Dr Morris described. But I noted a reference to the construction of a simple package as a design activity in itself. You go on to say that the student is encouraged, having built this, to have it tested and to market it to his contemporaries. This is an interesting philosophy, and I wondered whether you would like to elaborate on it?

A Johnson:

For the first time this year we ran some second-year undergraduate projects in the building of CAD packages and we gave the undergraduates who were interested a choice of two subjects to tackle. One was the presentation of vector fields and the other was the presentation of fourier analysis. Those were both suitable areas for bringing computer graphics to a place where they would be very useful to other undergraduates. And the groups who tackled these did them all in widely varying ways. On the vector field side, we have actually employed one of the undergraduates who did particularly well to come back this summer and finish off his package so that we can offer it ready-made as a package for undergraduates who might have a vector field that they would like to see physically realized. They can type the function in and a stereo picture of the field will be produced. On the fourier analysis side, we had again three completely different approaches. One group went in very carefully and discovered how many terms in a fourier series had to be applied to get the required fit of the series to the prime function. Another group actually wrote themselves a numerical integrating package which automatically fitted fourier series to a function entered by the user. And again, we feel that with a little work on our part we can put this together to form a very powerful and useful package. The minute we make useful packages, we get people interested because they realize that it takes less time for them to learn how to do it by computer than it does for them to go and do it by hand.

R B Morris:

The one thing which was totally unexpected, which came out of that latter exercise, was that one undergraduate produced a language processor for taking a line of commands and sorting out all the comments along the line. He spent two or three days at the end of the term just tying it up, so that we now have a line processor for general use.

S Bloor, Leeds University:

How much formal time do your undergraduates spend on computing, computer appreciation or programming, whatever the term you use?

A L Johnson:

Virtually none. The use of the mechanism synthesis package, which Dr Morris mentioned, is worked into our drawing and design course. Undergraduates select a mechanism and then select dimensions to meet a required phase plane plot; and they all use the simulator for the model aircraft engine testing. None of it is compulsory. None of it is examinable, so to that extent it is entirely voluntary. Everything else is essentially done in the undergraduate's own time because he is interested or because he thinks the use of the package will help him solve some other problem that he has met in some other area of his course.

R B Morris:

We actually noted this term, when there was no formal instruction on the subject at all, that the packages had been accessed over 400 times, and we weren't even asking them to do it. We have unleashed a lot of enthusiasm. They do have formal courses in programming and things like that, so that they have the background built in, but the vital thing we have found is to produce good manuals. It takes more effort to write a good manual than it does to produce papers.

A Guy, Hatfield Polytechnic:

I have noticed that engineering students often become so enthusiastic with computing and CAD that they often wish to transfer to computer science, or alternatively the lecturers in the other subjects become annoyed because they are working on these particular topics to the detriment of their subjects. Is this true in Cambridge?

R B Morris:

It is not universally true. In the particular case of two people this year who are transferring to computer science, they both got third classes in their first year's work. Partly, I suspect, because they spent a bit too much time playing with computing generally. It comes back to the fact that the engineering course is a very wide course and people do discover that they are not entirely suited to working out the intricacies of thermodynamics, or something like that, and prefer to concentrate on electrical matters or physics. One of the qualities of the Cambridge degree system is that you can change subjects fairly readily and in this sort of case, certainly in the case of these two who are going over this year, we have found that their reason for going is that they have discovered what they really want to do during the course of their first year at Cambridge.

Unidentified contributor:

I sensed that behind your question, there was the feeling that colleagues are getting cross because of their subjects, the traditional subjects, are being neglected while the student plays with computer graphics, terminals, programming, and so on, which are totally unrelated to his mainstream studies.

R B Morris:

We do have a problem here. There are certainly one or two members of our department who regularly glare at us when they think that their students are being diverted from what they should be doing. Computers are at a great disadvantage here because undergraduates' use of computers is logged while the amount of time another undergraduate might go and spend in the local pub is not. We do take the view that, if they are sitting at a computer, they are at least learning about something to do with engineering which we think is better than if they spend the time in the pub.

M Dooner, Lanchester Polytechnic:

You mention the packages DUCT and BUILD. Could you elaborate on how you plan to introduce these into the undergraduate course, and how important do you think it is for students to appreciate professional sophisticated systems?

A L Johnson:

We think it is important, but the problem is the amount of learning time that's required which is simply too great for the average student who isn't necessarily going to get involved with that kind of package when he goes out into industry. We are introducing DUCT and BUILD into our production engineering tripos, which is a four year course. Undergraduates specialize onto it after the first two years of general engineering, and there is a bit more time then for them to get acquainted with the package and follow a design process through from the initial conception through the design methods to the building and the testing.

D Main, Hull College of H E:

We are often accused by industrialists that we train students to look at the principles and that they have no real knowledge of the techniques. You have emphasized that you are looking at the principles rather than the techniques. Do you feel that there is a need for some knowledge at least of one technique – DUCT is a typical example of this – for the sake of the student going out into the industry and being able to say 'Yes, I have experienced and mastered this particular technique as well as having the basic principles in order to be able to master others if necessary'? My second question is, do you teach both BASIC and FORTRAN and if so, why?

R B Morris:

We do all our teaching in FORTRAN. We don't touch BASIC at all. Available on the Sigma are FORTRAN, Algol, Pascal and APL. All the work is done in FORTRAN from scratch, that is part of a computing course as well and by the end of the first year they have got a passing familiarity with the use of programmes, sub-routines, and so on.

With regard to packages one of the difficulties is that we would like to get hold of some packages and let people use packages as such. The trouble is they cost money and we haven't got any; it's as simple as that. We have devised one or two little bits of technique that the undergraduates can use but our general aim is to give the background because anybody can pick up the details afterwards. We have to cram so much in the three years of education that we are giving them the information so that, when they go out into industry after about five or six years, they will become the sponsors of the next generation of CAD software. That's the overall aim. To get them very familiar with the technique would be a jolly good idea, but we don't do it specifically at the moment, probably because we have had too much on our plate to keep us busy.

C S Wells, University of Liverpool:

I was interested in the Cambridge man mentioning the absence of screens. Does this not change the whole nature of the interactive CAD work?

A L Johnson:

Yes, I think it does, it puts different emphasis on different aspects of it. We find working without a screen to be very effective. The screen only really comes into its own when you have very high line speeds and preferably refresh display, and at the moment we simply do not have the computing power to support that. The small flat bed plotters will run at much lower line speeds and permit a much slower response time on the part of the computer. The advantages they have are that, as already mentioned, they sit in our well lit drawing office and there is no need to cut the light down; you can actually do graphical construction work on the paper in conjunction with the computer and you have hard copy at the end of the day to take away. The hard copy has two effects: one is the advertising because the graduates who do produce good drawings are understandably proud of them and can stick them on their walls so their friends can get interested; the second is that when undergraduates are learning to use our software, if they get into problems, they can bring us the pictures they have produced and dialogue which has come out on a teletype beside the plotter and we can see what they actually did which, in nine cases out of ten, is not actually what they thought they did, and we can help them to sort out their difficulties.

R B Morris:

It mightn't have been clear that the plotter is interactive; it's a 4662 plotter so that you can actually work interactively.

J Kinsler, Paisley College:

Dr Morris, why were the terminals sited in a drawing office rather than, say, a laboratory?

R B Morris:

The drawing office is where we teach design, and this is at the moment a changing field within the engineering department. One of the things which was always said about Cambridge engineers is, 'The first day they go into the department they teach them to sharpen a pencil', and that has always been held against us. At the moment the courses are under review; mechanical drawing is a necessary part of the qualifications in order to get exemption from the Institution of Mechanical Engineers' examinations, so that everybody sits a mechanical drawing examination. This has led on to the introduction of design as a subject for the second year and this seemed to us the right place to put the terminals. There is also the question of space and things like that. I certainly wouldn't put them in one of our laboratories, which are all specialized laboratories: thermo, electricity, etc. The drawing office did seem the right place for no other reason.

A Llewelyn, CADCAM Association:

Could you comment on the objective of the output of your course in the light of Dr David's division of expertise? I am particularly thinking of system builders. Do you think your course is sufficiently general to cover those various specialities, or are there other courses similar to yours which would fit them better?

Another observation you might like to comment on is this. A previous questioner talked about interactive working; experience in industry now suggests, I am talking of people who have had 15 or 20 years experience, that they are finding a difficulty which seems to emanate from that type of education, which perhaps Cambridge is pre-eminent in, which gives the man the idea he isn't a member of a project team.

Dr Morris/A L Johnson:

The work that we are doing and have been doing for the past four or five years is aimed at the first and second years of the degree course. This was a change that we made when we came back from Middlesbrough in 1977. Up to that time we had been dealing with third year projects and a few second year projects, and we got back to pulling CAD into the early stages of the curriculum so that people can use it in the later ones. The important thing to remember is that what we are teaching is a background for them to go on and do their production engineering tripos if they wish to do that. Alternatively, I might point out in the third year the undergraduate studies any four papers from about 30 available for Part II in which they deal with all sorts of subjects in depth, ranging from thermodynamics to surveying to soil mechanics (you name it, it's somewhere in our syllabus). The corollary to this lot is that an awful lot of our engineers end up ten years later as managers in some form and it is at that stage that we hope our initial teaching will come to fruition rather more, because nine times out of ten they won't have come across CAD since then but at least they will have some fundamentals and can speak the language.

14. New technology-based training and its role in CADCAM

R Martin

Abstract: CADCAM is only a tool and its effectiveness depends upon the skill of the operator. As with any tool it is easy to abuse it and hence the training given to its users must be of the highest and most effective nature. If CADCAM enables us to apply computers to the design and manufacturing process then can they not also be applied to the education of users of CADCAM. This paper outlines the background to the development by Control Data of their PLATO computer-based learning system.

The United Kingdom today exhibits both the best and the worst of the national traits for which we are well-known.

- We are a nation of shopkeepers; we expect the world today to come to our door.
- We are a nation of innovators, of ingenious and hard-working seekers after improvement.

In our first role we tend to retreat from the cold hard facts of modern life.

In our second role, we grasp the nettle of opportunity – the nettle stings, it is a painful process, but by applying patience and energy, great results can flow from our work.

In fact, we MUST look at new ways of doing better and more speedily those things which we have done indifferently in the past, or can no longer do as well as our partners in trade.

CADCAM, albeit a brilliant tool in its own right, is only a tool; a willingness and an ability to see its benefits, and the skills to use it productively are of even greater importance than the tool itself. Who has not seen an electric drill used as a hammer, or a chisel used as a screwdriver, or the similar abuse of a fine tool?

Appreciation of the potential of CADCAM, and the application of practical skills to achieve this potential, are prerequisites of the successful introduction of any CADCAM system. And this must be swiftly achieved if the economic benefits of CADCAM are to be reaped within an acceptable time-scale.

There is no point in our ignoring, though, the developments taking place even before the first or second wave of CADCAM equipment is installed and amortized. It is certain that CADCAM in ten years time will be quite different from the latest concepts available today. The availability of more sophisticated, powerful and speedy computing facilities, with aids such as voice recognition or access to immense data-bases of various kinds will inevitably revolutionize our thinking, even whilst we are implementing today's technology. These quantum jumps in capability are anticipatable facts.

The job which yesterday demanded the skills of a chief designer, acquired over 20 years or more, will tomorrow need little more than an appreciation of what is needed and an ability to interact with a computer system.

This is certainly not to say that human skills will have disappeared – far from it – they will be even more urgently required than is the case today. What *is* certain is that these skills will be in areas we cannot yet define, and that those possessing them will be even fewer and more valuable relative to the requirement of that not-very-distant future than is the case today.

I am sure there are few here today who will quarrel with this scenario, and I put it to you that the encapsulation of these skills, and their speedy distribution to a broad audience is a role which only new technology-based training can fulfil satisfactorily.

The problem is that of new technology; let us turn to new technology to resolve it. Control Data and its PLATO system stand ready to rise to this challenge.

We all know that there are two areas of need we must address: the managerial level and the operator level.

The question I ask is, can we address these problems and solve them in an integrated way?

The computer is a powerful aid in specific DP areas, but its use as an educational or training tool has only been realized on a large scale in the last few years: the advent of the 'mighty micro' has made available to schools and industry alike a resource whose power and potential remain in many areas largely unexplored.

It is over 20 years now since the University of Illinois commenced work on the application of a brand-new resource – the computer – to the problems of education. For several years the University laboured under a burden of inadequate funding and limited computer power; until the day that Bill Norris, chief executive of a relatively small company in the specialized field of very high power computers – Control Data Corporation – became interested in their work.

Bill Norris had some years previously been frustrated by the lack of trained computer specialists, and had decided to release this brake on expansion by training up people himself, ultimately setting up a quite separate division of his company to do this – the Control Data Institutes.

However, Bill's vision was broad enough to encompass a more general view of society's future needs, and he realized that training was the key to future prosperity, in particular given the very real difficulties which people and organizations had in mastering the 'information explosion'. Bill Norris realized that to tackle his own sector of industry alone was to ignore a future crisis, and to turn his back on a future opportunity: he grasped the nettle.

In 1966 Control Data, in partnership with the University of Illinois and the National Science Foundation began an investment programme in computer-based education which, up to the end of 1981, has led to an investment of over £400 million by Control Data alone, and which has given the world the PLATO system as it is today.

What is PLATO, then? First and foremost it is a tool, a method of creating, delivering and validating educational and training materials. It is an innovative method of providing individualized, self-paced, competency-based learning. It was designed from the start with the psychology of the learning process as a prime input – it is not a DP system which is being stretched. People speak glibly of the 'user-friendliness' of a particular terminal or system. How many computer systems do you know of which a child of ten years old with learning difficulties, or the chairman of a major corporation, could both operate and relate to within a few minutes of the first encounter? Where 'ownership' can develop within a few minutes?

This brings to mind a story relating to a project in which PLATO was being introduced to a group of secondary school children with learning problems and scant respect for State property. In the course of the project, several break-ins occurred at the school, during which no damage was done, and only small losses of material were noticed. The culprits were never apprehended. However, the Control Data monitoring staff noticed that certain children's sign-ons had been active for periods on those nights. The children's aim had actually been to get more time to work with PLATO!

'Friendliness' of this order can be the most powerful weapon we have in the fight to bring to full operating productivity the very computerized design and manufacture systems which my own company and others know well can solve the desperate shortage of skilled designers and draftsmen which, today and even more in the near future, hampers our economic progress.

That a crisis of awareness and skills exists in CADCAM, in industry world wide, we all know. Let us bring to bear on our present state of apparent powerlessness to affect our futures for the better this great resource which is PLATO. Let us use it to mould and motivate, to impart knowledge and skill, and this is a way which can relate very directly to the technology of CADCAM, but which uses that computer to draw on twenty years of experience in education and training.

The rewards for those who use the CADCAM resource efficiently are very great. The penalties for ignoring it or, worse still, for using it in inappropriate ways, and thus misappropriating a major investment, are equally severe in the long term.

The task before us is great. I say again: the problem is that of new technology. Let us turn to new technology to provide a solution. Computer-based training is available, it is appropriate, and most important, it is cost-effective.

The sunrise industries and CADCAM in particular ignore it at their peril.

R Martin, Control Data Ltd, Control Data House, 179-199 Shaftesbury Avenue, London WC2H 8AR

Discussion

P Cooley, University of Aston:

Mr Weeks made the point about the universities and polytechnics being in a pincer movement between the schools which are now acquiring desktop computers which normally operate in BASIC, and industry with its particular requirements. This is something that my own department is finding increasingly difficult. We began teaching students BASIC having taken over from the computer centre about seven years ago but we now find that it is almost unnecessary as so many students have at least a smattering of BASIC from their school. My question is, do we feel it entirely necessary to teach a particular language or should we not be thinking about the fundamentals of instructing computers in a higher-level language, because the number of fundamentally different instructions that one gives to the computer in the high-level language is quite small? My second question is directed to Dr Pollard, who quoted a figure which I would agree with of three hours per student to gain an initial competence with a drafting package. I would love to be able to offer this to my final year students but as there are usually between 50 and 60 of them it militates against any 'hands-on' experience of this nature. How does one get over this problem?

F Weeks, Newcastle Polytechnic:

On the question of the pincer movement, I think you are right, the particular language doesn't matter all that much. In general our students are being educated for the future. The important thing is to ensure that they understand the principles of instructing computers and that they leave higher education with an enthusiasm for continuing to learn how these principles may be applied in a changing technology.

I think that Jim Riley, the author who originally referred to the pincer movement, was addressing himself to the problem, which many of us experience, of encouraging our colleagues to become more involved in computer applications. In my paper, I refer to developments being driven by enthusiasts at present and suggest that there is a need for a considerable programme of staff development. If this is not achieved, many youngsters reaching higher education with significant computer literacy may be disappointed and, on the other hand, industry will be dissatisfied as new graduates fail to meet their expectations with regard to competence in this aspect of technology.

D Greatrex, EITB:

Obviously, when selecting software, you did a considerable amount of investigation and made your final choice. My question is related to the operating manuals of software. In your final selection were you happy with the standard of the manuals or did you have to modify them or supplement them with some additional material yourselves?

D Sheldon, Huddersfield Polytechnic:

The selection of software systems themselves is an interesting point. How would one react to the view that we as educators/trainers don't actually go through a selection process, because it is questionable whether the characteristics of systems really matter in the teaching environment? If you are a firm, you are concerned with certain products and therefore you can define and specify what solution will suit your own problem. In the teaching situation, it doesn't matter what systems there are for the students, it introduces them to the fundamentals, so we set about it that way. We had a certain amount of money and looked for who could give us the best deal for it. As for manuals the ideal thing would be for the lecturers together with programmers who do the application support on the systems to provide student manuals. Manuals supplied by standard suppliers are too extensive.

F Weeks:

The comment about commercial manuals being unsuitable for student use is very valid. They are designed, of course, for a different learning situation. At Newcastle Polytechnic we have produced a simple computer-aided drafting system based on software written in-house. The manual which is used with this was written as a programmed text. The students carry out an exercise on this drafting system working in groups of four or five for hands-on work at the terminal they work in pairs and the mutual support, which this provides, does seem to be beneficial. Our practice is to give the group a half-hour introduction to the system, provide them with the manual and then leave them to carry out defined tasks. This seems to work quite well.

J Gero, Sydney University:

Mr Weeks, you said in relation to the curriculum that, 'all were agreed that a fundamental requirement is a knowledge of computer programming'. Is there any rationale for this? I must say I am acting as the devil's advocate here, but why should engineers know anything about computer programming? Second, since most teachers started with BASIC, is the choice of BASIC a rational choice or is it chosen because it is the one they know? Computer science departments no longer teach FORTRAN or BASIC. They teach Pascal. In America a recent study showed that not a single computer science department in the accredited universities teaches FORTRAN or BASIC. I don't know the situation in the UK. FORTRAN, BASIC and Pascal are all fairly low-level languages for people who will end up in practice in four or five years' time. What function therefore does programming hold for engineers who are going to have a fairly low level of programming capability in the sort of scale that's been mentioned? Is BASIC chosen because that's the one the teachers know? It is a good choice if that's the reason? Is there any thought given to higher-level languages?

F Weeks:

Well, I can only speak for my own institution. Originally we taught FORTRAN, and when we decided to change that and introduce BASIC as a first step it was to encourage the students to become more proficient or to help them find programming more acceptable. At that time Pascal was not known: it certainly wasn't in my vocabulary anyway. Of course there are developments and we ought to bear these in mind, but to go back to your original question 'Do engineers need to know about programming at all?', it may well be that when they get into practice, certainly in the larger companies, they won't need to exercise that skill because they would be using packages produced by companies with specialist knowledge. On the other hand, there is a certain degree of awareness required and I would take the view that that is acquired only by getting involved in finding out how programs work; to do that you have to be able to program. How far you need to go is debatable and there is quite a marked variation of opinion from one establishment to another. In mine, we would say that we should be giving awareness, but that we would place more emphasis on the preparation of problems for solution by pre-written packages, either written in-house or bought from outside. We would think that that was more important than developing the basic skill in the individual.

Unidentified contributor:

There were others at the conference who are equally adamant that people ought to be able to program for quite complex tasks. Mr Riley, for example, from Kingston Polytechnic, was saying that his students have to be able to produce graphic/packages, they have to be able to program for NC machine tools. The question everybody was asking was: how do you manage to fit it all in to an already crowded curriculum? But he was saying that just as vehemently as I am saying that we ought to be using commercial packages.

F Weeks:

To be fair, the contributors to the Newcastle conference did make the point that these things were important. I referred in the paper to the fact that the students were given the marking scheme for the CAD projects and they knew from this how the marks were assigned: how many to the program, how many to the documentation and so on.

D J Williams, Plessey Co:

As a member of a company that has gone some way through CADCAM process already, I would like to make three brief observations. First, I agree that integration is essential and is largely achieved by having an integrated set of engineering data-bases preferably connected to the manufacturing data-base. Second, an aspect that is important is that most of the devices that are made, whether they be mechanical or electronic, are going to need testing at some stage, and we ought to be examining simulations of the particular product that we are making so that we can derive the test programmes which otherwise are going to be very long and costly to produce. Finally, our experience is that the teaching company associate scheme is a very good vehicle for getting academic institutions and manufacturing organizations working together. We are very pleased at the results.

D G Smith, Loughborough University:

Although this conference is concerned with CAD, which presumably means computer-aided design, we haven't heard very much about design. We have heard a lot about drafting, and manufacture, but very little about design as a total activity. The problem is that if we are not careful we will convey these attitudes to our students and not make them aware that design is an integrated activity. Mr Weeks, you said that people were talking about considering CADCAM in its own right. Is it intended to be taught against the activity of design itself? Because if it isn't, it can give a completely misleading idea of what design is to students.

F Weeks:

I entirely agree. It is clearly important that a subject which describes itself as computer-aided design should be related to the process of design. I made the point in my paper that for the computer-assisted technique, whether it be analysis or graphics, to be classified as computer-aided design it must form part of the feedback loop which characterizes the design process. An interesting point has been raised with regard to a specific topic of CADCAM being taught in parallel with a separate course in engineering design. How do we make sure that we get this incorporated within the design activity? I was reviewing what others had said in that particular context. I think we are at an early stage in their development of this separate topic, so I cannot really give a definite answer, but I do agree that its important.

A C Stuckey, Chelmer Institute of HE:

I am not an engineer, I am on the computer side of my own establishment and I am trying to get engineers interested in CADCAM. One of the ways we could promote the gospel, so to speak, is if the resources which have been developed elsewhere were available to other establishments like my own. I wonder whether anything has been done to provide a catalogue of resources which might be available?

F Weeks:

I am sure that wherever possible there ought to be procedures for exchanging software. I can't see any point in re-inventing the wheel. There are schemes: ACUCHE, the Association of Computer Units in Higher Education has a software exchange scheme; there is one based at Loughborough. The CADCAM Association to which Dr Abbas has referred is keen to encourage interaction between educational establishments to share experiences. And one of the things that we thought might arise out of that would be a scheme for sharing software. But having said that, there are difficulties of course; different machines, different languages may result in the problems of transfer being more difficult than the problems of writing one's own.

S Bennett, Ford Motor Co:

We run a number of courses in-house for our own draftsmen to be timed to fit in with our introduction of graphics tubes. We have something like 150-200 tubes coming in in the next five years, and for each system coming in we anticipate training three of our existing staff to use them. The courses we do are five-week intensive courses. Would any of the academic institutions be willing to help us in the task facing us?

Speaker:

I think it might be a good idea to approach the CADCAM Association. We have identified the need for a clearing-house or repository of information to which people could go to and which would know what's going on in various institutions in CADCAM and which would be able to put you and somebody else together or you and a number of people together to sort this kind of problem out.

K Ramsey, Paisley College:

If we do implement computer-aided design and manufacture and so on and incorporate them properly into courses, how do you examine this subject?

F Weeks:

I am not sure that I have an answer. I first came across this question five years or more ago. I was interested in computer-aided learning and was trying to use the computer and make engineering education more meaningful. One was able to get more information, go further into a particular problem than would be possible using a calculator. One's colleagues would say, 'Well, that's fine, but how do you go about examining it?' I remember telling that to our local HMI and his answer was 'Do you really mean to tell me that you are not going to put in train a worthwhile educational development because you can't think of a way of examining?'

Unidentified contributor:

I think there are two parts to the question. One is tied up with how you examine design, and if you can answer that one you have part of the answer to the other question. At Surrey we do it with a combination of oral exam and written course work. The other part of the question dealing with course examples with drawings, that is a relatively easy thing. We have a system at our disposal that enables us to change the data for every student and providing he doesn't know he has different data, even with the same object, the picture can look exactly the same to every candidate.

You can provide all the students doing their piece of course work with different data. How do you make sure that the course work has come from the individual rather than somebody else? All the students have their own UFDs: they log into that particular UFD, produce work in that UFD and we mark it in UFD. That's the way we propose to deal with the problem.

Unidentified contributor:

I have been involved in examining CADCAM at graduate and post-graduate level and it's perfectly possible. You would do it in two parts, one being the course work project assessment which we would assign 30 per cent of the final mark to, and one would be the formal examination. If you arrange for the course work to be practical-oriented projects which take a sufficient length of time to enable a good, hard, result to come through, and you choose specific topics which you can devise 45 minute questions to for the formal examination, we found it wasn't too difficult.

15. A review of computer graphics equipment for engineering applications

H Rippiner

Abstract: This paper looks at the development of computer graphics equipment from the 1960s up to the present time, and at the different technologies which have emerged to fill the needs of the engineering community.

History of graphics

The first commercial systems that became available towards the end of the 1960s from such companies as Sanders, Vector General, and Evans and Sutherland, had a price tag in excess of £100,000: sales were therefore restricted to organizations for whom computer graphics was essential (eg military aircraft/rocket movement simulation).

It was in the mid-1960s that Tektronix developed the dvst (direct-view bistable-storage tube) for use in oscilloscopes designed for recording transient signals. The inherent feature of all dvsts is that the writing beam only has to pass over the screen's phosphor surface once in order for the image to be recorded and displayed. As further developments were made to the dvst, it became possible to produce a cathode-ray tube with an 11in diagonal, and the availability of this device in a commercial display unit (the Tektronix 611) opened up new avenues of graphics display. Several manufacturers included this unit in their computer graphics terminals, and Tektronix itself entered this market in 1969 with the T4002 interactive graphics terminal. All these dvst-based products were significantly less expensive than other graphics terminals on the market, and thus were within the financial reach of more users.

In 1971, the computer-graphics market-place was really opened up with the introduction of a dvst-based terminal which was a fraction of the cost of competing products.

This meant that computer graphics were no longer restricted to government-funded military contracts, but were now within the reach of numerous commercial, scientific and educational establishments. From this small beginning, the computer graphics market has grown significantly, and it currently has an annual growth rate of between 30 and 40 per cent.

Computer graphics equipment for engineering applications can be divided into six categories: Non-intelligent terminals; Smart terminals; Intelligent terminals/desktop computers; Turnkey systems; Graphics input devices; and Graphics output devices.

Engineering computer graphic applications can be divided into five groups: computer-aided drafting; computer-aided design; computer-aided manufacturing; computer-aided test and scientific data analysis. These applications can be found in industries as diverse as vehicle manufacturing, ship-building, oil exploration, architecture/civil engineering, and garment manufacture.

Display techniques

Three types of display technology are used in computer graphics terminals: namely dvst, raster-scan and calligraphic (or directed beam).

Opponents of the dvst have for a long time been claiming that its days are numbered, but recent enhancements, and the development of new features have added another dimension to this technology.

The major elements of the dvst are the writing gun, flood guns and the phosphor target or storage screen. See Figure 1.

DVST Mechanics

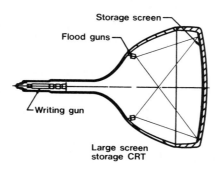

Figure 1 *Direct view storage tube*

The target includes a transparent conductive layer that provides a foundation for an array of collectors. The writing gun, in the neck of the crt is negative relative to the target, and the flood guns (four in a 19in tube) sit on the back wall of the crt funnel.

The dvst is normally used to display information stored on the screen. In that mode, the writing gun's beam scans (ie writes on) the target, creating an charge pattern that is positive relative to the flood-gun electrons, and thereby causing the phosphor to luminesce. The charge pattern is made positive as a result of secondary emission caused by the energy of the writing gun as it hits the target. The unwritten areas of the target remain at a much lower potential than the written areas, and hence do not luminesce to any great extent.

Once the information has been stored, there is no need to refresh it; it continues to luminesce by attracting electrons from the flood guns. In order to display new information in place of the existing image, it is necessary to apply a two-level erase pulse to the target. The first part of the pulse makes the whole target more positive, thereby attracting flood-gun electrons, and effectively writing the whole of the screen. The second part of the pulse lowers the target potential, thus erasing the screen.

In order to change even a small amount of displayed information, it is necessary to erase the whole screen (a process which takes about one second) and then rewrite the old as well as the new information. However, with the latest generation of dvst terminals, this is not a problem, since up to 26,000 short vectors can be displayed in less than half a second.

On earlier terminals, rewriting the whole screen of information at low data-transmission rates could take several seconds, and a technique called 'write-through' was developed to minimize much of the delay that users experienced in erasing and rewriting. In this technique, the beam writes the new information with just enough current to store the information. Continuously rewriting the new information (from a host computer or from the terminal's in-built microprocessor) maintains its luminance, and this process is called 'refresh'.

Because the target of a dvst is continuous and is not a discrete dot or stripe target, the resolution is limited only by the writing-beam spot size. The discrete nature of a raster-scan or shadow-mask type of target is not a limiting factor in the dvst; hence diagonal vectors, for example, do not show as jagged lines as they do on a raster-scan display.

The visual resolution of a dvst is similar to that of a 1000-line non-interlaced raster display on a 19in crt. There are high-resolution colour raster-scan displays on the market which approach dvsts for the number of lines per display, but that is in one direction only – their vertical resolution. They still show jagged lines, and they show them more prominently than the dvst because the raster-scan display is still a discrete dot target.

A very recent advancement of the dvst has been the introduction of colour write-through. As we saw earlier, write-through greatly increases the interactivity of the display. Adding a second colour enhances the viewability of that display (ie the user's ability to understand the information displayed, and to differentiate between stored and non-stored information).

When lower-cost dvst-based terminals were introduced in 1971, the only products using raster-scan technology were domestic/industrial televisions and monitors. However, in recent years, the cost of semiconductor memory has drastically fallen, and this has allowed the development of a wide range of raster-scan based terminals and graphic work stations.

In a raster-scan device, circuitry within the terminal or work station continuously generates a raster (irrespective of whether or not data is being transmitted to the screen) in the same manner as on a domestic television. On low- and medium-resolution devices, the number of lines generated will be 525 on US-manufactured units, and 625 on European manufactured units, although not all of these lines will be display on the screen.

Some systems use interlaced rasters, whilst others use non-interlaced rasters. In an interlaced display, the picture is divided into two fields (odd and even) and each field is made up of 312½ lines. So, starting from time zero, field one (odd) is displayed generating lines 1, 3, 5 etc, followed by field two (even) generating lines 2, 4, 6 etc, each field being repeated 25 times per second (ie creating 25 frames per second). See Figure 2.

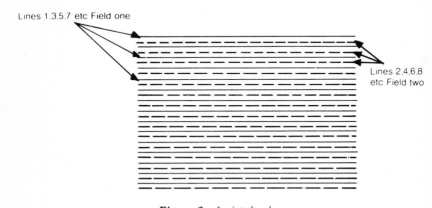

Figure 2a *An interlaced scan*

Lines 1,2,3,4 etc

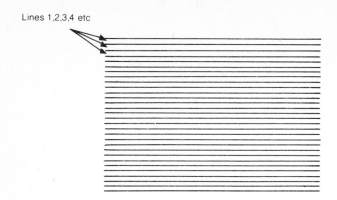

Figure 2b *A non-interlaced scan*

The reason for creating the picture in this way is to reduce flicker to an undetectable level (the eye integrates the information so that the brain does not detect the flicker), while at the same time allowing the use of a crt with a short-persistence phosphor, which permits full dynamic graphics.

In many graphics applications, such as computer-aided design, only limited dynamic graphics are required. In this case, a crt with a longer-persistence phosphor can be employed, allowing the use of a non-interlaced raster; this means that only one field is displayed, generating lines 1, 2, 3, 4 etc, the complete frame being repeated 25 times per second. For very detailed work, such as integrated-circuit mask design a 625-line display does not provide enough vertical resolution, and therefore more sophisticated (and more expensive) display units are used when there are more than 1024 lines per frame.

The horizontal resolution of the display is largely determined by the speed with which the crt beam (Z-axis) can be turned off and on, since it is the control of the Z-axis which determines what is actually displayed on the raster. This on/off speed is, in turn, a function of the display's video-amplifier bandwidth. Domestic televisions and low-cost raster-scan displays have bandwidths in the region of 4.4MHz, while the high-performance display units can have bandwidths as high as 24MHz. With this bandwidth, a raster-scan device can display about 1000 picture elements (pixels) per scan line, which is approximately the same optical resolution as that of a 19in dvst.

In any raster-scan computer graphics display, it is necessary to have a graphic bit-map memory to store the picture information which is being displayed. A typical medium-resolution display will have 640 pixels horizontally by 480 pixels vertically. Fortunately, it is not always necessary to have one bit of memory per pixel element, but only sufficient to store the actual pixels which are turned on (ie illuminated by the crt gun). A typical simple business graph displayed on a Tektronix 4025A terminal would require about 6kbyte of memory, while a complex engineering drawing would require 32kbyte. Whereas the architecture of the display device is simpler (using an absolute bit-map technique) it is necessary to have one bit per pixel on a monochrome display. For a display resolution of 640 × 480 pixels, this means 38.4kbyte of memory for displaying any graphics, whether simple or complex (more of this later).

The third type of display technology used in computer graphics is calligraphic, or directed beam. Relatively few commercially available products use this technology, because it is more expensive than both dvst and raster-scan technologies, but it does have some advantages in the area of dynamic simulation.

In a calligraphic display, the crt beam is directed from one end of a vector to the other directly by the driving computer. This is exactly the same procedure as used in the dvst device, and hence no jagged-lines are created. The difference between dvst and calligraphic devices is that, in a dvst device, the computer only has to direct the beam across each vector once for each displayed picture, whereas in a calligraphic device the computer must direct the beam across each vector at least 25 times per second, if flicker is to be avoided. This, of course, means that each display must have a dedicated computer associated with it for picture generation, in addition to the computer running the application program.

While this approach increases the cost of the display device, it does give the added benefit of allowing significant amounts of dynamic graphics to be presented, since every vector is redrawn 25 times per second. In practice, very few industrial or scientific applications need to display large amounts of dynamic information. Those that do are mainly confined to the simulation of real-life moving objects, such as aircraft, rockets, tanks etc. The vast majority of CAD applications require only limited amounts of refresh, as offered on raster-scan or dvst devices with write-through.

One drawback of the calligraphic technology, which the other two technologies do not suffer from, is a limit on the number of displayed vectors. Each vector takes a finite time to write; for short vectors, this will be in the region of 5 microseconds. Assuming that each vector must be overwritten 25 times per second (to avoid flicker), all vectors must be written in 40ms. This results in a maximum display capacity of 8000 vectors. If this number is exceeded, display flicker will result.

Graphics terminals

Having reviewed the display technologies, let us now consider the terminals. A non-intelligent terminal is one in which all the processing is done remotely in a micro, mini or main frame computer. As can be seen from Figure 3, the terminal architecture is fairly simple, since the storage tube eliminates the need for semi-conductor memory.

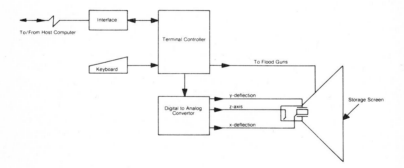

Figure 3 *Architecture of the Tektronix 4010 terminal*

In a raster-scan non-intelligent terminal, the architecture is slightly more complex (Figure 4), since the use of the raster-scan technique necessitates the addition of a bit map, memory controller and video controller.

In recent years, the advancement of semi-conductor technology has tended to change graphic terminals into 'smart' devices, containing one or more microprocessors which do routine manipulations such as drawing arcs and circles, filling panels with a colour or pattern, or re-designing the functions of the keyboard. A smart terminal does not allow the user to run a program within the terminal, but it does allow him to simplify the software algorithms on the host processor.

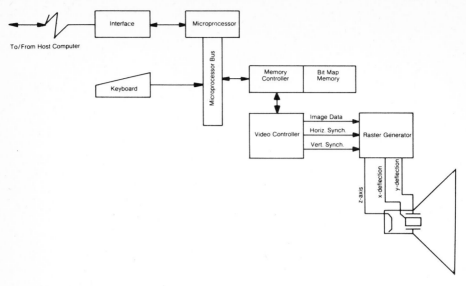

Figure 4 *Architecture of a raster-scan terminal (Counting House GT 1000)*

The Tektronx 4025A is typical of a smart raster-scan terminal, except that it employs a virtual bit map, whereas most raster-scan terminals employ an absolute bit map.

In a device employing the absolute bit-map technique, two totally independent memories are employed. One memory stores the alphanumeric text, with a window which can be displayed and moved to scroll the stored information, while the second memory stores the graphic information. As mentioned earlier, with this technique it is necessary to have one bit per pixel, and this results in inefficient storage of displayed graphics. In addition, with this technique the displayed graphics cannot be scrolled. This feature is not a problem in most engineering fields, but it is a serious drawback in business applications, where operating personnel expect to have terminals with full scrolling facilities.

The use of a virtual bit map overcomes this problem. This technique also uses two memories: one for the display list and alphanumeric characters, and one for the graphics (Figure 5).

The difference here is that the graphic memory works through the display list, and is not independently scanned before being displayed. This means that the graphics can be scrolled, and that the graphics memory only has to store information about those graphic characters which actually require illuminating. In addition, it allows multiple graphic areas in memory.

Initially, the display is alphanumeric only. To display graphics, a graphic area has to be programmed to cover a defined block of character cells. This action assigns a block of alphanumeric memory to the display list, and indicates that displayed in this area will be graphic characters. So far, no graphic memory has been allocated. As vectors are drawn within the graphic area, character cells are intersected, and the appropriate dot matrix of each cell is programmed into the graphic memory. Only character cells which are intersected by vectors or solid boundaries occupy graphic memory. Each virtual graphics memory character is formed within an 8 × 14 dot matrix and the sum of these characters make up the graphic area, and consequently the addressability of the graphic area (Figure 6).

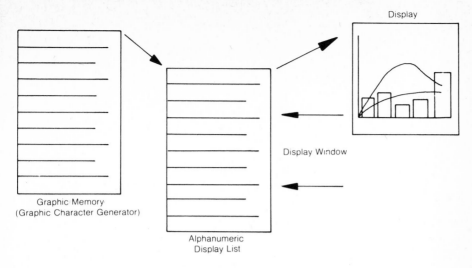

Figure 5 *Virtual bit map architecture*

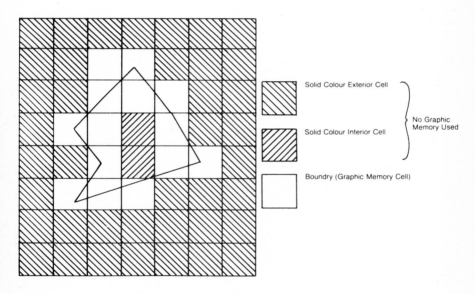

Figure 6 *Virtual bit map graphic memory*

The graphic requirements of engineering applications tend to differ from those of the business user in that engineers are typically working with far more complex graphics, and therefore require the smart terminal's microprocessor(s) to be used in a different way.

Newer terminals embody some extremely powerful intelligence features. Some terminals, such as the Tektronix 4110 series, possess a facility for storing local picture segments. A picture segment, as described in SIGGRAPH-ACM's proposed core system of computer graphics standard, is an ordered collection of output primitives (for example, vectors or text strings) defining an image which is part of the picture on a view surface. This concept is best illustrated if we study an example.

Suppose that an engineer is using such a terminal to design an electronic circuit. The host computer sends the graphic data to display representations of components, such as transistors, resistors and capacitors, on the terminal screen. Each component is stored in the terminal's memory as a separate segment, up to a maximum of 32,000. The segment is numbered, and therefore can be specified independently of the other segments. The designer uses the terminal keyboard controls, or a peripheral graphics input device, such as a graphics tablet, to select the desired component and position it on-screen. The component remains displayed at that position, while the designer selects and positions other components.

The capability to store and manipulate segments locally provides several benefits. Host communications traffic is substantially reduced, as it is no longer necessary to re-transmit the entire sequence of graphics data to the terminal. The host need send only a command that says, in effect, display segment 'N'.

In addition, there is a marked increase in repaint speed (the rate at which an erased display can be re-drawn). This capability is particularly valuable when working with dvst displays containing large numbers of vectors or alphanumerics. As indicated earlier, a diagram containing 26,000 short vectors, stored in ram can be re-drawn on the screen in less than half a second.

Another powerful feature is the ability to perform two-dimensional image transformations. A segment may be rotated, translated (shifted in position) and scaled (made larger or smaller). The user has the flexibility to define a picture segment of a standard size and shape and, after it is displayed, modify its image to fit the application.

The majority of graphics terminals are optimized for graphics work, and the dialogue between the operator and host computer is given a much lower priority; in some cases, it may well be written on top of an already complex graphics image. However, the new terminals are equipped with a dialogue area. The display of alphanumeric text is restricted to a defined area of the screen (Figure 7). Communications between the host graphics information is displayed on the rest of the screen.

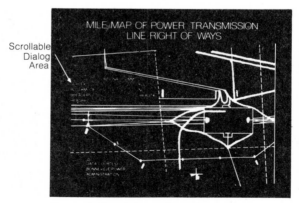

Communications between the operator and the host computer are displayed in the dialog area. The size and location of the dialog area can be defined, to avoid interfering with displayed graphic information.

Figure 7 *Dialogue area on the 4110 series*

The dialogue area is scrollable when filled with text, and the top line is automatically moved up and out of view to make room for the next line of text at the bottom. Each line of text that is scrolled out of the dialogue area is stored locally in ram. This enables the operator to read previously displayed text by scrolling backwards with the keyboard thumbwheels.

The user can define the size of the memory buffer used to contain scrolled text, and can tailor the on-screen dialogue area in terms of number of characters per line, number of displayed lines of characters, and the position of the area within the screen boundaries. In addition, the dialogue area can be turned on or off.

The last few years have seen an explosive growth in the number of desktop microcomputers with graphics facilities. In 1975, this field was pioneered with the introduction of the 4051 desktop graphic computer system, which has since been joined by the much faster (but fully compatible) 4052A and 4054A. All these systems are based on dvst technology and, as such, offer the highest resolution of any desktop computer on the market. Since 1975, this product range has been joined by those of other manufacturers – from Hewlett-Packard's 9845C at one end of the scale to Clive Sinclair's ZX81 at the other, with many more in between.

At the heart of all desktop microcomputers is a central processing unit. Unlike smart terminals, in which the microprocessor is only used for graphic manipulation, panel filling etc, the desktop microcomputer allows the user to construct and run his own programs within the desktop unit. Practically all desktop microcomputers use BASIC as their programming language, with one notable exception: ICL's Perq, which runs Pascal. In the last 12 months or so, CP/M operating systems have become available for some desktop microcomputers, particularly Commodore's PET.

Apart from the BASIC language, most desktop microcomputers have another common feature. Communication with the 'outside world' (ie other peripheral devices) is usually achieved by either the well-established IEEE 488 interface (more commonly known as gpib or general purpose instrumentation bus), an 8-bit wide, bit-parallel, byte-serial interface, or the RS-232-C standard data-communications interface. These interfaces give the user the freedom to select peripherals from different suppliers, so that he can choose the best one for the job. If peripherals are chosen from manufacturers other than the supplier of the microcomputer, a software overhead is often incurred, but in most cases this is fairly small.

By the addition of a send/receive data-communications interface (usually RS-232-C), many microcomputers can be converted into intelligent terminals. In such a terminal, an operational program in the desktop unit is normally controlling a bigger program in the associated host computer, and accessing the host's data-base via this program. Although the microcomputer is being used as a stand-alone device, or as an intelligent graphics terminal, it will almost certainly have some form of magnetic storage medium. This can range from a small data cartridge containing 100kbyte of data and/or program up to file-management systems, which can support multiple users, and have hundreds of megabytes of online storage capacity.

With the addition of peripherals such as digitizers and plotters, the modern desktop graphic computer system can rival many small minicomputer-based systems in terms of cost and performance.

The top end of the display device spectrum is taken by the turnkey graphic systems offered by such companies as ComputerVision, Applicon and Intergraph. These systems are based on powerful minicomputers and comprehensive software. They can use one or more of the display technologies we have already discussed and, in some cases, where two screens are required, both dvst and raster technologies are used.

Turnkey systems are usually designed for use in CAD areas such as ic mask and pcb applications, in which complex graphics displays are required, and the software associated with the system is optimised for the application in question.

Graphical interactivity

In many areas of computer graphics (particularly in engineering), graphical interactivity is essential, and there are currently three methods of graphic input in common use.

On dvst-based terminals and desktop microcomputers, graphic input via the screen is achieved using a cross-hair cursor. The cursor, controlled by thumbwheels or a joystick, is generated in the same physical plane as the stored image. This has the advantage that there is no parallax error, as there often can be with a light pen (because of the thickness of the glass). Also, because the cursor is electronically generated, its cross-hair point is much finer than the tip of a light pen, and therefore allows more accurate placement.

On raster-scan based display units, the graphic input via the screen can either be by means of a cross-hair cursor controlled by thumbwheels/joystick, or by means of a cursor controlled by a light pen. The majority of thumbwheels and joysticks use precision potentiometers, whose output is fed to an accurate A/D converter. The converter's digital output is then fed back to the computer as the operator's input of cursor position, and is also fed to the digital input of the display in order to determine the position of the cursor on the screen.

A more recent technique for thumbwheel operation is to have a toothed wheel connected to the thumbwheel. A light beam is passed across the wheel's teeth, and is broken up into digital pulses. After processing, these pulses are used as the operator's input.

A light pen works by measuring the time taken for the raster lines to reach the point at which the pen is touching the screen. In practice, it is difficult for the light pen to pick up only one raster line and, as a result, the accuracy of this technique is inferior to that of a cross-hair controlled by thumbwheels or joystick.

Many graphic displays can have digitizing tablets attached to them. In such a device, the interactivity is by means of either a stylus or a cursor (commonly called a 'mouse') which is passed over the surface of the tablet. The pen is usually arranged to control an electronic cursor of some kind (cross-hair or symbol) which is generated on the associated display. The tablet itself is used to digitize existing drawings/sketches, or to send specific commands to the host by means of a menu attached to the tablet. In many cases, it is used for both activities.

The majority of tablets operate on the principle of magnetostrictive ranging. Below the surface of the tablet is a mesh of closely-spaced parallel wires in both the X and Y directions. These wires are made of nickel-cobalt and exhibit magnetostrictive properties. Current is pulsed along a send wire that lies perpendicular to the mesh of magnetrostrictive wires. This current pulse changes, slightly, the dimensions of the magnetostrictive material, and a strain wave propagates down all the wires in one direction simultaneously.

The receive coil in the stylus or cursor provides an electrical signal from the flux change. The time delay between the pulse leaving the send wire and reaching the receive coil is directly proportional to the distance between them. By pulsing from both sides (or top and bottom) of the tablet sequentially, two sets of distances are recorded, resulting in a highly accurate digital co-ordinate.

In some application areas, particularly cartography, manual digitizing is slow for the vast amount of data which requires digitizing. Two techniques are available to assist in this area. The first is the raster-scan technique, whereby the whole document is scanned by a laser, and the resultant massive file of data is then cleaned up, compiled into features (symbols, vectors, etc) and coded in either a separate processing or editing stage. While at first glance this might seem to be an ideal solution, any interaction during the rasterization sequence is difficult, and changes can mean lengthy re-runs or extended editing.

The second technique is that used by Laser-Scan's 'Fastrak', which automates line following by using a local raster-scan technique. A very fine laser-beam probe is steered at high speed, and with good resolution, in both X and Y axes. The beam is used to interrogate a film, which is a reduced photographic negative of the original map or document. The line co-ordinate information is obtained by automatically scanning a small area along a selected feature on that film negative. This local raster-scan is called a scan pattern, and the line data are derived from it (Figure 8).

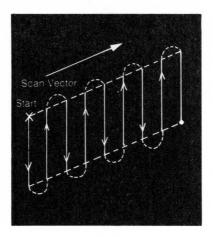

Figure 8 *Vertical scan pattern showing start point and scan vector*

Hard copy

Hard copy output from computer graphic systems can be generated in one of four ways: digital plotter; matrix printer/plotter; electrophotography/thermal/electrostatic copier; and computer output to microfilm.

Digital plotters are designed for applications requiring only limited solid colours, but otherwise very high line quality (eg engineering drawings). Both flat bed and drum plotters are available, and pen types can vary from ball-point to fibre-tip and wet-ink (eg Rotring) varieties.

Pen plotters draw vectors and characters directly from the computer output or from a magnetic storage device, and many have considerable amounts of intelligence built into them.

Functions included are arc/circle generation, variable-size character generation, character rotation, local scaling, mirror imaging etc. Some can also function as digitizers, with the pen cursor being manipulated by a joystick on the plotter.

Flat bed plotters have the advantage that it is easier to see the plot while it is in progress, and also various output media, such as paper, Mylar and transparent film, can be used. Accuracy is also greater, for a given picture size than on drum plotters. However, A0-size flat bed plotters are both physically large and expensive; hence the majority of large plots are made on drum plotters.

One drawback with drum plotters is the difficulty of using pre-printed media, since the paper is normally sprocketed and supplied in continuous blank rolls. However, the recent introduction by Hewlett-Packard of the 7580A A1-size drum plotter has overcome this problem by the introduction of a micro-grip drive using small grit-covered wheels to move individual sheets.

In some applications, the graphical output is a relatively small proportion of the total computer output, and for such applications a digital dot-matrix impact printer/plotter will often suffice. With these products low copy cost is a major advantage, but low image resolution can be a critical restriction. Some impact printer/plotters, such as the

Ramtek 4100, have four colours available, with separate cartridge ribbons and print heads for each primary colour, plus black.

The electrophotographic technique uses a beam of light to expose a light-sensitive medium (silvered paper) which is then developed by heat (Figure 9).

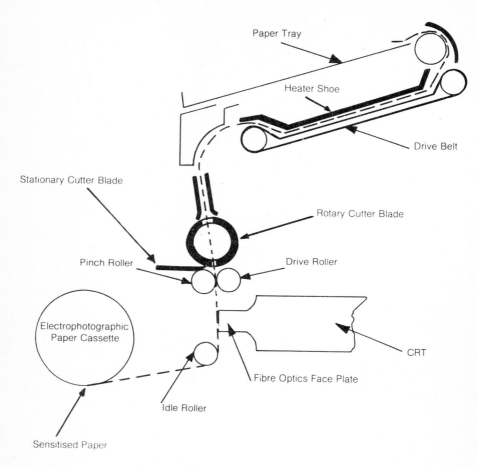

Paper Tray

Heater Shoe

Drive Belt

Stationary Cutter Blade

Rotary Cutter Blade

Pinch Roller

Drive Roller

Electrophotographic Paper Cassette

CRT

Fibre Optics Face Plate

Idle Roller

Sensitised Paper

Figure 9 *Electrophotographic copier operation*

The beam of light is produced by a crt with a fibre-optic face plate, which maintains a very fine light spot on the paper being exposed. By varying the intensity of the light, it is possible to produce fine grey-scale, as well as pure black-and-white images.

The image information for electrophotographic copiers can either be read directly from the graphic display crt, in the case of dvst's, or from the associated ram, in the case of raster-scan devices. The main advantage of the electrophotographic technique, apart from its grey-scale capability, is speed. Regardless of complexity, any screen image can be copied in 15 seconds.

Electrostatic technology is similar to the electrophotographic technique but, in place of the light beam, wires and electrodes are used to put down a charge on dielectric paper, which is then passed through a toner. The toner is attracted to the charged areas, and is fused to the paper.

In most electrostatic copiers, charge transfer is achieved using a matrix of fixed-wire styli as wide as the paper being imaged. With up to 200 styli per inch, the manufacturing cost of this matrix and the electronics necessary to drive each stylus is high. On some new electrostatic copiers, this problem is overcome by using a moving stainless-steel belt with raised styli in contact with the paper. Because this belt and the associated drive electronics are much less complex than the wire matrix and its drive circuitry, it is less costly to manufacture. Another innovation in these two new copiers is the use of a dry toner. Most electrostatic copiers use liquid toning systems. Liquid toner consists of a suspension fluid (usually Isopar, a refined form of kerosene) containing suspended black charged particles that are attracted to imaged areas when this liquid is pumped against the paper. These systems have several disadvantages: the liquids are messy and inconvenient and, as copies are made, the black particles are used up, resulting in progressively lighter images. Both concentrate and suspension fluid must be added periodically to maintain proper image density. Also, if the paper sits still for any length of time in contact with the liquid applicator, the toner seeps into the paper and leaves a grey smudge.

The use of a dry, single-component magnetic powder toner overcomes these problems. The powder is inert, non-toxic and consists of particles that are a uniform mix of carbon wax and magnetite. The most significant benefit of single-component toner is consistent image quality. In addition, since there are no liquids or separate elements to mix, it is far more convenient to use.

Thermal technology closely resembles electrostatic technology, except that wires heat a thermally sensitive paper. The heat breaks open microcapsules in the paper, releasing a liquid that creates the graphic information.

Although not yet in widespread use, graphical computer output on to microfilm (com) is available. This technique uses a laser to plot the resultant image on microfilm in much the same way as a crt beam plots the image on a storage-tube screen.

In recent years, colour graphics terminals and displays have become more widespread. Where colour hard copy is required, this is achieved in one of five ways. The first is the multipen flat bed plotter; these devices have a choice of up to eight pens, and are able to use a variety of plotting media. The second is the colour matrix pattern/plotter mentioned earlier.

The third is a conventional Polaroid camera attached to a purpose-built hood which is held against the display screen, and the image photographed directly from the crt. Although this technique is convenient and low in cost, it does have the drawback that, since all colour crt's have a curved screen surface, the finished photograph will exhibit 'pincushion' distortion.

The fourth (and most interesting) colour-copying technique is the directly coupled colour camera system. In this device (Figure 10), a duplicate image of the picture seen on the display screen is produced on a high-resolution flat-screen monochrome monitor. The monitor's image is focused on to the film plane of a Polaroid or 35mm camera.

By intercepting the light path between the monitor and the camera with a red, green or blue filter, each primary colour can be exposed in turn for the correct amount of time. The finished photograph exhibits little, if any, pincushion distortion, and has excellent colour rendition.

The fifth and final method is the ink-jet printer. These devices, such as the Applicon colour-plotting systems, have three ink jets of primary colours (red, yellow, and blue). A constant-size spot is created by the ink jets, producing a uniform resolution in the region of five points per millimetre. A single picture element may contain any of the three colours, alone, or superimposed on top of one another, thus providing a wide choice of colours and shades. Since there is no mechanical contact with the writing surface, a wide choice of recording media can be used.

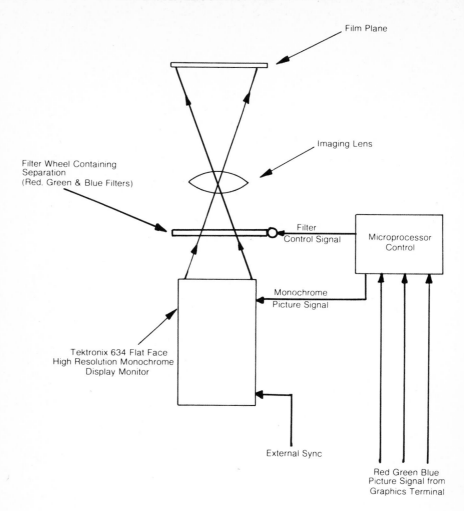

Figure 10 *Colour camera operation*

The future

There are several developments yet to be commercially exploited on dvst-based displays. Raster-scan graphics are unlikely to go much beyond 1000-line non-interlaced resolution. In industry it is expected that many more companies will invest in computer-aided-design tools, and that manufacturing will be more closely linked to design. In hardware terms, the one technological barrier yet to be broken is the design and manufacture of medium-cost colour high-resolution hard-copying devices. Once it is possible to produce A0-size colour prints at a reasonable cost, the days of the dyeline printer are numbered, and colour graphic terminals will be the 'norm' in much the same way as the domestic colour TV.

H Rippiner, Product Manager, Tektronix UK Ltd, PO Box 69, 36-38 Coldharbour Lane, Harpenden, Herts AL5 4UP

Discussion

J S Gero, University of Sydney:

A brief comment on one statement that you made about 1,000 line raster systems. The Sony company is just about to release its 5000 × 3000 line three bit deep bit-map display.

H Rippiner, Tektronix:

That represents quite an incredible thing. If you work out the mathematics, once you get above 1,000 lines, the video bandwidth shoots up tremendously. The scan rate is so fast.

J S Gero:

It doesn't use one beam.

H Rippiner:

Oh, so that is how they are getting round it. Thanks for that information.

J S Gero:

It's, relatively speaking, cheap. £12,000 in Japan.

16. Educational software for CAD teaching

I C Wright

Abstract: Good quality engineering software suitable for educational purposes is in short supply. Most of the development is done within the educational establishment where it will eventually be used. The work is time consuming, and is frequently considered to be an improper use of an academic's time. Two types of educational software are identified and suggestions are put forward to make them more available for academics.

Introduction

In the summer of 1982, the second SEED* seminar was hosted by the Department of Mechanical Engineering at Loughborough University of Technology. The seminars are annual events which are organized with the intention of providing a forum for the discussion of topics relating to the teaching of engineering design in undergraduate courses. The general format reserves the morning session for presentations describing the methods of engineering design teaching in the host Institution, whilst in the afternoon, discussion groups are formed to deal with specialized topics in depth. At the Loughborough seminar, one of the discussion groups considered the problems associated with the teaching of computer-aided design (CAD). Among the many problems which were identified was the difficulty encountered in writing or otherwise acquiring good software for use by undergraduates.

Types of software

Two types of software were identified as being required for CAD teaching. These were:

1. Professional standard software
2. Comprehension demanding software

Type 1 software is typified by the CAD package which may be used in an industrial environment by a practising design engineer. To be properly used, the engineer would be aware of the engineering science upon which the program is based, together with the assumptions made therein. Furthermore, the engineer would have to be capable of interpreting the results presented to him in the form of output from the program. In general, however, the program user would not be required to make decisions during a program run which would test his understanding of the theory in any depth. For example, a user of a commercial finite element (FE) package such as PAFEC would not need a detailed knowledge of FE theory to obtain output, but he would be unwise to accept that output at face value without knowing the limitations of both the program and the theory. A similar understanding is also required for other type 1 software such as STEM (Shell and Tube Heat Exchanger Design, marketed by the Computer Aided Design Centre and NEL).

*SEED is an acronym for Sharing Experience in Engineering Design.

When type 1 software is used in an undergraduate engineering course, it is used to:

(a) provide a demonstration to students of the type of commercial package that they may use in industry, and/or

(b) provide a useful tool for students looking for an optimum solution to a design problem.

In both of these uses, the student's knowledge of underlying theoretical relationships may not be tested as a *direct* result of using the program, although his understanding of the interrelationship between design parameters can be improved if the programs are iterative in nature. For these reasons, type 1 software, when it is used in undergraduate courses, is usually reserved for more advanced students in the final or penultimate years of their course.

In contrast to the 'professional' standard software just mentioned, type 2 programs are designed so that the user must be familiar with the engineering science basis of the program before progress can be made. In essence, this type of software provides a means of solving tedious, iterative mathematics without giving assistance in the form which would remove the need for ongoing understanding. The engineer would therefore be required to make professional judgements during the program run, and the wisdom of such judgements would be reflected in the suitability of the final solution. Programs of this nature are produced in some quantity in educational establishments as an aid to student learning. They have particular relevance when it is desirable for students to become aware of the relationships between the variables in a design problem. Two examples of this type of software may help to identify the general form:

(a) The design of helical compression springs is a problem which is often given to engineering undergraduates. A typical problem may have around ten variables, with most of the variables being subjected to regional or discrete limits. If an optimum solution is required, an analytical approach is usually impossible or, at least, difficult. An alternative means of obtaining the solution may be to use standard optimization techniques such as those that are available via the NAG library programs. Even so, the problem is difficult to formulate correctly, and the method does little to advance the student's understanding of the relationships between the variables. A better approach is for the student to be given access to an interactive program which will allow him to learn about the problem as he feels his way to the optimum by changing the variables in a series of small steps.

(b) Projects incorporating the design of spur or helical gears to BS 436 are found in several undergraduate engineering course. Values obtained from graphs within BS 436 are substituted into equations which yield the maximum power and torque which may be carried by a gear from the point of view of both wear and strength. There are several commercial gear design programs available which automate the process to an extent where a satisfactory or even optimum solution may be obtained with only minimal understanding of the underlying relationships. An interactive program which requires the student to refer to BS 436 during a run may be used to ensure that the student is fully conversant with the standard.

A large number of practising engineers write and use their own software as an aid to the solution of design problems. The proportion of engineers writing their own software has been suggested by Clifton (Clifton, 1982) to be as high as 75 per cent and it seems to be desirable that the necessary skills are taught. Type 2 software can assist in this requirement by being similar in nature to the type of program produced by designers in industry, particularly when it is coded to run on a desktop micro.

Availability of type 1 software

Professional standard software is available from a large number of commercial software houses and finds a ready market in some sectors of manufacturing and engineering service industries. The software is made available to the user in one of the following ways:

(a)   purchase
(b)   lease
(c)   bureau service.

The way in which a particular piece of software is made available often depends upon its complexity and its development cost. The smaller the program or less the development cost, the more likely it is to be made available by direct purchase with the availability of an ongoing maintenance contract if required. Software with a high development cost is more likely to be leased to the user and/or offered as a bureau service via on-line terminals or by post.

In general, the pricing strategy adopted by British software houses makes the cost of leasing engineering application programs prohibitive for educational establishments. Similarly, bureau services are also expensive if a large number of students are to use the facility as part of their undergraduate course. The practice followed by most USA software houses, of making considerable discounts available to their universities, has by no means been accepted in the UK. The only area where the discount scheme has been accepted is that of computer-aided drafting where the predominance of American companies has influenced marketing methods. For engineering application packages then, educational establishments are restricted to the cheaper end of the spectrum if they wish to expose their students to professional software. As has already been stated, the smaller programs which are available in the lower cost range are more likely to be acquired by straight purchase. This method of acquiring software is, in any case, better suited to the financial structure of universities and polytechnics, where ongoing costs are difficult to sustain, particularly with a contracting budget. Table 1 gives information on the cost of some commercial application packages, their areas of use, and their portability.

Availability of type 2 software

The production of type 2 software by academics is a necessary but time-consuming activity. Good educational software is just as difficult to produce as software for the practising engineer. It always demands from the author a clear understanding of the purpose for which the software will be used, as well as professional engineering competence. The fact that these stringent requirements are placed upon a writer who is usually an engineer rather than a computer software specialist makes the task even more difficult.

The incentive to become involved in educational software writing are small. The activity is, in general, considered an unrespectable pastime for an academic engineer. Outlets for publication of such endeavours are at present inadequate, and much good work therefore goes unnoticed by other workers. The Loughborough SEED seminar concluded that an effort should be made to provide a forum where writers of educational software could make their work known to other academics with similar interests. The outcome of this resolve was the 'Computer Aided Design Software Exchange' newsletter (CADSE) which was first issued in January 1981. The initial purpose of the newsletter was to foster contacts between academics with common interests in software development. This was done by enabling software authors to enter details of their programs in CADSE which was then circulated to departmental heads and known CAD activists in departments of mechanical, and production engineering in British universities and polytechnics. The services of the newsletter were made free of charge to

Table 1 *Type 1 software – commercial engineering application programs*

Sales Organization	Software Details	Cost
Engineering Sciences Data Unit	COMpacs. Originally designed for use on Tektronix 4050 series desktop graphic computers. Applications covered include buckling of flat plates, stresses and deflections in plates, bending of curved beams, stresses of cylindrical shells and circular plates under pressure, and fatigue damage estimation.	COMpacs and ESDUpacs are leased at £250 pa: this figure includes associated hard copy Data Items.
	ESDUpacs. FORTRAN or BASIC programs to provide computerized versions of some of the hard-copy ESDU Data Items. The software is for incorporation by the user in his or her own computer system.	
Robinson Ford Associates	TRAIN. A low cost interactive finite element system for education. Written in ANS166, FORTRAN 4 for use on a Micro Nova. Can be implemented on other types of computer.	£5,000 for licence if a Micro Nova is purchased. Otherwise the licence is £7,500.
	ROSE. Post processes finite element results.	Same as TRAIN.
	PVE5. Pressure vessel design to BS 5500. Written in ANSI FORTRAN.	£10,000 pa for five year leasing arrangement.
Whessoe Technical and Computing Systems Limited	PSA5. Pipe stress analysis. Written in ANSI FORTRAN, and can be implemented upon many computers. There are two versions of the program, one taking account of dynamic effects, whilst the other is for statics only.	Static analysis, £16,000 pa licensing arrangement. Dynamic analysis, £24,000 pa licensing arrangement. Both of these are on a five year basis.

contributors and recipients. To ensure some degree of uniformity between the entries in CADSE, a proforma was used which required the following information:

(a) program title
(b) programming language and size
(c) the type of computer(s) on which the program has been mounted
(d) whether the program was interactive or batch
(e) what peripherals were required

(f)   the name and address of the program author

(g)   a description of the program.

It seemed obvious from the outset that a service which was provided free of charge should not become a vehicle for authors wishing to advertise software which had been written primarily for commercial purposes. Authors of this type of software could be expected to pay for their advertising in other media. However, producing good quality software is a skilful and time-consuming task, and the results are rarely something that the author will wish to give away. The return to an author for contribution in CADSE could be in two ways. First, he could charge for a listing or tape of his program, the only restriction being that the software must not have been produced with commercial marketing being the primary concern. Second, he could exchange the software for a program produced elsewhere.

The first issue of CADSE (CADSE 1) was circulated to approximately 150 people. It contained details of eight programs from authors in universities and polytechnics. This original circulation, plus some editorial comment which the newsletter attracted in engineering journals resulted in an encouraging response from that sector of higher education concerned with CAD teaching. A response which had not been foreseen came from the schools. The rapidly growing availability of microcomputers in schools has produced a situation where there is a serious 'gap' in the development of software which is suitable for teaching purposes at that level. As a result of a short editorial comment about CADSE 1 in *Educational Computing* there were close on 100 requests for copies from comprehensive and grammar schools. However, a large majority of the programs which are produced for higher education are totally unsuitable for school use, and the degree of difference between the requirements make it difficult to visualize a publication which could meet the needs of both. The main thrust of CADSE will remain in engineering software, and the demand for development information in this area was reflected in the CADSE 2 circulation list reaching 250. In addition to these, 50 copies were sent to manufacturing companies to gauge their interest in this field. Over half of the companies circulated have expressed their support and requested to be placed on the mailing list for CADSE 3. CADSE 2 contained details of 11 programs, but returns for CADSE 3 suggest that it will contain at least twice this number. The circulation list for CADSE 3 will be restricted to those making personal requests for a copy. There will be no general circulation to departmental heads as in the first two issues because it is important to minimize cost if the service is to remain free of charge. The result of this change, together with the increase in the number of personal requests from academics and industrialists will probably bring the CADSE 3 circulation to around 300.

A compacted version of the programs described in CADSE 2 is presented in Table 2.

Future developments

The difficulties involved in acquiring good engineering software for educational purposes are, if anything, greater than those associated with obtaining the necessary hardware. At least, a wide range of hardware is on the market if the necessary finance is available for its purchase. The marketing policies of companies selling type 1 software make it an unattractive proposition for higher education, whilst good type 2 software is in short supply due to the lack of reward for authors.

A promising trend is the recognition by a few software houses that there is a potential market for type 2 software if realistic price levels can be set. Several such companies have recently commenced the marketing of engineering software written specifically for education purposes. Another interesting development is the setting up of a software development and marketing facility by the Engineering Sciences Data Unit Limited. Many authors of otherwise good software are let down by their lack of marketing knowledge. ESDU Limited intend to use their experience in both of these areas to assist authors to develop their programs to a stage where they are more acceptable as

Table 2 *Type 2 software – extracted from CADSE 2, January 1982*

Author	Details
Wilson, R North Staffs Polytechnic	Combines graphics and text on a 48K Apple II. Positioning is by means of a cursor, and diagrams can be saved.
	Designs gears and shafts, and aids the selection of bearings and shaft keys for spur and helical gear boxes. Runs on 48K Apple II.
	Aids the selection of a pair of bearings to support a simple shaft with specified loading and speed. Runs on a Hewlett-Packard 2000E.
Pressnell, M Hatfield Polytechnic	Evaluation of area, centroid position, second moment of area, and stresses for complicated solid or hollow sections. FORTRAN IV. Dec 10/81.
Spicer, H R Bristol Polytechnic	Aids the design of helical compression springs, taking cyclical stresses into account. Based upon ESDU 65005. BASIC. Apple II.
Clarke, D University of Strathclyde	A library of programs used to determine the bending design requirements of simply supported prestressed concrete beams. BASIC. H6060.
ESPE Exchange Co-ordinator Queen Mary College London	Models linear phased array antennas by superimposing point source radiators. FORTRAN. Nova 2. Determines gas generator equilibrium running lines. FORTRAN IV plus GINO. PDP – 11/40.
London	Facilities parametric study of the influence of various independent variables upon uniform and critical depth in an open channel.
Wright, I C Loughborough University of Technology	Produces dimensioned drawings of turned parts with input from keyboard of digitizer. Stores part geometry to disc. BASIC. Tektronix 4051.

marketable products. Profit on sales would be shared by the author and ESDU Limited. Whilst there is no intention to restrict this service to educational software, it would seem likely that the scheme will prove useful to engineering academics.

An important step forward would be a recognition by the grant-awarding bodies that work on the production of good educational software is acceptable as an area for sponsored research. Whilst a limited amount of work is already being done, the general level of funding is too low. An increase in the amount of funded research in this topic would not only directly support work which is badly needed, but would also tend to give respectability to the developments which are already underway.

From the response to the first two issues of CADSE, it would seem clear that there is demand for a vehicle by which information on software development can be disseminated. Existing journals are, in general, concerned with publishing work which concentrates on the mathematical developments associated with new software. Important though this aspect is, it is often not the most relevant criteria by which educational CAD software can be assessed.

It has often been proposed that national software bank would benefit both producers and users of engineering CAD programs. The idea of such a scheme has, in the past, attracted support from many august engineering bodies. The problems associated with the organization and funding of a software bank have always been found to be intractable, although the advantages which would accrue from its presence would be substantial.

References

Clifton, C J (30 March-1 April 1982) The economics of micro computer use in the engineering design office. Proceedings CAD 82 5th international conference and exhibition on computers in design engineering: Brighton, UK: 494-500

I C Wright, University of Technology, Department of Mechanical Engineering, Loughborough, Leicestershire LE11 3TU

Discussion

S A Abbas, Teesside Polytechnic:

From what I understand, you don't put a stamp of approval or comment or critical analysis on software that's submitted to you and then disseminated to members.

I C Wright, Loughborough University of Technology:

None whatsoever. The point of the interchange of information is to encourage workers in similar areas to come together. If a reader of CADSE sees that there is a particular program that looks interesting to him or is something along a similar line of development to what he is doing, then his contact would be with the author of that software who had entered it into the newsletter. Having established contact with the author of the software, the reader can see whether it would suit his own particular application.

S A Abbas:

Do you think there might be a need in future for some sort of categorization of levels of software? This would depend, I suppose, upon the extent of support that there might be for it. For example, there might be no support at all where it's totally at the risk of the user, or at the other extreme there might be support for the computer centre from the originating polytechnics or universities.

I C Wright:

We have been in something of a dilemma in producing the newsletter about what should go into it. We decided earlier on that we shouldn't put commercial software into it. If a piece of software is being developed commercially for sale then we would expect that organization to find their advertising somewhere else and not use the newsletter as a free vehicle for advertising. So we would limit entries to software which were developed in the first place primarily for undergraduates CAD teaching which may perhaps at a later stage be developed into commercial software.

S A Abbas:

I wondered whether in certain instances a computer support group, or a computer centre, or a unit in a polytechnic or university might provide the documentation which year by year would get updated or the package withdrawn or scratched. That's a level of support I would suggest.

I C Wright:

We already have quite a large range of software. We have that sort of software already in the newsletter. From very large packages to the other end of the scale with people working on PETS and Apples. I take your point that as this develops, I think some division between programme types and complexities may be necessary.

M Dooner, Lanchester Polytechnic:

Do you have any experience of software that has been developed under SERC grants for example and what are the problems in trying to exchange this sort of software between institutions?

I C Wright:

I don't have any experience personally of that type of software, although I am aware that some of the software published in the newsletter has been developed in that way.

M Dooner:

What are your feelings about publishing software that has been produced under a research contract?

I C Wright:

I would leave that purely to the conscience of the people who were putting it into the newsletter, but it would have to come from the department rather than the student. If the department feels it fit to publish details of that software then we wouldn't put any restriction on it.

17. Implementing graphics in design, process and manufacturing industries

P Daniel and A Pinnington

Abstract: CARBS is a modelling system that allows the user to model in either two or three dimensions at a number of levels of sophistication. This paper outlines the various ways in which the system can be used and points out the benefits to be gained in the training of engineers in universities and colleges by the use of this system within their courses.

Introduction

CARBS is a suite of computer programs which enables a designer or draftsman to define a two- or three-dimensional computer image or model of a building and structure, a ground surface, equipment and piping units making up a process plant complex or virtually any combination of groups of physical objects.

From the model the user can obtain from any viewpoint orthographic, perspective or isometric drawings with all necessary hidden lines removed.

The system allows the user to develop his model so that related physical properties and dimensional parameters can be identified and used to form schedules of materials and costs or numerical data for manufacturing purposes.

The creation and manipulation of the model is carried out by a range of command statements. These may be initiated at an interactive level, giving immediate response, or by preparing an ordered set which may be executed in sequence. As the system contains a powerful programming facility, the latter process enables designs to be done from which automatic dimensionally variables models can be created together with final drawings and actual production documentation. These processes can be controlled, that is, the user can intervene by using commands which prompt the user for information should certain constraints be violated.

Most of the data input to the system for pure drafting purposes is done by cursor in conjunction with menu facilities on the tablet or digitizer. Whilst this can be still used for 3D input, a more powerful facility is the 'design language' which can be combined with commands and program statements but is more readily used in the definition of the 3D models particularly in say, the piping field where traditionally the designer thinks in 3D.

CARBS has a powerful data-base management capability which enables an organization to set up generalized data-bases for access during modelling or to transmit data from a model developed in one design discipline to a user creating a related model in another discipline.

The input to the system is via keyboard, function keys, cursor on graphics screen or digitizer and menus on the VDU or tablet/digitizer. The menus and function keys are programmable from within the CARBS system, hence can be changed by the user to suit his particular requirements.

Output of drawings can be to screen them to a hard copy unit or similar. This form of drawing would be for project evaluation purposes and for small details. Larger study and final drawings would normally be to an incremental plotter. Text output can be initially to VDU and then to hard copy or printer as required.

Basic modelling

Objects to be modelled may be two- or three-dimensional, simple or complex. Objects are described using four basic graphic elements, lines, faces, solids and pipes.

To define a two-dimensional 'wire' diagram, lines are used. Figure 2.1 shows a simple square made up from lines. By adding more lines a cube may be constructed, as illustrated in Figure 2.2; the transition from a two- to a three-dimensional object has occurred. However, this simple lattice is not an opaque object, to define such an object, faces need to be used.

A face is a polygon with no thickness; a sheet of paper. It is opaque when hidden line removal is carried out. Figure 2.3 illustrates a cube made up from six faces, viewed in perspective with hidden lines removed. This type of prismatic shape can be extruded from a base polygon, thus short-cutting the definition to one polygon only. However, an object made up from faces is hollow; when sectioned it has not cut surface. To obtain an object that will show a cut surface a solid must be defined.

Prismatic solid objects are defined by a polygon and an extrusion. When sectioned, a cut surface is revealed which may be automatically cross-hatched. Figure 2.4 shows a cut section of a solid cube. Figures 2.5, 2.6, 2.10, 2.12, 2.13 and 2.14 show examples of solid shapes that have been defined by extruding a polygon. All are perspective views with hidden lines removed.

Non-prismatic shapes can be formed by defining several faces that are adjacent to each other. Figures 2.7, 2.9 and 2.11 illustrate objects made from combinations of faces and solids (extruded polygons), again hidden lines have been removed from all the views. Figure 2.11 also shows an example of pipe definition, the taps.

Figure 2.8 shows the use of the curve fitting routines available. At present, these are restricted to line graphics only and consist of bi-cubic and cubic spline curves as well as conventional circles and circular arcs.

Complex modelling

Simple objects can be built up from the basic graphic elements, as two- or three-dimensional images. These objects can be stored away on disc, by the modelling system, and are known as sub-models or components. Components can be retrieved at any time and used as many times as the user requires. A component can be as simple (a rectangular column) or as complex (a washbasin or a door leaf) as required. There is no limit on this complexity.

As well as adding components to the model, it is possible to erase (or kill) components that are no longer required, replace objects by others and duplicate items. When the model in the work space has been changed to your satisfaction it can be saved away as a new model or can overwrite an existing model, provided you have the authority (security) to do this.

Thus, all the component parts of, for example, a building, can be built up and stored away and then recalled and assembled to form the whole building. This three dimensional model (if its component parts are described in 3D), can then be viewed from any angle to give conventional drawings. Plans, elevations and sections are all 'taken off' the same 3D model built up in the model work space. Figures 3.1 and 3.2 show examples of orthographic views taken off a model of a room. Figure 3.3 is a perspective view of the same model.

Models with curved surfaces

The operation of the system to produce the models shown in 2 and 3 is applicable over the whole of the system. An obvious extension of *faces* and *solids* is *pipes* where by using the directional phrase facilities complex process equipment and piping configurations can be modelled.

The pipes command contains extra facilities for defining diameters, bend radii, reducers and other piping parts, by their inclusion, a wire frame model can be transformed into a piping model.

The basic fittings in a line may be defined entirely by the user to his discretion. In process plant piping valves and strainers for example, are normally shown in a symbolic form. In ship-building, a more realistic representation is required as the objective is to minimize the space occupied by say, ballast, pipework, etc. Hence, accessibility to control equipment is important and it is necessary to be able to model obstructions more accurately.

The pipework shown in Figure 4.1 demonstrates quite adequately the capability of the CARBS system to model up this type of pipework. The valves, with the exception of the handles, are shown in fairly conventional manner. The valve chest is detailed to give a more realistic representation of space occupied.

All the equipment is related to suppiers' data and command files were written which contained questions like: 'Give supplier X's dimensions A, B, C and J', which when answered with the appropriate dimensions would form the graphic component which was then named and stored in the system, ready for inclusion in the pipe model at a later stage.

On completion of the model, line references can be assigned to it and final dimensioned isometric or plan and elevation drawings can be produced automatically together with material lists for fabrication and erection.

A further example of modelling curved surfaces is shown in Figure 4.2. The vane or screw thread on the device has been obtained by writing a command file which when initiated, asks the user to input the necessary parameters. The embedded program determines the surface geometry and the 3D model is produced automatically. When the design is proven, final dimensioned working drawings can also be automatically output.

Model storage and control

The information and data files used by the modelling system allow great flexibility in the method of working on a model. There are several different types of files available for holding graphic information (components), component descriptions (as text), sets of co-ordinates and global data (data files), 'frozen' views of the model (plot files) and predefined sequences of commands (command files) which can be simple 'user programs'. The latter is discussed later in this document.

As a component is recalled and placed in the current model (work space), there are commands that enable you to shift it laterally or vertically, rotate it about any axis, tilt it, invert it, mirror it or scale it. Control of the location of components and the model is maintained by using a conventional co-ordinated axis system with x, y and z axes. However, positions need not be referred to by absolute co-ordinates, which can be tedious and confusing, but can be relative to the current position. This is done by reference to direction and distance (ie a 'vector description'). The directions used are north, south, east, west, up and down. (North-south is the y axis, east-west is the x axis and up-down is the z axis.) Thus, E5,000 means East 5,000 units of distance, relative to your current position. Non-orthogonal directions and angles are also allowed in this 'vector description language'.

eg   E5,000 & N3,000
        a diagonal line, east 5,000 and north 3,000
        E1,000 & N@60
        a diagonal line, east 1,000 and north at 60 degrees

All objects are built-up and stored as 'life size'. The units of measurement usually used are millimetres. Distances and positions can be held with up to four places of decimals, (eg 21345.6073 mm). There is no upper limit on distances. The modelling system automatically scales objects as necessary when a view is requested on the graphics screen. Thus, a command SHIFT E4000 (which means move the current object eastwards by 4 metres) moves the object a scaled distance on the graphics screen. (This may, in fact, move it off the screen if the current view does not look upon a large enough area. However, this does not invalidate the information held in the model work space.)

CARBS as an educational aid

One of the traditional problems facing universities and technical colleges is the training of engineering graduates in the truly practical aspects of design and design realization.

Using CARBS the education establishments could acquire or create realistic practical models for demonstration purposes. These models could convey to the students far more information that the equivalent manual system, and in a much shorter time-scale. The other advantages to the student are as follows:

(a) He can create complex models of the conceptual stages of design.
(b) He can embed design/analysis routines in command files, with user prompts, to see the effect of change on the geometry of the model.
(c) He can set up mechanisms within command files which will enable parameters to be extracted from output files of external design/analysis programs to aid in the formulation of models.
(d) The user interrupt facility enables models to be changed at any stage.
(e) From the final model he can obtain final working drawings to which can be added any necessary text for clarification purposes.

CARBS are prepared to make the system available to universities/technical colleges subject to suitable agreements.

Acknowledgement

The authors wish to thank Building Design Partners for their contribution to the parts of this paper referring to building modelling.

P Daniel, CARBS Ltd, c/o Architects Department, Clwyd County Council, Shire Hall, Mold, Clwyd

A Pinnington, CARBS Ltd, c/o Architects Department, Clwyd County Council, Shire Hall, Mold, Clwyd

Discussion

S B Kamath, CMTI, India:

Approximately what is the cost of this system? Can it do cinematics and dynamic representations?

P Daniel, CARBS Ltd:

Answering the second part first, yes. Cost? It is very flexible; I am the one who determines what to sell it for and a target price for a single user station would be round about £25,000 if the application was fairly general.

If it was a small organization, with a couple of designers, a couple of draftsmen with a very narrow product range, we would cut that down to between £5,000 and £10,000 according to what we thought they could support. At the other end of the scale, putting it into a very large organization with around eight terminals, we would probably expect around £80,000. Basically we are aiming at a figure per display. You can't avoid the initial cost being fairly high because a lot of the maintenance still goes in at the early stages.

18. PELICAM: An interactive educational software for training students to the finite element method

M Blanc, G Coffignal and G Ishiomin

Abstract: PELICAM is an interactive educational software designed for the initiation of students into the finite element method applied to the mechanical vibrations of deformable bodies. It offers the student the opportunity of building up and running his own example during a working session.

Introduction

At the present time, the industrial applications of the finite element method (FEM) are occurring more and more frequently in the field of computer-aided design (CAD) and a large number of programs exist for evaluating displacements and stresses within complex structural systems. These computer programs may be used as 'black boxes' without a complete knowledge of underlying assumptions and theories which have been incorporated into them. Such a use may cause many problems owing to the inability of an unqualified user to get a critical view of the results. Pre- and post-processors connected to the kernel of these softwares are now making an extensive use of interactive graphical displays.

The Ecole Nationale Supérieure d'Arts et Métiers (ENSAM) is an engineering school, oriented towards mechanics, where the students have to learn FEM. The theoretical aspects of the method are developed along with the courses in vibration mechanics, structural mechanics and mathematics. Usually, a large number of examples are needed to explain and exhibit the different aspects of a complete analysis by the FEM. These examples have two major weaknesses:

1. They only allow a very crude idealization of a real structure if a computer program is excluded.
2. They are imposed on the student as the set of solved problems is not very numerous.

The need for a specialized computer-aided learning tool, in the sense developed in *Interactive Computer Graphics in Science Teaching* (McKenzie, Elton and Lewis 1978), arose naturally while using standard batch-oriented packages to show real life examples to our students. Some important work has been carried out in this area, see for instance, CAL described in 'CAL a computer analysis language for teaching structural analysis' (Wilson, 1978); or 'A teaching package for computer solution of engineering problems' (Brebbia and Ferrante, 1977). But as far as we know, these CAL programs are not directed towards the representation of motion on a convenient display.

In industrial packages, the description of the finite element model in the first stage of the analysis, and the aspect of the results in the last one, are concerned with interactivity, but the kernal itself is a (complex) 'black-box' process where interactivity has not been included.

PELICAM is a fully interactive software designed for training students in FEM applied to mechanical vibrations of deformable bodies.

A main advantage of the FEM from the viewpoint of CAL is that the model offered to the student is not chosen by the program designer excepting, of course, the choices related to the FEM itself.

Educational aims

As mentioned above, PELICAM has been developed to enhance fundamental courses by computer-aided application sessions. The computer is essentially used to provide opportunities for learning to take place, in a field where one of the final fundamental facts is motion.

The main education aims of PELICAM are to:

1. develop the student's understanding of
   - the concepts and methods presented in theoretical courses;
   - modelling by FEM in dynamics;
   - the dynamic characteristics of finite element models;
   - the results available from simulation by models;
   - the numerical algorithms involved in the solution procedures;
2. familiarize students with:
   - the basic features of industrial structural analysis computer packages;
   - CAD working techniques.

As a consequence, because the aims are directed towards understanding FEM, there is no requirement that the student should be able to program the computer: while using PELICAM, the student needs no particular knowledge of any programming language.

Structural modelling

In practice, the mechanical engineer has to extract, from the actual structure, the pertinent parameters which will allow him to study its behaviour. A frame of hypotheses is selected which describes the space of displacement fields and constitutes the basic theory. The three-dimensional continuum idealization, the plate idealization, the beam idealization and the rigid body idealization are examples of such frames of hypotheses.

The complete definition of the basic engineer's model is achieved, within the theory, by a description of:

- structure geometry
- boundary conditions on displacements
- the set of given applied loads
- constitutive laws of materials (elasticity, viscoelasticity . . .)

Usually, such a basic model leads to a mathematical problem, involving partial differential equations. In most industrial situations it is not possible to obtain an explicit solution. However, such a model is supposed to be the best one, so the associated problem can be considered as the 'reference problem'. Among approximated solutions, the FEM offers a versatile, wide range of modelling capabilities. From the mathematical viewpoint, a new problem for which a solution can 'easily' be found is substituted to the reference problem.

In the particular case of the displacement method, a space of displacement fields of finite dimension is substituted to the initial one. This operation can be thought of as a modelling of the displacement fields.

From the engineer's viewpoint, the basic model has to be approximated by a finite element model, whose behaviour is supposed to be a good model of the actual structure's behaviour.

Practically, the engineer must take into account the softwares which are available to him. Each software package offers a set of elements. An approximate displacement field is associated to each type of element, according to some mathematical requirements, during the design of the package. So, the user builds up his model by assembling compatible elements together, in order to give back the geometry of his basic model. Simultaneously, an approximate solution space of finite size is automatically built, according to the type of element selected. Very often, that approximate space is not understood by the beginner: this is one of the major reasons for mistakes.

A finite number of ordinary differential equations can be automatically derived from the finite element model and is substituted into the original set of partial differential equations.

The current version of PELICAM (82.0) deals with planar structures that can be built up with beam elements, linear springs and dampers, concentrated masses and pendulums. The constitutive law of the material may be either elastic or viscoelastic, the shear flexibility may be neglected, if need be, and the kinetic energy may be evaluated using either a concentrated or a consistent mass matrix for the beam element. The displacement approach has been used.

It must be noticed that the approximate solution depends, for a large amount, on the finite element model built up by the user, but it also depends on the algorithm used to solve the associated set of equations. Some parameters associated with these algorithms have to be chosen with a good knowledge of their action: the time step in integration schemes or the tolerance value in iterative algorithms are good examples of such parameters.

Algorithms

Under classical assumptions, (Coffignal, 1982); (Bathe and Wilson, 1976), the finite element model leads to a set of ordinary differential equations which can be conveniently expressed in the following matrix-form:

$$M\ddot{u} + C\dot{u} + Ku = F(t), \text{ where}$$

M, C and K are, respectively, the mass matrix, the damping matrix and the stiffness matrix,
u and F are, respectively, the unknown displacement vector and the applied force vector
$\dot{u}$ and $\ddot{u}$ denote, respectively, the first and second derivatives of u with respect to time.

The solution algorithm depends on the results that are expected from the model:

- modal properties ie mode shapes and natural frequencies as solutions of an eigen value problem which can be solved using different algorithms (Jacobi, Householder, Inverse Power . . .)
- simulations ie transient response and frequency response are obtained through the use of direct integration schemes, normal mode analysis or complex analysis.

The current version of PELICAM allows the determination of mode shapes and natural frequencies using Jacobi's method, or the inverse power the shift method. The simulation of a transient response can be performed using the central difference method.

Program design and implementation

PELICAM is implemented on a 4054 Tektronix desktop microcomputer which has been selected for its dynamic graphics capabilities associated to a high resolution display. The working station includes a hard copy. The programming language is a graphic enhanced BASIC. PELICAM represents, at the present time, about 7,000 instructions.

The amount of guidance or decision given to the student by the program was an important decision. In order to allow a quick, meaningful output for the beginner, many default decisions or values of parameters have been chosen. On the other hand, no definite finite element models have been chosen: this choice is the first active decision of the student. Practically, PELICAM follows the logical flow of a dynamic analysis, to manage the dialogue with the user. At each stage the program asks questions when a decision or data are needed. At any moment it is possible for the student to come back to a previous stage using a set of special function keys (see Figure 1): so a change can be made in a previous decision or data input. This permits modifications to the model itself, modifications to parameters, changes of algorithm . . .

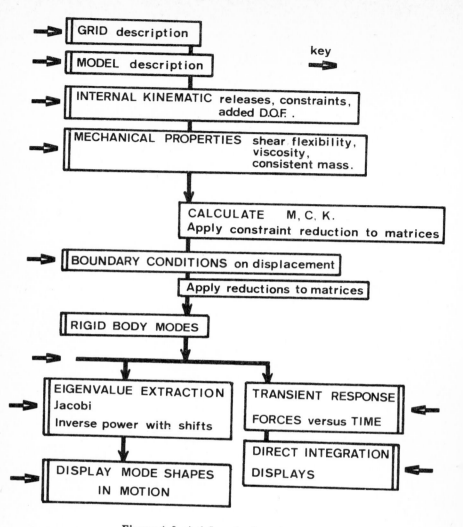

Figure 1 *Logical flow of a dynamic analysis*

To satisfy these requirements, PELICAM is organized into independent functional blocks which are automatically linked according to the logical flow of the analysis and/or to the (licit) requests of the user. Data are exchanged between these blocks via the memory and/or files.

Of course it was decided to protect the program against as many errors as possible, and this was really a very difficult task. It was also decided that the screen will be, as far as possible, always showing something about the current computations when a dialogue is not engaged. So matrix filling, number of iterations, . . . are visualized in order to show that the machine is working for the user.

Working session and dialogue

As mentioned above, the dialogue is based on the logical flow of a dynamic analysis (see Figure 1), but at any moment the student may alter natural flow, returning to a previous stage. The form of the dialogue depends, essentially, on the kind of decision or data needed: when the student has to take the initiative in doing something, an extensive use is made of graphic interactivity, otherwise when a simple choice has to be made, a simple question appears on the display. Two examples of such situations are given below.

The student is supposed to start the session with a precise idea of the model he wants to build up, and he is strongly advised to draw it on paper. Then the dialogue is engaged and the student has to define the dimensions of the domain in which he wishes to set up his model. A grid of points is then presented to him according to the previous choices (see Figure 2), and commands are recalled on the right side of the display. The model geometry is then entered by direct designation using the cursor and some selected key to decide whether a node or a beam or a spring, . . . is to be added at this location. Simultaneously, the model is drawn on the grid allowing immediate control and/or correction. In that phase the dialogue is limited to remarks concerning the use of a key or the logical impossibility to define one thing before another.

Figure 2 *(see page 9)*

On the other hand, when the student is entering the mechanical characteristics associated with each beam family, previous choices that have been made, such as whether shear flexibility has been neglected, and viscous damping. These choices are taken into account when printing the question: the computer only asks the numerical values which are needed, recalling their present default value if any.

So the working session begins by the definition of the finite element model, regardless of boundary conditions which will be described at a later stage. Many possibilities are included in addition to those previously mentioned such as constraints between degrees of freedom, joint releases local changes of co-ordinate systems. Their selection is simply initiated by answering yes to a question. Mechanical characteristics of elements being organized in families, a small amount of numerical data is needed.

These characteristics being entered, PELICAM builds up the K M and C matrices and while performing this task, the filling of each matrix is dynamically drawn on the display. Then displacement boundary conditions are needed and it is possible to command a static condensation. If model properties are needed, the eigenvalue extraction block is entered and after a dialogue concerning the choice of the algorithm and its related parameters, the computer extracts eigenvalues and eigenvectors. Mode shapes being available, the student may ask for their display (in motion) on the screen. If a transient response is wished, a dialogue is engaged to define where loads are to be applied and to describe their time dependencies. An extensive use of graphical possibilities is used during this dialogue (see Figure 3), to show simultaneously the entire model, the region where a load (concentrated or distributed) is to be applied (ie a node or a beam) and its associated function of time. Some parameter values are then given to the central difference scheme and the simulation of the transient response is entered. Two visualizations are available: the animation of the deformed shape or the plot of a response curve, as the calculations are done (see Figures 4 and 5).

Figure 3   *(see page 9)*

Figure 4a *(see page 9)*

Figure 4b *(see page 9)*

Figure 4c *(see page 9)*

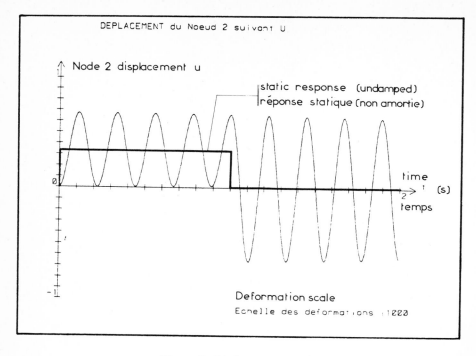

Figure 5 *Displacement versus time*

Educational material

To gain maximum benefit from PELICAM, the student should understand the
elementary theory of structures and the basic elements of the theory of vibrations. To
help the understanding and mastering of PELICAM, a documentation has been written
which consists of students' notes and a theoretical manual. Students' notes are
subdivided into user's manual and a collection of examples.

In addition to these manuals, two video tapes (12mm each) are available. The first
one allows the student to get an overall view of what is possible to do using PELICAM,
the second one exhibits the influence of the finite element model complexity and
boundary conditions on modal properties, showing animated mode shapes obtained
using the program.

Conclusions

The use of graphics alone is not new, but its combination with interaction and moving
displays, at a reasonable cost, now allows a wide range of investigations in the field of
CAL tools. PELICAM is one of these new powerful learning tools which forces the
student to become active, and which allows him to get qualitative and quantitative
responses to a real life problem that he chooses himself. Moreover, the animated
visualization of motion provides learning opportunities that can not be provided as easily
by any other educational medium.

Experiments with PELICAM have been carried out on small groups of students and
the first results are encouraging. A working session seems to be interesting if about two
hours long.

The feasibility of the system being now exhibited on a single work station, we are
now concerned with extension to a large amount. The choice of a new hardware

configuration must be studied in order to allow a standard FORTRAN version to assure the portability of the package. That is what we are doing now.

Acknowledgements

The authors wish to thank the following people for their help in preparing this paper:
Paul Johnson, BSc (Hons), lecteur d'anglais
Paul Fera, Guy-B. Mazet, Marc Williame, Alain Nobre, Florence Vassalo, students in the Structures Departments at ENSAM who implemented the BASIC version on our microcomputer.

References

Bathe, K J and Wilson, E (1976) *Numerical methods in finite element analysis.* Prentice Hall: Englewood Cliffs, New Jersey
Brebbia, C A and Ferrante, A (1977) *A teaching package for computer solution of engineering problems.* Pentech Press Computational methods for the solution of engineering problems
Coffignal, G (1982) PELICAM manuel theorique. ENSAM (in French)
McKenzie, J, Elton, L R B and Lewis, R (1978) *Interactive computer graphics in science teaching.* Ellis Morwood Limited: Chichester
Wilson, E L (September 1978) CAL a computer analysis language for teaching structural analysis. IPSI; Paris lecture notes for finite element analysis

M Blanc, Ecole Nationale Supérieure, d'Arts et Métiers, 151 Boulevard de l'Hôpital, 75640 Paris Cedex 13, France

G Coffignal, Ecole Nationale Supérieure, d'Arts et Métiers, 151 Boulevard de l'Hôpital, 75640 Paris Cedex 13, France

G Ishiomin, Ecole Nationale Supérieure, d'Arts et Métiers, 151 Boulevard de l'Hôpital, 75640 Paris Cedex 13, France

19. Computer-aided design for design and craft students

J H Frazer

Abstract: This paper attempts to identify some of the characteristics of hardware and software which facilitate the introduction of computer-aided design.

The paper begins by relating the background and experience of introducing computer-aided design techniques into an art and design college. Problems of suitable hardware and software are discussed and a justification is made for the adoption of a low-cost graphics work station featuring a small dedicated flat bed plotter instead of a high resolution screen. Approaches to improving the user interface by both hardware and software techniques are described.

The paper concludes by discussing not only some of the advantages accruing from the introduction of computer-aided design techniques, but also by highlighting some of the problems. Three critical questions are posed for the future of computer-aided design education.

1. How to choose suitable hardware and software.
2. How to encourage students to accept CAD.
3. How to overcome the damages and problems associated with CADED.

Suitable hardware and software

Introduction

Although schools of architecture and engineering started introducing the teaching of computer-aided design techniques over a decade ago, the application of computer techniques in courses in other areas of design, such as industrial design, graphic design and textiles, has been more recent. There has been a change of emphasis, computing style, programming language and hardware. A typical architectural program in the early 1970s was written in FORTRAN for a main frame or large mini and was accessed from a remote terminal. By contrast the relative latecomers to other areas of design (and second generation architects and engineers) have written programs in Pascal or BASIC for microprocessors with immediate graphic output to small flat bed plotters.

Background

I suspect our experience at the Art and Design Centre of Ulster Polytechnic is typical of the difficulties encountered by design schools attempting to introduce computer-aided design techniques. Enthusiasts on the staff of our foundation course had been struggling since about 1975 to extract computer graphics from an ICL main frame. The main frame was in a central location remote from the art college, had no suitable peripherals, had only a very slow plotter which had to run-on line, there was no really suitable software, and no staff in Computer Services who had any knowledge of, or interest in, computer graphics. Nevertheless, as often happens under these circumstances and despite having to enter all data points on punch cards, the staff of the foundation school did succeed in getting drawings out of the machine of sufficient interest to whet the appetites of the art and design staff. The main Art and Design Centre is even more remote from the main Polytechnic campus and the main frame machine and before I arrived in 1977 it had no computing facilities whatsoever – not even a terminal. I was able to negotiate some funds for providing a graphics terminal and plotter and whilst worrying about the problems of trying to work with the ICL machine, almost by luck, I came across the Tektronix 4051 intelligent terminal which had just appeared on the UK market. Everyone (including myself) assumed that we would be able to achieve very little using the 32K local processing power and that most serious computer-aided design work would have to be done using the main frame and the 4051 as a remote terminal. The local processing power of the 4051 was envisaged as being a bonus and a possible advantage for teaching purposes.

Thus, for Christmas 1977, there arrived a Tektronix 32K 4051 with data communications interface, a large digitizing tablet, an A3 plotter and hard copy unit. (We subsequently added a disc system, printer and second tape unit.) In the event we were absolutely delighted with the equipment and with the aid of a temporary research officer were able to write our own graphics software and computer-aided design programs and managed without the use of the main frame at all.

Students and staff trickled in and began to use the machine. Over three years the trickle became a flood, and in 1980 it was proposed that we should give a full introductory course on computing to all design areas of the Art and Design Centre. This included three-dimensional design, visual communication, textiles/fashion. This generated a new problem which was the numbers of students, typically 30 in each year of each school, who would be unable to get adequate access to the single microprocessor. Finances were very tight and whilst, in the meantime, it had been possible to provide the long-suffering foundation school with a separate 4051 terminal, it was clearly going to be impossible in the foreseeable future to provide adequate systems of the Tektronix type for all students.

The problem of providing an adequate number of work stations

We looked very hard at alternative equipment, always assuming that we needed a high resolution screen approaching the quality of the 4051. To this end we were looking at medium resolution raster systems such as the research machine but I remained unconvinced that the quality of the resolution was adequate.

Throughout this period we have been in close contact with a colleague, Paul Coates, at Liverpool Polytechnic School of Architecture. Liverpool invested in PETs for their architectural students, having had a similar experience to ourselves with the difficulties with their main frame. They also wanted graphics which they could not possibly obtain on the PET, and so they attached an A4 Hewlett-Packard plotter. Using just a small plotter as a high resolution graphic output instead of the screen, proved remarkably successful. To cut a long story short, we also subsequently installed two PET work stations driving A3 interactive flat-bed plotter/digitizers and we have just obtained four more similar work stations with disc drives and plotters. This configuration gives students immediate access, they can fully comprehend what is going on and can see it happening, they can themselves put special paper in the machine, can change the pens and colours of inks and interfere in a number of other ways which would be impossible if the computation and plotting were taking place remotely. Perhaps most significantly, these new work stations (complete with disc systems and plotters) are costing for all four, no more than one new intelligent high resolution graphics terminal to the remote main frame.

Software problems

The Polytechnic was able to provide a research officer for a total of two years, which allowed us to introduce computer graphics and computer-aided design in the first place and provide all our standard graphics packages for the Tektronix. It was clearly going to be impossible for the Polytechnic to fund the level of input required to establish really sophisticated computer-aided design and graphics programs on a long-term basis. We tried, unsuccessfully, to obtain funds for this purpose from the Science Research Council (who told us that others were doing the same thing – which, of course, they have not done) and so we established, in collaboration with our colleagues at Liverpool, a small company for the purpose of developing really low-cost microprocessor-based graphics systems. We had also become conscious over this period, of the extraordinary dilemma of small firms of engineers and architects. The profile of these professions is that over 75 per cent are offices of four or less staff and for them the amount of investment generally assumed to be necessary for computer-aided design equipment (a minimum of £50,000 per work station) was out of their reach. We believed we could develop a microprocessor-based system which would answer their needs, for approximately £5,000.

Problems to be overcome with low cost graphics work stations

Three problems to be overcome, were the speed of the graphics transformations, the limitations of data storage using a small floppy disc system, and the poor raster graphics screen resolution.

To deal with the problem of resolution first, we have essentially stuck to our philosophy of avoiding the problem by doing all drawing direct on the plotter. Using 400mm per second plotters this does not present a serious speed limitation and has the enormous advantage that users can see the drawing evolving as they work. This enables the draftsman or designer to work in a very natural way sketching in guidelines, trying a component in one location, re-locating it, and so on. If they change their mind about a location on the plotter they simply leave it there until changing the paper and re-running a final perfect plot whilst sipping their coffee at the end of the session.

The problem of high-speed graphics manipulations in BASIC was solved by developing a ROM which performs all the graphics matrix manipulations in machine code. This ROM was written to our specification by John Stout and after some initial problems of compatability between the ROM and the BASIC compiler had been solved, it proved exceptionally fast and successful.

This leaves the question of disc storage. Clearly storing every vector of a complex engineering drawing was going to consume considerable disc space. What we have done is as far as possible avoid the problem by storing separately the individual shapes from which a drawing is constructed and filing the command lines which reconstruct these shapes into the total drawing. Apart from allowing a very dense form of storage by separating these two functions, it also avoids the problem of storing more than one shape when it is repeated and, of course, avoids storing any superfluous information such as hatching lines. In the compiled version of the program the vector instructions for the hatching lines can be re-computed at a comparable time to that which a 400mm per second plotter can draw the lines. One advantage of storing a command line is that it makes it very simple to edit a complete drawing. The shape can be deleted or moved by a simple edit to the command line, similarly, global edits can be rapidly performed – a circle of hard chairs can be changed to a circle of soft chairs by simply changing the shape name. What is a little more awkward is to change one of a ring of hard chairs to one soft chair leaving the rest hard – but this problem is also true of a number of very expensive and supposedly more sophisticated drafting systems (Coates, Frazer, Frazer and Scott, 1982).

Student acceptance of CAD

Initial acceptance

The students' acceptance of computer-aided design techniques divides between first, problems of their initial acceptance – that is, if they cannot rapidly come to terms with the equipment at all, then they are unlikely to get any further – and, second, the longer-term question of whether the student continues to use computer-aided design as a natural part of their design activity – or indeed as a replacement for it.

There is only one problem more difficult to overcome than initial acceptance, and that is total addiction. We have a small proportion of students whose use of the computer becomes addictive and a substitute for creative decision-making.

There are many factors contributing to immediate acceptability.

1. Immediacy of results. Drawings must appear immediately on the screen or better still on an adjacent plotter so that students can see the results of their labours and take them away at the end of the session.
2. The command system must be exceptionally simple, yet it must be capable of producing complex results. Programs only capable of producing trivial results rapidly lead to boredom.
3. The system must be robust and able to cope with human error without failing. It should be capable of making reasonable guesses at what the student intended to do and if remotely possible it should always draw something – preferably within the visible plotting area!

Continued use of CAD techniques

Having overcome the problem of initial confrontation with the equipment, two major issues have emerged as controlling the further use of CAD techniques.

1. How easy is it to change the data-base?
2. How easy is it to get complex output?

In other words, assuming that the student can successfully input data in the first place, does the system encourage the student to explore alternative ideas and perform comparative analysis by making it easy to change the data-base? Second, whilst it may be easy to get some output, how easy is it to get complex output?

At this point I propose to limit myself to describing how we approached these problems.

Simplification of changes to the data-base

The two-dimensional drafting and shape manipulation which we use – the Shape Processor – consists of defined shapes and command lines. To facilitate changes to the drawing we have facilitated changes to both the shape definitions and the command lines themselves. Changes to the shape are done with a simple editing program which enables the user to change the two-dimensional data-base rapidly by deleting points, adding points or inserting points and, additionally, has the facility for smoothing curves through points to simplify radiusing and free curves and the facility for adding automatic dimensions and dimensioning lines. These shapes can also be deformed and distorted and the automatic dimensioning routines take care of correcting the dimensioning information.

Any particular drawing will consist of one or more shapes repeated, deformed and so forth. A command line which controls this is also saved and can also be simply edited. It is possible to change a shape name in the command line so that the same repetition or drawing occurs with an addition or modification to a shape. It is also possible to change the location of drawings, the number of repetitions and so forth.

Quicker routes to complexity

One way that we have tried for getting students to achieve reasonable complexity at an early age is by devising a special syntax for defining user defined commands (Coates, Frazer and Scott, 1981). A series of simple commands to control the matrix transformations such as rotate, shift, shear, etc, are structured to operate cumulatively, that is, the matrix does not reset between operations unless specifically cancelled. The syntax devised allows anything inside brackets to be repeated the number of times subscribed outside the bracket and that any number of transformations can be built into one complex line inside or outside the brackets. This becomes a user-defined command which can be called by a user's chosen name.

Furthermore, items can be included or omitted in the command line such as shape names or numbers of repeats or parameters to rotation angle etc. This means that the same user-defined command can be applied to a different shape or that the command could be generalized to vary the number of repetitions or extent of rotation and so forth.

Even more powerfully, further command lines can be structured consisting in part or whole of previously defined user commands so the user-defined commands can be combined into complex supercommands and nested many layers deep.

This allows users to move to a high degree of complexity very quickly. In the first instance we supply them with a few user-defined commands which produce surprising results as a sort of catalyst or primer. The student then tries these commands with their own shape, then tries some previously defined commands with other parameters missing and is then encouraged to build up their own user-defined commands. As standard we supplied, for example, all possible two-dimensional symmetry repeats including the seven frieze groups and 17 wallpaper patterns.

We are now experimenting with the use of an Intelligent Menu in an attempt to further improve the man/machine interface (Coates and Frazer, 1982).

Changing a three-dimensional data-base

As far as possible we have tried to make it just as simple to change the three-dimensional data-base but the particular problems associated with this have led us to experiment with alternative forms of input devices for three dimensions. In particular, we have been experimenting with building physical models which contain electronics within the parts. As the parts are manipulated and changed so the three-dimensional data base is changed automatically (Frazer, 1980; 1981.)

Advantages and dangers of CADED

It is perhaps worth pausing for a moment to reconsider why we are trying to teach students computer-aided design.

I am enthusiastic about the use of computer techniques in design education not merely in order to prepare students for the use of computers in the real world and make them more employable, but because I believe that the use of computer techniques will in the long run completely change the design process, will open up new possibilities for creativity that do not exist at the moment and will encourage students to take a more disciplined approach to design. This enthusiasm is tempered by serious concern about some problems with the application of CAD techniques.

Amongst the many advantages and disadvantages for the use of computers in design education I would like to highlight the following:

Possible advantages

The use of CAD programs should enable the student to:

1. Obtain more synoptic feedback about proposed solutions in terms of performance, appearance and cost.
2. Explore a wider range of design solutions.
3. Move more rapidly to the prototype stage.

4. Explore alternative routes to design solutions such as public participation in design.
5. Extend their creativity by making possible design processes impossible by non-computer techniques.
6. Consider alternative design techniques including automatic design.

Problems and dangers

There is a tendency for the use of CAD techniques to:

1. Dull critical faculties.
2. Induce a false sense of having optimized a design which is fundamentally ill-conceived.
3. Produce an atmosphere where any utterances from the computer are regarded as having divine significance.
4. Distort the design process to fit the limitations of the most easily available computer program.
5. Distort criticism to the end product of the design process rather than examining the value of the initial concept.
6. Concentrate criticism and feedback on aspects of the problem which can be easily quantified.

To take up some of these points: There are now a number of readily accessible programs which, particularly in the architectural field, enable students to rapidly obtain feedback on a range of different design ideas. The potential advantages of this are, first, that obtaining such accurate information will provide a better basis for critical judgement and design decisions and, second, the facility of doing this easily, will encourage students to explore a wider range of ideas. It is notoriously difficult in schools of architecture to persuade students to do even a partial environmental or performance evaluation of a design because of the tedium of the task and if they do succeed in doing a partial analysis it is virtually impossible to persuade the student to change the idea on the basis of the analysis, let alone re-perform the calculations on a new design to see whether it had actually been improved. The introduction of computer techniques has facilitated this process by making it quicker and easier to obtain an accurate analysis and within certain limits, to change the design and re-run the program (Frazer, 1982).

My concern about this is, first, that limitations in the program may produce irrelevant design constraints such as, for example, some architectural programs limit students to rectangular rooms or level ceilings or flat roofs; second, even with some systems where it is simple to change the data-base to explore alternative ideas, they are often only minor alterations to what may be a fundamentally inappropriate idea. Using the computer seems to induce a false sense of having optimized an idea and avoids questioning the original concept. Similarly this affects criticisms, first by tending to concentrate critical attention on the end result and the comparative evaluation and, second, by tending to concentrate on those aspects of the design which can be easily quantified. There is also a tendency, more commonly observed in the popular press of either blaming the computer for everything or imbuing it with divine powers leading to statements like 'The computer predicts that . . .'. It is essential that students are continuously reminded of the assumptions on which programs are based and their intrinsic limitations. This requires a certain transparency of operation of a computer program and raises other aspects of ethics and professional responsibility of which one would hope staff and students are aware.

In an attempt to use the computer to explore alternative approaches at a pre-design stage Liverpool Polytechnic and Ulster Polytechnic obtained joint funding from the Royal Institute of British Architects to write a problem worrying program (see Scott, Coates and Frazer, 1982). This has proved valuable as a teaching aid in terms of making explicit different approaches to a design problem, but we would be the first to admit that at the moment the dialogue with the machine is too long and tedious to be of use to any but the most persevering of students, and the program is thus not yet a useful design aid. At the risk of damning our own work I perhaps might add that students who do persevere through these programs often display a tenacity bordering on computer addiction.

With regard to exploring alternative approaches to design such as the use of computers to encourage public participation in design, there have been a number of interesting experiments of which those at Strathclyde are the best known (see Maver in Proceedings of CAD 76). The problem with most of these programs is the difficulty of use of the equipment by members of the public with no computing experience. We have been trying to make a modest contribution to this with another example of our data input devices employing the logical scanning of physical models (Frazer, Frazer and Frazer, 1982).

Examples of students using the computer to extend their creativity significantly or to achieve results which would have been impossible by non-computer techniques are so far less common. I had this experience with my own work as a student and that provided the motivation to attempt to make such computational techniques available to all designers (Frazer and Connor, 1979). The most immediately available examples are in the field of animation where particularly the use of simple inbetweening programs has led not just to the speeding-up of the tedious process of hand animation but also to the production of extraordinary new images (McCrum, 1982).

There is a curious assumption that there will be something necessarily technical and precise about computer graphics, which has a tendency to be confused with some technological image and hence with realism. There is not necessarily any such connection and whilst traditional animation has been concerned with obtaining a level of reality, computer animation has become concerned with surreality (Hayward, 1974).

A nice example is recounted by Brian Smyth, research fellow in Computer Graphics at the Royal College of Art who describes a jewellery student who used an animation inbetweening program for the design of some new cutlery. The student started with a drawing of an ordinary fork and a drawing of a pig and went one quarter of the way in an inbetweening sequence between the pig and the fork until he arrived at a strange piece of cutlery – of course for eating pork!

It would be rash of me to suggest that the introduction of computer-aided-design techniques has yet improved the quality of design education in our college, but there are secondary effects which are also important. The use of computers has led to interdisciplinary discussions – jewellery and ceramics students using the same curve fitting program as marine engineering students opened an interesting debate on how a ship designer evaluated the fairness of a ship's lines. Staff courses on computer-aided design have also stimulated interesting discussion amongst staff of varied disciplinary backgrounds and different faculties. The introduction of computer techniques has certainly encouraged students, particularly with simple examples like textile design, to explore a wider range of pattern repeats than they would have tried by hand and, similarly in graphic design, there have been some interesting experiments in animation which the students would have been unlikely to have tackled by traditional means. So far the most valuable side effect of computing has been its action as a mirror to question tacit assumptions about noncomputer design techniques. In lectures on artificial intelligence techniques and demonstrating programs using heuristic aesthetics we have provoked directly questions from students about the nature of their own creativity.

Conclusions

The future of CADED

I believe it is easier to train designers to be programmers than to turn programmers into designers. This is not to imply that it is necessary to teach a designer to be a programmer in order to use CAD techniques – there is nothing more disastrous than sending an enthusiastic sculpture student on a 15-week FORTRAN course – but a designer clearly needs to know a lot about computing if they are not to be limited to existing packages. Computer-aided design will not be able to make a significant contribution to the quality of design until more talented designers become involved in the creation of design programs. The computers/graphics in the Building Process conference in Washington earlier this year ended with an impassioned speech from Marc Goldstein, a partner in Skidmore Owings and Merrill, one of the world's most prestigious architectural practices. He took for granted the use of computer drafting along with the use of computer environmental evaluation; he seduced his audience with breathtaking computer images of their buildings, and then launched an attack on the complacency of the CAD specialists and system purveyors, exhorting them to set their sights higher and apply their efforts to the real problem of CAD – the problem of design itself.

I think there is a very real risk that design might get worse rather than better unless CAD takes a leap out its current stagnation. Better that designers should reject the use of CAD techniques entirely than that CAD should attract only those who will sit mindlessly watching a plotter turn out banal repetitions, or those designers who actually need the aids to design visualization that the system purveyors are promoting as representing design aids. Better still would be if we could provide computer aids which would attract the very best designers to use them and influence their development.

Postscript

Questions to ask of a CAD program

1. Is it assisting with the most important design problems?
2. Is it extending the designer's creativity?
3. Is it actually improving the quality of the product being designed?

Acknowledgements

The author would like to thank the Royal Institute of British Architects for joint funding for Liverpool Polytechnic School of Architecture and Ulster Polytechnic School of Art and Design Research history and criticism for the work on the problem worrying program and thank the Awards Research and Consultancy Committee at Ulster Polytechnic for funding for the work on the use of computers with design students.

References

Coates, P S and Frazer, J H (1982) The intelligent menu – A proposal for improving the man machine interface. National Academy of Sciences Proceedings of computer/graphics in the building process

Coates, P S, Frazer, J H, Frazer, J M and Scott, A E (1982) Low cost micro processor based draughting systems. Butterworth Scientific Proceedings CAD 82: Brighton

Coates, P S, Frazer, J H, Frazer, J M and Scott, A E (1981) Commercial and educational impact of shape processors. Proceedings computer graphics 81: London Online conferences

Frazer and Connor, J M (1979) A conceptual seeding technique for architectural design. AMK: Berlin Proceedings PArC'79

Frazer, J H, Frazer, J M and Frazer, P A (1981) New developments in intelligent modelling. Proceedings of computer graphics 81: London Online conferences

Frazer, J H, Frazer, J M and Frazer, P A (1982) Three dimensional data input devices. National Academy of Sciences Proceedings of computers/graphics in the building process: Washington, USA

Frazer, J H, Frazer, J M and Frazer, P A (1980) Intelligent physical three-dimensional modelling systems. Proceedings of computer graphics 80: Brighton Online conferences

Frazer, J H, Frazer, J M and Frazer, P A (1982) Use of simplified three-dimensional input devices to encourage public participation in design. Butterworth Scientific Proceedings CAD 82: Brighton

Goldstein, M (1982) Keynote speech. National Academy of Sciences Proceedings of computer/graphics in the building process: Washington, USA

Hayward, S (1974) *The computerised studio*. Focal Press: New York Computer Animation

Maver, T M (1976) Democracy in decision making. Proceedings of CAD 76

McCrum, J (1982) Computer graphics and the application of artificial intelligence. Ulster Polytechnic

Scott, A E, Coates, P S and Frazer, J H (1982) Problem worrying program. Princelet editions Proceedings of levels and boundaries conference: Amsterdam (April 1981)

J Frazer, Ulster Polytechnic, Art and Design Centre, York Street, Belfast, N Ireland

Discussion

J Geo, University of Sydney:

Optimization programmes only optimize quantifiable aspects of the design and in the kind of design I am interested in and the kind of design that John Frazer is talking about, the important things aren't really quantifiable. I think that anybody who really does believe in an optimization programme is making a very simplistic mistake. The way that we do use multicriteria optimization is to satisfy minimum criteria for certain fundamental design criteria and then know that any design we work up incorporating those basic criteria meet our fixed formal calculable requirements. But it's important to see the computer not as an end in itself but as an extra tool in the kitbag of the designer.

J Frazer, Ulster Polytechnic:

My concern is that the difficulty of changing a data-base may lead to only minor changes to a design being made in response to synoptic feedback, and that as a result, a false sense of optimization may be induced when in fact the fundamental starting point was ill considered or inappropriate. For example, with an architectural student's presentation there is a tendency for them to pin up their work with accompanying computer-generated analysis and then show a few minor alterations to the building which have led to some improvement to the thermal performance of whatever. The conversation with the critics and tutors then becomes concentrated on those aspects of the scheme which are emphasized in the presentation, instead of concentrating on the fundamental approach that the student has taken to the design and early strategic decisions. There is a disincentive to making such major changes as a result of feedback from computer analysis if this will require a major change to the data-base. However useful the quantitative information from a computer analysis may be in preventing students kidding themselves about the technical performance of their schemes, a small improvement in performance is of no great value if concentration on that aspect leads to a failure to see a radical alternative and better approach to the whole problem.

A Bridges, University of Strathclyde:

The actual consequences of using computer systems in our student criticisms are exactly the opposite to what you just surmised they might be. The traditional form of criticism for architectural designs is that the student puts up drawings and all the students and tutors come and discuss the drawings and say how well they answer the particular problem. This is in the nature of a very, very conjectural discussion, and they say, 'Well, I think that's a bit bigger than it should be in the brief' or 'I don't think you would really get it for the kind of cost limits' and it really is a desultory debate about whether the student has really designed within the brief. The interesting thing using the computer is that it answers those questions. There is no getting away from what's on it, what size it is, what it costs and so we ignore those things and say, 'Right, we can see what you have done in that respect, let's go on and discuss the architectural aspects and merits of the scheme', and we can concentrate much more on the non-numeric aspects of the design because all the formal things have been taken care of.

Part 5:
Training – the introduction of CADCAM into particular disciplines

20. Training requirements for architects: a view from an experienced user

S F Race

Abstract: Many organization in the past have not appreciated that for a successful implementation of a CAD system it is necessary not only to try and staff at all levels in the organization but to ensure that adequate training time is made available and that it is undertaken away from the trainees and office. From experiences in the installation and training of users in architectural practice into the use of CAD the author identifies the timing and level of training appropriate together with the level of personnel concerned and uses case studies to illustrate his points.

The D'Arcy Race Partnership is an architectural practice formed in April 1978, specializing in CAD.

The practice has expanded and diversified and is now applying its expertise in three main areas:

1. Conventional design services where computers are used exclusively for the preparation of all finished drawings leaving the practice.
2. A computerized drafting service which is used by other organizations either to ease a resource problem or to explore the use of 2D drafting systems, or both.
3. Specialists services based on the promotion, development, and implementation of computer aids in the construction industry.

This paper will concentrate on a single but important aspect of this last area of our work training.

Training is a crucial but often neglected factor in acquiring a system: installation is the beginning of the story and not the end. Training is neglected by system vendor and purchaser alike. Without exception, the amount of training provided by the system vendor as part of the system purchase price will be inadequate, and in some cases may be available only to ease the transition from a period of high customer attention before a sale to a period of relatively low customer attention afterwards. Similarly the amount of capital made available by the system purchaser will reflect what he believes to be the amount of training required to get a system started and will usually be seriously underestimated.

The meaning of the word 'training' has a wide range. It can be used to mean instruction in a number of commands to a specified standard of competence, usually for the middle and lower levels of an organizational structure. Alternatively, it can be used to envelop the wider issues of senior management education, and the reorganization of traditional work patterns and job organization. The kind of training system vendors give will be closer to the former end of the spectrum. In many cases they are unqualified to advise on the wider issues of training, and therefore the effective use of a system usually becomes the responsibility of a small number of committed individuals.

Expectation, problem and provision

Senior members of an organization who are currently shopping around for a CAD system see computer usage advancing at a very fast rate. System vendors are saying these are the products of the future. A new pace has been set by people who have already acquired a system and those few who have shown that spectacular results can be achieved. There are now new pressures on winning jobs in the international market place and systems seem to offer the potential for retaining a competitive edge. Systems promise to ease resource problems either by providing that extra bit of horsepower at the crucial moment or by eliminating the difficulties of employing and dismissing staff. There is also a feeling that a better service could be provided and possibly therefore, better architecture; profit margins could certainly be enhanced.

The problem facing people taking decisions on investment is how to proceed from their current state towards a new *modus operandi*. This may be difficult to achieve from a position of little or no knowledge. New and unfamiliar skills need to be defined, and existing skills need to be fitted into changing organizational structures.

The reality of buying a system is different. An expensive mini computer-based system is going to make little impact on offices with 50 to 150 or more drawing board staff. Large systems are expensive for small firms, and small systems for small firms lack the horsepower and lose some of the benefits. Management will be under great pressure to allocate resources to maximize the benefits the system can allow. This problem will be particularly acute initially where no experience exists to guide management decisions. Disillusionment can grow very quickly. Very often what is provided in a system falls far short of the standard expected and which sales techniques imply. It soon becomes evident that life is much more complicated than the one portrayed in a sales demonstration, and that achieving spectacular results requires a monumental effort and is not simply a button-pushing exercise.

In the very broadest sense, training begins with the educational system, which in recent years has been overtaken by events and is failing to provide the necessary skills to match technological developments. Course structures need a complete overhaul. CAD needs to be established and accepted as a respectable part of training for all the disciplines in the construction industry. System design and problem-solving with a system are disciplines in their own right and should be taught as such. Because things are changing so fast at the moment, there is a pressing need for refresher courses and in-job training courses and these should be given the necessary help and support. In other words, we should be much more aware of what we are looking for in a system, be able to use a system when we leave our further educational establishments and those who are caught mid-career should be given the opportunity to adjust.

However, given the fact that most people at a junior or senior level in an organization do not have the necessary skills, they must face the problem in the best way possible.

When?

Training should start pre-purchase, when it is a process of developing an appreciation of what is available. People from all levels in an organization should be nominated to undertake this process. This fact-finding group should consist not only of people who are responsible for the initial strategic decisions, but also of people who are going to operate the system day-to-day, and who will be expected to achieve results. It is important that these activities are incorporated into the forward predictions of an organization and that as with any project, financial and manpower resources are allocated. At this stage, it is a question of gathering intelligence, reading reports and explanatory articles in the press, and visiting system vendors and users. A familiarity with common terms and techniques will grow. A further possibility is to run alongside an already established user for a few months. There is no better way to gain experience than on a live project achieving real results, under the protection of an experienced user who can ensure that there are no disasters. Even if this is not the system which is eventually purchased, the experience gained will serve as a datum against which other systems can be measured. It may also be thought advantageous to seek the advice of an independent consultant.

Eventually a final decision will be made on the system to be purchased. At this pre-delivery stage, I believe there are substantial benefits to be gained by training staff before installation takes place. It is very important both technically and politically for some members of staff to be confident and proficient by the time the system is installed and commissioned. Undoubtedly, the people involved in acquiring the system will want to demonstrate it as quickly as possible before interest from the rest of the organization dissipates. Similarly, the capital expense of the system necessitates its early introduction to production work where it can begin earning.

Post-delivery training should continue on a gradually expanding basis. Existing knowledge should be consolidated and disseminated. Disseminated because it is important that the system is not seen to be operated by an élite computer corps. It will also be necessary to ensure that trained staff are capable of stepping into someone else's shoes.

Who?

As already mentioned, it is important for a hierarchical cross-section of people to be involved in training. Senior people are being charged with a new resource which they must manage effectively, and one or two of them should gain some practical experience at a work station. A system must be driven from the top and senior people have a responsibility not only to take the initial strategic decisions, but also to ensure that the system continues to run effectively. A new resource of this kind must be fully understood so that new pressures can be met and different decisions made.

Project leaders will have to contend with major changes in the pattern of work. Individual projects will still need project management but there will also be a need for data-base management and for the co-ordination of data from different projects with company standards and procedures. A project team can potentially interact through a CAD system and use a common data-base. They will no longer operate as independent professionals with clearly defined interfaces. Resources will have to be allocated in different quantities and at different times than previously. People who have hitherto spent their working day at a drawing board will now become the new work station operators. They will be at their most effective if they are given the responsibility to make decisions and if they are committed to using a CAD system. It has been our experience that people can be taken from any of the traditional disciplines and job descriptions and training to use a system properly. Indeed, much work can be done by people without any formal qualification. There are many aspects of data capture which can be done by people who are simply conscientious. Out of the people who are selected, one or two will have to develop an interest which is nearer to the computer end of things. The system will need some attention in its own right. It will need maintenance contracts, insurance policies, consumables, and internal resources will need to be allocated and reallocated. Its performance should be monitored, and job costing will be an integral part of the system's operations. Eventually the system will need expansion and at that time it will be necessary to understand which computer resource needs extra power, and which pieces of equipment can be added in an appropriate and compatible way. For some of these tasks therefore, a knowledge of some computer jargon is essential.

Finance

Budget figures should be incorporated into cash flow predictions for pre-purchase, pre-delivery and post-delivery training. Budgets should be realistic; it is false economy to believe that people will 'get by' or 'sort it out for themselves'. This is a certain recipe for disaster and disillusionment. Perhaps, as a rule of thumb, it may be appropriate to allocate an amount roughly equivalent to annual maintenance costs, which are usually fixed at 10 per cent or 11 per cent of purchase price. Allocate resources for a continuous programme of training, because systems will not remain static nor will be operators.

(*Note*: in the short term, slightly higher salary levels may be anticipated. For the time being there will only be a relatively small pool of expertise capable of operating such systems, and when that pool realizes it is a valuable commodity on the job market, salaries will rise.)

Make due allowance for advanced training on the first few jobs. Every office has its more relaxed project and this is the one which should be chosen first time round, probably irrespective of size or building type. By the time a trainee is ready to tackle a real job, they should have a basic technical competence in knowing the commands on a system and their effects. However the assistance and supervision of someone with experience is needed to undertake the strategic planning of a major data-base which has to work over a period of time, alternatively be prepared to make mistakes. This initial start-up period should be regarded as supervised training, and it may be better not to expect a profit on the first two or three projects through the system. When new organizational procedures have been established, it will become much easier to pass on knowledge to new trainees and new projects ought to find their way onto the system in a more comfortable manner. It may be prudent to allow three non-fee earning months per trainee which should cover the initial stages of training when no real work is done, through to the supervised beginnings of live projects.

Where?

In our experience, there are substantial benefits to be gained by training staff outside their own organization before, during and after installation. This applies particularly to the technical competence type of training referred to earlier. There are several benefits. First, staff are released from their day-to-day responsibilities and no conflicting activities will distract them from their training; total concentration is therefore possible. Second, staff will be assisted by, indeed totally surrounded, by CAD experts who are using the same computer system to produce drawings under great pressure. Much experience and technique will be gained by working in an environment where a CAD system is the norm rather than a special activity.

Advanced training and the gradual transition on to project work can be done either in the original training organization or in the trainee's organization. For the people whose special interest will be in the management of the computer system, system vendors almost always hold training courses on their own premises.

What?

Training falls loosely into four categories – technical, strategic, operational and managerial.

Technical training is concerned with learning each command on the system and realizing what effect it has. On most systems it is a case of first learning the command individually, and then progressing on to the use of powerful combinations of command. With regular use a trainee will soon commit to memory many of the commonplace commands. It is not necessary for him to plough through commands gradually; the use of simple exercises is more helpful. It is most important to get the person who has just left the drawing board to draw something familiar on the computer as quickly as possible. Technical training, whilst appearing formidable on the surface because of the requirements to communicate with a computer in a new language, is in fact less problematical than strategic training.

Strategic training gives the trainee the ability to structure his computer information in a way appropriate to his project and possible other factors. Every system has its unit of information to which symbols we recognize on our traditional drawing sheets must belong. It is the way these familiar symbols are represented in terms of the systems units which determines how easy or difficult it is to extract from the system the appropriate scale and combination of information which project team members and others will require. This is the key issue in determining the ease and speed of response when amendments are necessary.

Operational training is for trainees who have been nominated to take a special interest in the computer system. The tasks to become familiar with are: the appropriate allocation of resources for particular projects, routine copying of information for security purposes, long-term archiving of information, prevention of the accumulation of unwanted data, simple diagnostics when things go wrong, the ordering of consumables and the general administration of associated paperwork.

These are the functions which need to be carried out on a day-to-day basis so that the system and the data it holds are maintained properly. Many organizations may already possess a computing division which could take on these functions. This can be convenient, but occasionally computing departments full of programmers and system managers are not sympathetic to or cognizant of the requirements of a CAD system user. If a ready-made computing department does not exist, it does not necessarily mean that a computer specialist must be taken on to the staff. The duties outlined above are ones which a willing person, recruited from existing ranks, could easily carry out.

Managerial training. It is worth re-emphasizing that it is imperative that higher management levels know enough about the system to manage it effectively and exploit its benefits to the full. Training here should really include some 'hands-on' experience, but may take the form of seminars designed to give an appreciation of the limitations and powers of the system. It is sometimes surprising to find that what was thought to be easy for the computer to do is in fact difficult and vice versa. It will be tempting for senior people to encourage potential clients to use their services by emphasizing the use of a CAD system, but their bluff may be called at a later date if they cannot assess the implications of a client's requirements.

STUDY 1 is a firm of private architects employing more than 50 people. They were prepared to invest time and money in training and decided that they would nominate eight people to spend four man days per week over a six week period in our offices. Those chosen represented a cross-section of expertise and experience. They were two junior architects, one technician, two interior designers, two senior architects and one other architect. None of them had any previous experience of computers. They used the ARC training manual and split themselves into four groups of two. Using two work stations, two groups of two did some computing while the other two pairs read the training manual and prepared for further processing. This was interspersed with seminars and informal sessions. Progress was good and by the end of the fourth week a reasonable level of technical competence had been reached, including some aspects of system management. In the last two weeks of training, the first live project was started and advice given on the structuring of the data base. The project was eventually transferred to their own system after installation.

This was an almost ideal situation from the training point of view; none of the trainees gave up or proved to be unsuitable. Two people needed a little one-to-one tuition but managed to master the system in the end. One of the older architects was particularly successful and was later promoted to a position of being responsible for the system. Senior management looked favourably on training and made an effort to ensure that the system got off to a good start. The exercise cost in the region of £8,000, but it was money well spent.

STUDY 2 involves a nationalized industry. Two people were nominated for a week of training in our offices while at the same time the system was being installed in theirs. Despite talks with senior management to try and persuade them for their own good that a longer period would be better, the single week of training remained. After that period, the two trainees were expected to start two large 6 million refurbishment jobs the week after. The two trainees luckily were well suited to the system and made the best of their week's training.

Conclusion

Training should be viewed as an essential and integral part of buying and installing a system; it determines the success or failure of a system within an organization, irrespective of the capabilities of a system. If training is given the status it deserves, this will ensure that CAD does not get a bad name.

Discussion

J Gero, University of Sydney:

Most architectural schools are now introducing into their curricula something to do with computers, although it's not always clear exactly what. Do you see this as having an important effect on the ease with which you can train architects in such systems as GDS?

S F Race, D'Arcy Race Partnership:

I hope so. At the moment things which have been done are fairly small scale, with some exceptions, but I hope that the people who come out of those establishments have a better appreciation of what systems can do. I would also like to see people like myself, who trained basically as an architect, being able to make comments about even designing chips, and designing software, and I think that that ought to be an established integral part of their training. I hope very much that this will lead to people being much better informed.

D Hodges, Hull School of Architecture:

How can these things be incorporated in a course of architecture? Do you feel it should be a hands-on process or should it be a course of lectures on theory and structure of drafting?

S F Race:

It should be very much a hands-on experience. We are beginning to see new hardware concepts and new operating systems which mean that a larger number of powerful self-contained units can be installed in higher education establishments allowing more people access to them. I think the first priority must be for hands-on use, backed by seminars and lectures.

D Hodges:

Do you feel that fairly simple programmes are worth using to get the students to use the computer, or is it necessary to go for the fairly complex, very high resolution graphics, or all the sophistication of the very expensive packages?

S F Race:

Ideally, I would like to see a variety of systems from simple to complicated, but I think there is certainly no harm in first installing the simple system as long as somebody on the staff knows about the more complicated systems and can say 'These are the limits of the simple system, another system could do "A", "B" and "C".'

F Weeks, Newcastle Polytechnic:

Derek Appleton referred to using component catalogues in the data-base. It must be difficult to make sure that this information is up to date. How much of an overhead is this?

S F Race:

Surprisingly enough, the processing part of putting a large data base together is small. We can put a fairly large, 2,000-3,000-5,000, piece data-base together in five or six days. That's not the main problem. The problem is getting the information out of the people involved to make the data-base in the first place.

D J Williams, Plessey Office Systems:

With CAD, we are giving design engineers an opportunity of designing and updating their designs regularly so that there will be issues 1, 2, 3, 4 of their designs. However, that design eventually has to be frozen and a particular set of parts, the data-base, has to be frozen because at some point you are going to be ordering the components from your suppliers. I think it very important therefore that the penalty of making a subsequent change is appreciated and that cost awareness be another part of the training required.

S F Race:

I wouldn't like to view amendments as necessarily being penalties. I think if you have the right system, the system could score heavily on amendments. I suppose on the drafting side of our business I spend most of my time just doing amendments, and I can still produce a sheet price which is competitive with contract draftsmen.

21. Education and training in computer-aided building design

A H Bridges

Abstract: In this paper the need for all architects to be given an appreciation and understanding of the capabilities of CAD is put forward. As in other subjects the problem of availability of suitable systems and the means by which it is integrated into existing syllabuses are discussed. Because architectural design is less scientific and more philosophical than engineering may be an argument for not using a CAD structured system, but there are many advantages to be gained from the use of a system to set against the possible disadvantages.

Introduction

There is, currently, a widespread dissatisfaction with the state of architecture – not only with the quality of design, but also with the physical realization. One approach to remedying the situation is based on the accumulation of a body of knowledge derived from architectural research. Approaches to design are affected by a number of things, not least among them being the currently prevailing philosophical attitudes. The shift in approach with changing ideas is not, of course, exclusive to architecture, but, lacking any fundamental theory, the subject is subjected as much to fashion as to philosophy. A scientific approach to design may establish underlying theoretical principles so that models might be established and understood, and the validity and usefulness of results arising from this theory recognized and correctly applied in architectural practice. This paper discusses the use of computer applications in architecture to achieve some of these objectives.

Background

The ratio of research and development expenditure to all other input costs is lower in building than virtually any other industry. A number of reasons may contribute to this statistic but it is a matter of fact that architects do not generally feel the need to carry out fundamental research, and consequent upon this, do not utilize the results of such research as is done. There is, therefore, no demand for research, and, accepting lack of demand implies no need for research, Architectural practice continues to rely on intuitive selection and the processing of personalized information for decision-making.

In the short term it is quite possible that 'experience' or 'design flair' or 'intuition' may work as well as, if not better than, a systematic but imperfectly formulated methodology. But systematically organized knowledge is at least communicable; it is possible to identify sources of error if anything goes wrong; but more importantly, it enables fundamental issues to be discussed with less scope for misunderstanding. By making design processes explicit the possibility for progress and improvement in the quality of the design product is made more feasible.

Design theory

To enable the creation of a design theory it is helpful to regard architecture as a 'problem-solving' activity. The term 'problem-solving' is used here in a particular way.

In its literal sense a problem-solving activity is one in which problems can be defined and solutions obtained, but this would only apply to a small range of highly regular and predictable phenomena. Using this interpretation the great discoveries of science (not to mention architecture) would fall outside problem-solving, even though scientific method is consistently adopted as the problem-solving paradigm. A wider definition of problem-solving must, therefore, be adopted, and we are led to consider problem-solving as a style of discovery, a process and a method.

The pure sciences are concerned with explanation. The physical world is explained or accounted for by theories, and, if a theory is shown to be unsatisfactory then a better explanation replaces it. Since repetition of what had already existed cannot solve the problem, science calls for continuous innovation. Applied science (architecture) is not so defined, and would not appear to meet the criteria of problem-solving in science, although the methodologies have many important aspects in common. To define those aspects of architecture which do have direct comparability it is necessary to distinguish between architectural process and architectural product. Process may be considered as being the educational and research work building a body of architectural theory, and product may be considered as the built artefact resulting from architectural practice. The important point is that the product needs the support of the process and the process can be considered as a problem-solving activity.

Architectural design theory has only come to the fore over the last 20 or so years. The first design methods relied heavily on operations research techniques, cybernetics and tightly formulated mathematical descriptions. As a reaction against this self-imposed straitjacket a new generation of techniques developed around theories of public participation. These methods, too, have been rather less than entirely satisfactory in practice. A so-called third generation of design methods now attempts to provide the best of both worlds – those first generation techniques which do work, augmented by the attitudes from the second generation.

One source of successful first generation models is Systems Theory. At its simplest level a system may be considered as a set of logical operations acting upon, and acted upon by, one or more inputs. These inputs lead to the production of outputs from the system and this process of throughput is capable of either sustaining the operational structure of the system or of transforming it in some way. Inputs, throughputs and outputs of systems can involve flows of mass, energy, information or ideas depending on the manner in which the system is defined. Figure 1 illustrates the basic elements of a general systems model and its adaptation to architectural design and learning.

The problem-solving approach to architecture is exemplified by a 'third generation' design method based on Popperian falsification: if one is interested in a problem then attempt to solve it by making a tentative assertion and devise a test to appraise that assertion critically. A number of different models form the basis for data generation, hypothesis testing and theory generation, but within an architectural context, design practice may be seen as a matter of generating design proposals and then testing them – and modifying and improving them where necessary. Design education is then a matter of learning how to generate and test the design proposals. Figure 2 shows the design process may be considered to proceed by conjecture and analysis rather than the previous model of analysis and synthesis. The 'conjecture and refutation' centre of the design method may be elaborated as shown in Figure 3. From the initial briefing and analysis the design problem is defined (P_1). A number of tentative solutions (TS_n) to this problem are postulated. These solutions may arise by any means, but are all subject to critical testing or error elimination (EE). This evaluation may identify unsatisfactory aspects of the design which will lead to a reformulating of the original problem into a further problem (P_2) which is then subjected to further testing, and so on.

This conjecture and refutation model, together with the general systems model, provide the necessary background to consider the use of computer models in architectural education.

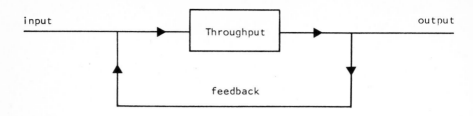

Architectural System

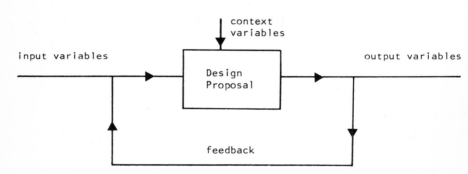

Figure 1 *General systems model and its adaptation to architectural design*

Figure 2 *Design by conjecture – analysis; rather than analysis – synthesis*

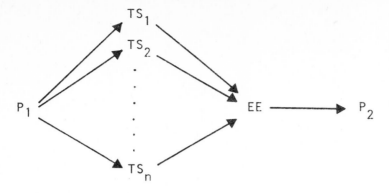

Figure 3 *Conjecture and refutation model*

CAD education

Computer education in architecture must fulfil several different roles. For students of architecture it may be a learning resource in itself, serving specific educational objectives. Alternatively, it may be the vehicle for introducing new techniques into practice, or, as its simplest, be training in basic computer applications for practising architects.

Computer-based education exists in one of two principal modes which may be classified as 'structured learning' and 'learning by discovery' (see Figure 4). In the structured learning mode the tutor exercises control over the learning environment by selecting and arranging the computer-assisted learning material to elicit the required responses through the use of reinforcing stimuli or remedial routing. There is an implication of well-defined objectives associated with discrete concepts or skills and the procedures tend to be strongly dialogue-oriented. In the discovery learning mode control is exercised by the learner who builds on a prepared knowledge base by 'exploring' a computer-based model of a particular problem.

Computer-assisted learning procedures tend to gravitate to one or other of these modes and, in architecture, it is the latter – the discovery learning mode. The objective is to give students an awareness of the essential complexity and multivariate nature of design. Typically, these programs present to the student a computer-based model of a system, offering the student interactive access to the principal parameters and the opportunity to calculate and display the effects of their variation. The educational objectives are a mixture of the generalized – such as obtaining some understanding of the system responses, and an awareness of the magnitude and sensitivity of the system parameters – and the specific, which may be to determine a given optimum condition or achieve a specific design objective.

A simple example

A limited time project from the second year course in 'Systems Approach to Design' at Strathclyde University Department of Architecture and Building Science provides a simple example of learning by discovery and multivariate design decision-making. Figure 5 shows a general systems model for design and the particular system used for the project. The purpose of the project was to give the students an awareness of the relative importance of each of the design variables and to illustrate how a general relationship can be found to, in this case at least, provide an optimum solution.

The problem is based around the responsibility of the architect to provide a design solution which is efficient in its use of expensive, and increasingly scarce, energy

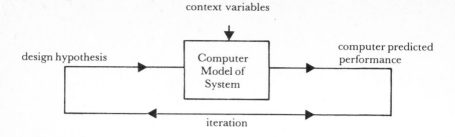

Learning by discovery

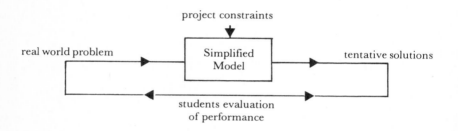

Figure 4 *Systems view of design and learning*

resources. For this simple example energy consumption is determined by 'steady-state' heat flow, such that the level of consumption is dependent upon the shape of the building and its construction. The task set the students was to establish this general relationship for a limited class of building forms and to find the optimum solution for an individually set problem. Figure 6 shows part of the program output and search sequence followed by one student.

There are obviously some limitations to this approach (such as the need for carefully defined projects) and this type of educational facility is complementary to the traditional studio teaching reinforcement of building physics lectures. The learning by discovery acts as a counterpoint to the learning by doing. The systems approach does, however, have the advantages of individualizing the learning process, enabling the student to progress at his or her own speed, and to make their mistakes in relative privacy.

A slightly different educational use of CAD is to incorporate the techniques into projects which can provide the student with a level of realism and depth that is otherwise impossible within the confines of a traditional studio-based design education. ABACUS have developed a number of computer programs for the evaluation of architectural design proposals, following the 'generate and test' model. Amongst these programs is GOAL (Sussock, 1978) which, from the input of geometry and construction data provides information on size, cost, and energy use. In the studio design project the student firstly derives a brief from a specially modified linear programming model (Maver, Fleming and McGeorge, 1978). This sets the accommodation and cost 'targets' to be met in the design. Then, following the conjecture and refutation model, as design proposals are developed they may be evaluated by computer to compare accommodation areas, environmental conditions and costs with the requirements of the brief. Figure 7 shows a summary of design alternatives investigated in one particular example.

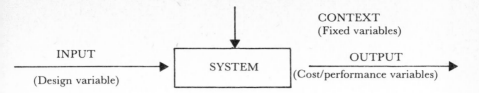

The relationship between the design variables (over which the designer has control) and the cost/performance variables (which represent the quality of the solution) are, for any set of context variables, complex and ill-understood. The purpose of this limited project is to illustrate how, in one case at least, a general relationship can be established from an exploratory investigation of specific examples.

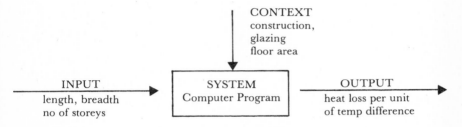

A Bridges, ABACUS, University of Strathclyde, Department of Architecture, 131 Rotten Row, Glasgow G4 0NG

Figure 5 *Systems model used in a simple project*

Used imaginatively, CAD can be seen as offering an enrichment to the educational process and meeting an educational need. It provides an experience which is not otherwise available to the student; it extends the range of systems which can be studied; and it consolidates points of theory and design in a more 'realistic' manner than is otherwise possible.

Education in CAD

The design and development of teaching packages such as those described demands considerable effort and the majority of existing software has been developed in those schools of architecture with sizeable CAD research groups. Few of the other schools have the staff with the expertise to write such packages or even to modify or adapt software written elsewhere to either suit their own requirements or to run on their own hardware. The result is that many CAD courses tend to be based around the local availability of software and the particular interests of the teaching staff. As a result there is little overall coverage of the subject or academic structuring. Again, all too often, CAD is treated as a specialist or elective subject rather than just one aspect of the broad subject of design in general. The educational objective should be for all designers to have the knowledge and skills to use computer aids effectively – but with the possibility for a small category of designers, who, taking a deeper interest in the subject, could go on to more advanced courses and then participate themselves in CAD research and teaching, or act as links between designers and computing specialists.

A number of authors have published course syllabuses for computer-aided building design (Gero, 1980); (Amkreutz, 1977). These courses include large amounts of technical detail – about computer hardware and software, peripheral devices, languages

TYPE IN YOUR STUDENT NUMBER FOR THIS EXERCISE
? 19
THE VALUES FOR YOUR PROBLEM ARE:-
TOTAL FLOOR AREA   = 2026.00
PERCENTAGE GLAZING   WALLS 1-4 = 22   43   88 8
U - VALUES   WALLS 1-4 =1.7 1.7 1.7 1.7
TYPE IN NUMBER OF STOREYS - INCLUDING GROUND
                              FLOOR
AREA PER FLOOR = 289.43
TYPE IN LENGTH OF WALLS 1 & 3
? 15

LENGTH (1 & 3) = 15.00   BREADTH (2 & 4) = 19.30
          HEAT LOSS (W/DEG C)

WALL	GLASS	SOLID
1	323.4	348.07
2	813.1	327.20
3	1293.6	53.55
4	151.27	528.11
5	0.00	463.09

TOTAL HEAT LOSS = 4301.40 WATTS/DEG C
DO YOU WISH TO RUN PROGRAM AGAIN 1=YES 0=NO
? 1
TYPE IN NUMBER OF STOREYS - INCLUDING GROUND
                              FLOOR
AREA PER FLOOR = 289.43
ETC.

Figure 6 *Hard copy of part of the simple project*

and computational algorithms, etc – only really necessary for the intending specialist. There is a need for a 'basic awareness' course whose contents would be based around the structure of the design process; design methodology; computer applications specific to the particular discipline being studied; and the social and economic implications of computer-aided design.

Computer-aided design teaching at the University of Strathclyde, Department of Architecture and Building Science combines elements of both these approaches. The main undergraduate design course includes substantial amounts of mathematics, systems modelling and design methods teaching. Special studio projects link systems concepts and computer applications to the general studio design work. The specific computing elements are covered only in an elective option course which may be taken by those students wishing to develop the more technical aspects of the subject. The projects and class teaching subjects are described more fully by Maver (Maver, 1979). The real benefits of this approach arise from its modularity. The design projects have been run successfully at other schools throughout Europe (Schijf, 1982); sections of the courses have been used in post-graduate teaching and in mid-career professional development courses; special seminars drawing on the same material have been organized for individual design offices; and so on.

In order to develop the future specialists it is necessary for more advanced architectural computing courses to be made available by the centres of excellence in this field. Gero (1980) describes one such course. Another example, at the post-graduate level is at the University of Strathclyde which offers a Master of Science in Building Science instructional degree course in Computer-aided Building Design. This course is a one year full time course for architects, engineers, surveyors or computing scientists, wishing to develop a high level of expertise in CABD techniques. The course contains three main themes – Design Methods, Computing Methods and CABD Applications. The general framework is that the course consists of two-thirds course units and one-third project work. The course units themselves follow a two to one split between lectures and assignments or short projects specifically related to the course unit. The course outline structure is summarized in Table 1.

Conclusions

CAD is a subject which needs careful integration into the overall design education syllabus. The time has passed for arguing about the place of CAD in architecture: it is now inevitable that it will be an essential tool of the architectural practice and it behoves schools of architecture to train their students to make effective use of CAD. In the current transition stage of adopting CAD techniques there is a large demand for 'basic awareness' courses for practising architects to familiarize themselves with the new technology. As CAD is reliant upon a range of sophisticated hardware and software there is a need for architectural computing specialists. This paper has described one approach to provide the range of educational services needed to meet these demands.

References

Amkreutz, J H A E (1977) Educational implications of CAD. IPC Science and Technology Press Proceedings of CAD Ed

Gero, J S (1980) The diploma in architectural computing at the University of Sydney. IPC Science and Technology Press Proceedings of CAD 80: 293-6

Maver, T W (1979) Methods and models: alive and well at Strathclyde. *Design Methods and Theories* **13** 1: 18-22

Maver, T W, Fleming, J and McGeorge, W D (1978) INVEST: a programme for analysis in hotel design. University of Strathclyde ABACUS Occasional Paper 63

Schijf, R (1982) Modular CAAD courses – a vehicle to discuss CAAD education. Butterworths Proceedings CAD 82: 287-95

Sussock, H (1978) GOAL general outline appraisal of layouts. University of Strathclyde ABACUS Occasional Paper 62

Table 1 *Outline structure of the University of Strathclyde's MSc in Building Science (computer-aided design)*

Terms 1 and 2

Course Modules	Unit Length
Design Methods	
– Introduction to design methodology	2
– Systems and models in building design	1
– Formal design methods	1
– Brief analysis and layout planning	3
– Operations research applications in design	3
Computing Methods	
– Introduction to computing	2
– Advanced introduction to computing	2
– Mathematical methods	3
– Computer graphics for architecture	3
Applications	
– Computational methods in building design	2
– Computer application in architecture	5
– Computer applications in environmental analysis	3

Term 3

Design Method Project (15 units)
  – Practical application of advanced or experimental techniques to a
     specific design problem
or
CABD Project (15 units)
  – Implementation or modification of piece of applications software

Figure 7 *The use of a general appraisal model (see page 9)*

Discussion

Mr Hafner, University of Massachusetts:

You mentioned that you have what appears to be single objective linear programming application in the building, such as maximizing profits, and I was wondering if there was any thought in the future to using multi-criteria programming such as goal programming?

A Bridges, University of Strathclyde:

That was deliberately a very trivial example to try and get over the point of how these educational packages might be structured.

R Guy, Hatfield Polytechnic:

As I see it, architecture is multi-disciplinary, so how do you link the architectural design with, for instance, the structural engineer or the ventilation engineer, and acoustic engineer and so on?

A Bridges:

I think that as soon as you set up a computer model, formal professional boundaries almost disappear or merge into each other. The computer program mentioned which was measuring the building to work out the cost and everything else was really doing very much what the quantity surveyor would be doing. There are other modules of that same programme which look at the structural analysis problems and do all the beam and column sizing which is the traditional structural engineering job. The thermal model in that particular programme was a simple sort of degree day modification of a steady state system, but we have sophisticated analysis that environmental engineers can get without access to similar models. When computers were first introduced into architecture, they were as great unifiers. One of the problems in practice is getting up-to-date information from the different specialists. The architect might make a design proposal and by the time he found out how much it would cost from his quantity surveyor, he has already discovered from structural engineers that there is a fundamental error and it can't be built. By this time the QS has wasted his time and everyone else has moved on a pace. The computer can not only give those quick rough answers straight away, but using some of the drafting systems linked to the computer model can ensure that everyone gets up-to-date information as well. The way it really works is that the architect troubles the specialist rather less early on; he plays off his ideas on the computer to get a design that he knows is roughly right and only then goes on and talks from that knowledge base to his consultants to refine his ideas.

22. Introducing CAD into the design office

M R Prince

Abstract: This paper describes the implementation of CAD techniques in two major design areas within the W S Atkins Group, namely, building design and highways design. The characteristics of the two systems, BDS and MOSS, are described briefly and the circumstances necessary to their successful use are set out. The paper concludes with some more general comments about the impact of CAD techniques on the design process based upon four years practical experience within the Group.

The W S Atkins Group

Activities and Organization

The W S Atkins Group is a large multi-disciplinary engineering and management consultancy with headquarters in Epsom, Surrey and regional offices in five cities in the UK, currently employing some 1,600 professional staff. The main constituent company, W S Atkins and Partners, offers consultancy services in the building design disciplines, process engineering, oil and gas engineering, transportation engineering and project management. Atkins Planning offers expertise in transportation planning, management planning and more recently computer consultancy. Atkins Research and Development provides specialist services in the areas of advanced stress analysis, dynamics and fluid mechanics, environmental assessment, and additionally supports the growing use of computer techniques in other parts of the Group.

The Group is principally organized by divisions, each comprising a single discipline or technology, but this structure is strongly overlaid by a project organization in order to deliver an appropriate service to clients, particularly on multi-discipline projects.

Historical use of computers

Like many similar organizations, the Group first became aware of the potential of computing techniques in the 1960s, and in 1967 began to make significant use of a time-sharing and batch bureau service. A number of teletypes were installed in design offices, a series of training courses were provided by Research and Development and a concentrated effort to develop application programs, both by R&D staff and engineers in the design divisions was put in hand. The objective was to spread computing expertise as widely as possible whilst co-ordinating development by influence and advice, rather than by edict from a central computing department. Early applications were principally in the areas of structural and civil engineering, transportation engineering and project management, and typically mechanized existing manual analysis procedures contained within one discipliner. Such programs therefore had no radical effect on design office procedures or interdiscipline communication, but they did enable existing calculations to be carried out more quickly, more accurately and more frequently, thus improving the quality of analysis and, indirectly, design.

The introduction of CAD

The development during the early 1970s of less expensive interactive graphics hardware, notably the storage tube terminal, and the explosion of investment in CAD software led to a re-examination of the potential of computer techniques in the consultancy. The first generation of applications, described above, had enhanced the quality of the analysis phase of the design process. The advent of CAD techniques held out the prospect of significant enhancements in productivity in the areas of design, detailing and drafting, traditionally the more labour-intensive elements of the design process. However, it was clear that the implementation of CAD on any significant scale would have a considerable impact on the nature of the design process and would radically affect interrelationships between disciplines in a multi-discipline project. Thus, unlike analysis applications, the successful introduction of CAD would only be achieved if an appropriate organization for management, administration, training and support could be established either partially in replacement of, or in addition to the existing structure.

In an organization with such a diversity of interests and a number of regional locations it was apparent that the introduction of CAD on a broad front across the Group would require an unacceptably high investment of resources both financial and personnel, and that the consequent organizational changes would be extremely disruptive. It was therefore decided to adopt a policy of gradual introduction of CAD techniques on a step-by-step basis, dealing with one area of interest at a time. Because of the investment required it was also felt desirable to time the steps to coincide with growth in work load, so that full advantage could be taken of increased productivity without the need to redeploy existing staff. In practice this policy has meant that new implementations have been geared to meet the needs of major new design projects, thus ensuring the interest and support of the project director and manager.

At the time of writing, two major areas of consultancy activity have been tackled in this way, and CAD techniques are now in full production use by the Building Design Disciplines and the Transportation Engineering and Land Survey Divisions. These two applications are described in detail below, and some observations about their success in practice are made on p. 202.

CAD applications

Building design

In 1978 the Group was commissioned to provide architectural and engineering services for a new teaching hospital, university and halls of residence in Algeria. The total project has a floor area exceeding 250,000m^2 and it was estimated that 4,000-5,000 drawings would be required to be produced during a 23 month design period. Even without the use of CAD techniques it was clear that a high degree of rationalization would be necessary if the programme dates were to be achieved. It was felt that this particular project would provide the incentive necessary to the successful introduction of CAD into design office practice, and consequently a small study team was set up, comprising staff from Research and Development and the relevant Design Divisions, to evaluate the need and recommend a solution.

Time constraints determined that the chosen system should be commercially available, well supported and with a good track record. There would be no opportunity for software development to remedy deficiencies or put right bugs. The immediate requirement of the project manager was to improve drawing productivity and it was therefore natural that attention was first focused on drafting systems. However, consideration was also given to 3-dimensional modelling systems, and their advantages in terms of design evaluations, co-ordination and the longer-term possibility of integration with analysis programs were seen to be important. One example of each type of system was selected following a desk study, and evaluated in detail by processing part of an existing design project through the systems. Although the decision was finely balanced, it was decided to adopt the modelling system solution because of the greater long-term potential described above. The selected system was BDS (Building Design System) marketed by Applied Research of Cambridge, and this package was installed on a dedicated PRIME mini-computer in mid-1978. A summary of the salient features of BDS and a description of the current hardware configuration are given in Appendices A and C respectively.

From the outset a policy was established that the facility should be seen to be under the control of the users, and a Computer-Aided Design Manager was appointed at a senior level in the architectural practice to introduce the system into design office use, who was able to call on technical support from Research and Development as and when required. It is clearly desirable to spread expertise in the use of the system as widely as possible throughout the user community, but time constraints on the Algerian project militated against this approach, and it proved necessary to establish a 'core team' of expert users to service this project. The core team initially comprised the CAD manager and an architectural assistant and was later strengthened by the addition of a further architect and a structural engineer.

Relationships between the core team and the project team were generally good, but stress situations inevitably arose from time to time, particularly at critical stages of the project when design changes had to be incorporated. It is felt that these occurrences were partly due to a lack of knowledge in the project team members of the capabilities and limitations of the system, and partly due to a natural tendency for the core team to oversell the versatility of BDS. Now that the Algerian project is complete it has proved possible to extend a working knowledge of the system into the design offices by on the job training, and the core team has a more mature knowledge of its capabilities which can only be achieved through experience.

Surface modelling and highway design

The Transportation Engineering Division has been using computerized design techniques for many years, traditionally through the BIPS suite of optimization, alignment and earthworks programs on a bureau basis. A steady change in work load pattern from predominantly UK projects towards overseas contracts prompted a re-examination of design procedures during the latter half of 1980. Increasing pressure on margins brought about by fierce competition for work suggested a move towards an in-house solution, and an examination of design requirements, particularly on urban projects, indicated that a more appropriate solution might be provided by the adoption of the MOSS system. MOSS (Modelling of Surfaces with Strings) was originally developed in the Highway Departments of three county councils, and was made commercially available to private practice through the MOSS Consortium. The major capabilities of MOSS are described in Appendix B.

Following a desk study and various informal discussions with existing users, it was decided to evaluate MOSS by processing a small live project using MOSS on a bureau service in London. Whilst the facilities provided by MOSS fulfilled expectations, the operational and financial constraints of using a bureau service, even with time-sharing access reinforced the belief that full benefits could only be obtained from the system if it were installed in-house. After some deliberation it was decided that the most cost-effective in-house solution would be obtained by upgrading the existing PRIME computer so that it could support both BDS and MOSS, and the hardware configuration shown in Appendix C reflects this decision. The upgrade was completed and MOSS installed by mid-1981, and production work commenced immediately, capitalizing on the operational experience gained with the bureau service.

The policy for the introduction of MOSS into the design office was essentially the same as that employed with respect to BDS, although the changes in the highway design process were much less radical, since considerable expertise in the use of computational methods already existed. However, it was still considered essential to appoint a senior engineer to the position of Computer-Aided Design Manager and prudent to restrict the initial use of MOSS to a core team of experts. The introduction of MOSS into more general design office use has proceeded faster than was the case with BDS for two reasons. First, as already stated, considerable experience in the use of other computer programs for highway design already existed, and second, the work load in Transportation Engineering comprised a number of smaller projects on short time-scales, thus permitting more flexibility in staffing.

One of the more useful facilities provided in MOSS is its ability to accept direct input of land survey data recorded on a variety of data logging instruments. Whilst this facility is potentially useful and extremely cost effective, it has required an organizational change which is symptomatic of a class of interdisciplinary problem areas likely to be thrown up by the introduction of CAD techniques. Traditionally, the Survey Department would produce fully finished mapping, probably including contours, for subsequent use by the highway designers, and these documents represented a visible and fully defined limit of responsibility. The use of the MOSS system provides the possibility of highway designers to have much earlier access to survey data before it is finally plotted and checked, and requires closer co-operation and interaction between the two disciplines. Whilst this facility creates the possibility of improved working practices it also raises the question of changed limits of responsibility which require careful definition.

Organization, management and costing

The previous section described in some detail the introduction of CAD systems into two major design areas with the Group. This section expands on the organizational and management philosophy adopted and comments on its success. The question of costing and charging is also addressed.

Management of applications

It is essential that the responsibility for the management of the use of a CAD system should rest with staff in the end user departments. They are most highly motivated to ensure the successful acceptance of CAD techniques, and they have a detailed understanding of the design process in its traditional form. However, in order to discharge this responsibility and to derive the greatest benefit from the new facility, it is necessary to have a focus of interest in the design department in the form of a Computer-Aided Design Manager, whose role is to introduce and regulate the use of CAD. The choice of the right candidate for the position of CAD Manager is critically important to the successful uptake of CAD. The position itself needs to be at a relatively senior level at least equivalent in status to that of project manager, and the holder must have comprehensive design experience, an unshakeable faith in the viability of the CAD system and a diplomatic personality. Indeed, a paragon of virtue, but the lynch pin of success.

Organizational requirements

The CAD manager will require support in at least three areas. In the period following the introduction of a CAD system, it is highly desirable to adopt the core team approach in order to give rapid credibility for the system, but this should be seen as an expedient with the main objective being direct use of CAD techniques by project staff, with assistance and advice where necessary. The CAD manager will almost certainly require technical support, since he is unlikely to be a computer expert, and this support may be provided by an in-house service department or externally by the system vendor.

Education and training is best provided by the CAD manager and his core team, but again back-up may be required from a computing service department or the vendor. As the system develops maturity through project use it will spawn standards and necessitate new disciplines, possibly requiring co-ordination and documentation. This new role of data management would naturally fall to the CAD manager, and thus create a new set of interactions with other managers concerned with rationalization of design office practice.

Costing

Financial management of a CAD investment requires careful consideration and an agreed policy ahead of the event. The costing of a CAD system should be full and comprehensive, and in addition to the obvious costs, including hardware, software and support staff, attention should also be given to such items as consumables, power consumption and lost floor space. For example, the electricity used by the system described in this paper costs some £5,000 pa.

Having determined a true figure for the annual running cost, then attention can be given to the question of how to recover this cost, either through company overhead or by direct charges to users. This decision must depend upon individual company policy, but it is important to ensure by one means or another that individual users are aware of the value of the resources they are using. In the case of the installations described in this paper it has been decided to pass the total system cost back to the users by utilizing a comprehensive accounting program on the PRIME thus ensuring that the Group's investment may be recovered fairly from individual projects benefitting from the application of CAD techniques, and enabling project managers to make a rational judgement on the financial benefit to be gained.

Conclusion

Experience over four years had demonstrated that CAD techniques can be successfully introduced into traditional design office practice, given a number of factors. Implementation should be timed wherever possible to coincide with a situation of rising work load so that the benefits of enhanced productivity can be directly realized and the need to redeploy staff avoided. It is clearly necessary to recognize and resolve the technical problems associated with the introduction of CAD, but it is vitally important to assess the organizational impact of these new techniques and to set up a suitable structure ahead of the event. In practice it has been shown that the single most critical step is the appointment of a CAD manager at a senior level. The manner in which investment and recurrent costs are to be recovered must also be agreed in advance and to the satisfaction both of the company management and project staff.

Acknowledgement

The author is grateful to the Directors of the W S Atkins Group for permission to publish the material contained in this paper.

Appendix A: Description of BDS

BDS (Building Design System) is a modular software package for three dimensional modelling of buildings marketed by Applied Research of Cambridge Ltd. Its salient features are as follows:

1. Planning grids may be created over the extent of the site, and these may be regular, irregular or tartan. The grid definition is a trivial task and alternatives may easily be evaluated.
2. Building zones may be created to represent functional requirements, fire performance, cost etc. Main zones may be subdivided through four levels which may represent departments, suites, rooms and subdivisions of rooms. Zones may have a variety of properties associated with them such as names, areas, costs, equipment sets etc. Zones may be located throughout the site grid in an orthogonal manner correctly to 1mm in three dimensions. Any overlap is reported as a clash. Drawings of the resulting building representation may be produced either on the Tektronix screen or plotter as plans and elevations or perspectives with or without hidden line removal.
3. A library (or Codex) of building components may be created, containing details of any structural, building or fitting element required. Each component is contained within an orthogonal three dimensional box, and may have a variety of properties associated with it including full graphical representation, name, cost or any physical property. Once a component is created it may be used repeatedly in one or more projects, but may at any time be revised.
4. Individual components may be placed on the planning grid to an accuracy of 1mm in three dimensions, and comprehensive checking for clashes with previously placed components takes place. Components may be moved, copied or deleted either singly or in groups up to 2,000 maximum. In practice whole buildings may be copied, deleted or moved, enabling the very rapid creation of a project containing repetitive features.
5. Drawings may be produced at any scale and as plans, elevations, sections or perspectives. Any combination of components may be selected for drafting enabling the production of drawings precisely matching individual needs. The drawing style may be varied but maintained consistently throughout a particular project.
6. Schedules of zones or components may be produced for selected parts of the building and for specified ranges of components within the Codex.

Appendix B: Description of MOSS

MOSS stands for modelling system and is the title of a computer program which provides a range of modelling facilities for civil engineering work.

Information defining a surface model is the starting point for most MOSS work. This information is collected by standard land or air survey methods or from existing maps. MOSS will accept data from all types of land survey and will process the surveyor's readings directly – no manual reduction being required.

MOSS was originally written for use by highway engineers and many of the features within MOSS relate specifically to highway design but other features are of more general application. A typical sequence of operations for the design of a highway using MOSS would be as follows:

1. *Carry out a survey and prepare a digital ground model*
   Digital modelling is a method of recording the shape of surfaces in numerical form suitable for storage in a computer.
2. *Prepare alignment designs*
   Alignments are geometrically specified lines which are used in highway and railway design to define the skeleton shape of a project. MOSS provides two methods for designing horizontal alignments and three methods for designing vertical alignments.
3. *Design main highway features*
   Once the highway alignment has been designed, other components may be added. These will include carriageway edges and verges top and bottom of

embankments or top and bottom of cuttings. These can be designed by simple horizontal and vertical off-setting from the main alignments. Cross falls super-elevation and any other features may also be added.

4. *Plotting*
    Three types of drawing are usually produced. They are plans, long sections and cross sections. MOSS provides the facilities to produce all three types of drawing. Any data contained in a MOSS model may be plotted.

5. *Area and volume calculations*
    Calculation of volumes is a requirement of most civil engineering projects and MOSS provides two methods. In the first method volumes are determined relative to a defined line permitting the calculation of running volumes for evaluation of mass haul problems. The second method is more general and calculates the volume enclosed between any two surfaces.

Two additional facilities are provided in MOSS which do not necessarily form part of a typical highway design sequence, but which are nevertheless used extensively in the planning and enquiry stages of civil engineering project design.

The first facility is that of contouring. The data stored in the computer can be used to produce contour plans of models. Normally these would be topographical contour plans but data such as population density or noise intensity can also be displayed in contour form.

The second facility is the ability to create perspective views. Techniques of superimposing perspective views of proposed constructions onto photographs of existing landscape (known as photomontage) have been used increasingly at public enquiries to give a better impression of the impact that new works will have on their surroundings.

Appendix C: The hardware

Both software packages run on a PRIME mini-computer having a P400 processor, 1 Mb memory, 600 Mb disc storage, a 1600 bpi, 75 ips tape deck and 32 asynchronous lines not all of which are in use.

Access to BDS is provided through three work stations, each comprising a Tektronix 4014 storage tube display and an alphanumeric VDU. One work station has an A0 digitizer attached. Two further alphanumeric terminals are used for data entry and file management. Drawing output is directed to a Benson 1322 drum plotter and printing is achieved on a 120 cps character printer. A Diablo daisy wheel printer is also used for contract documentation.

MOSS is used in a remote office linked to the PRIME by two high-speed multiplexed telephone lines, and access is provided by three alphanumeric VDU's and a Tektronix 4016 with A0 digitizer. A second Benson 1322 is provided in the remote office for production plots and a Versatec V80 electrostatic printer/plotter is used for printed output and local hard copy from the Tektronix.

M R Prince, Atkins R & D, Woodcote Grove, Ashley Road, Epsom, Surrey KT18 5BW

Discussion

P de Santos, Preston Polytechnic:

I would like to ask first, whether Mr Prince's firm tried to identify the cost benefits of using the system, and second, whether they used that in order to do some of the traditional investment analysis of pay back periods, etc?

M R Prince, Atkins Research and Development:

We did not do a full cost benefit analysis to show that we can pay back all the investment by guaranteed benefits in a certain period of time. I think that in our kind of business it would always be impossible. There has to be some kind of act of faith, speculative investment, call it what you like. However, I think it would be foolish and probably a route to computer-aided bankruptcy to base your investment entirely on that approach. What we have tried to do is to identify times in our business development where we can be reasonably safe in making investments and expect to get a reasonable proportion of that investment back in reasonable time. So it's a mixture of accountancy and confident speculation. What I think is important is that we recognize the full cost of what we are doing. It is not sufficient to add up the hardware costs and add a bit on for luck. We do try to take account of all costs, including maintenance, training, staff support, etc, and we have a policy to recover those costs from production. What we have done in the case of hardware is to amortize capital costs over five years, which is perhaps a bit long but it does give us some opportunity to get off the ground and if we have perhaps 18 months work load ahead of us then we think we are well on the way to getting a satisfactory return on investment.

C W Allen, British Aerospace:

In our case, the investment in the first instance was £350,000 and it was our idea to get the pay back over an 18-month period. We did in fact get it down to less than that by strictly controlling the way that we applied the use of CAD into the pure drafting activities in the first instance. We were able to cut down our sub-contracts from over 80 to 55 in the first 15 months of operation, so on that sort of basis we were able to prove that it did work and were able to get some more money 18 months later to buy some more equipment. We proved it, but we have kept a very strict control of the use of CAD.

T Maver, ABACUS:

One has to take a rather more daring stand when involving the application of computers in the conceptual stage of design, and that's a trickier calculation, as it were, and it is very difficult to evaluate how much extra business you are going to get because of increasing client satisfaction.

23. Training and learning during the introduction of an interactive computer-aided building design system into government design offices

B G J Thompson

Abstract: The tasks and levels of achievement involved in learning to use an integrated, interactive computer-aided building design (CABD) system are described.

During the introduction of the system into multi-professional design offices situated in different parts of the country, alternative methods of training were considered. These ranged from *ad hoc* individualized tuition and intensive design exercises to the investigation of prototype computer-aided learning 'System Tutor' software. Several levels of user support were provided including advice over the telephone, provision of job aids and documentation together with diagnostics built into the system itself.

A preliminary evaluation of learning performance is made from retrospective examination of hard copy records of user interaction with the system. Errors per command typed, and measures of 'mental set' and of 'fluent use of the input language' seem, on the basis of the limited data available, to discriminate adequately between experienced users and trainees. Such measures appear suitable for automatic monitoring of the lower levels of learning achievement. Efficient strategic use of design systems is much more difficult to teach or assess, however.

Only *ad hoc* tuition and practice were monitored and it is recommended strongly that systematic comparisons be undertaken, in real life conditions, of alternative methods of training and support. Heuristics should be sought that can be used to develop improved computer-based teaching aids for use with CAD.

Introduction

Background

This paper describes work carried out as part of an overall programme of development, evaluation and production use of a computer-aided building design system (CEDAR 3) in the Department of Environment, Property Services Agency, during the period 1978-80. The resources required to train and support users of such systems are generally thought to be quite large and of the order of cost involved in the development phase itself (Annett, 1976). Many factors may affect learning in the 'real life' design office environment and little guidance seemed to be available, at the time, for the choice and implementation of appropriate training techniques. Hence it seemed advisable to investigate the learning process and to develop a training programme on the basis of empirical study during the introductory use of the system. Also, at that time, it seemed appropriate that central government should not only encourage the use of CAD in building but also provide fundamental information to those wishing to develop user support for man-computer interactive systems generally. Consequently, related research work into man-computer interaction (MCI), which had been initiated by DoE during 1975, proposed an examination of the feasibility of computer-based adaptive teaching machines for CAD training, by means of the on-line System Tutor concept (Goumain, 1979).

The CAD system

CEDAR 3 (Thompson and Webster, 1978; Thompson, Lera, Beeston and Coldwell, 1979; Thompson and Young, 1980) is an interactive, integrated system intended primarily for use at the early or 'sketch plan' stages of building design. A user command language is employed together with both two-dimensional and three-dimensional graphics to enter and manipulate descriptions of orthogonal building shapes and their non-geometric properties such as cost rates, U-value and reflectance, for example.

The principal aim of the system is to enable individual designers or multi-professional teams to reach a better overall solution to the client's requirements by analysing and comparing alternative designs. To do this the performance of each design version can be assessed by one or more of the suite of eight applications programs which use the data from the 3D building model within CEDAR.

Design and drawing

The use of a design system should be clearly distinguished from the application of computer-aided *drafting* systems to realize a chosen design concept in the form of co-ordinated drawings and production documentation. As Table 1 shows, the properties of design systems and drawing systems are different and hence the learning tasks and appropriate training methods are likely to be different, also. In particular, the pattern of use by design professionals 'hands-on' will be intermittent, occurring in short periods of intense activity as each project passes through the early design phase. This contrasts with the almost continuous use of a drawing system by (usually) technician grades.

Table 1 *Comparison of characteristics of CAD and drawing systems*

Design System	Drawing System
No well-defined solution – many alternatives should be explored.	Relatively well-defined end point to task from a given design input.
Assessment of benefits difficult objectively.	Assessment straightforward in terms of savings of time and manpower to produce a set of drawings that have a manual equivalent.
Essentially 3D. Geometric model supporting both graphics and other applications.	Essentially 2D. Graphics only.
Group or individual use by professional designers.	Usually single technician user.
Both regular and intermittent use patterns.	Regular use on a daily basis maintains skills of user.
Requires manipulation of large quantities of attribute data for analyses.	Input almost entirely geometrical information.

Changes in design practice

The introduction of CAD necessitates (ideally) the adoption of new ways of working in the design organization. For example, involvement of all disciplines in joint decision-making at the outline design stage, and the management of more precise analysis data and results. Hence the eventual performance of the design team will depend on:

(a)   their innate ability to *design*
(b)   their ability to use the CAD system effectively
(c)   their response to opportunities resulting from changes in design practice.

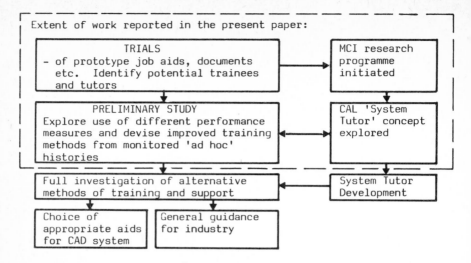

Figure 1 *The structure of the DoE investigation of training and user support techniques for CAD systems*

The present paper is concerned primarily with (b), although a full study would consider the interactions between all three aspects.

Learning tasks

To carry out a typical design run using CEDAR, the user must know how to:

(a) input a valid building description
(b) manipulate both 2D and 3D graphics
(c) analyse the performance of the design by means of the CEDAR 3 applications programs, and to
(d) modify the building model either to correct errors or to introduce deliberate design changes in preparation for further analyses.

In addition, every user should also learn how to:

(e) enter and leave the system and hence recover from computer or communications line failure, and to
(f) operate the work station equipment, consisting of a teletype and a Tektronix 4014 storage tube display.

Training and user support aspects of the DoE programme of work in CAD

The overall programme outlined in Figure 1 consisted of three phases: field trials, the preliminary empirical study whose results are reported in the present paper and a full comparison of training methods envisaged for 1981-2 that was subsequently not carried out because of policy changes within the PSA.

The trials

During 1978, as reported by Thompson *et al* (Thompson, Lea, Beeston and Coldwell, 1979), field trials of the pre-production system were held in several multi-professional design offices situated in different parts of the country. Two intensive design exercises (IDEs) were also held. CEDAR 3 was run by experienced operators from the development team on 'live' projects in each office.

Training of design staff was not attempted during this phase but potential trainees were identified.

The preliminary investigation of learning and training

The trials had been monitored extensively using a number of recording techniques. The feedback obtained enabled the prototype user documentation, system messages and other aids to be improved on the basis of users' reactions rather than from the opinions of the system development team.

The system was then introduced to two design offices on a regular basis during 1979 as described by Thompson and Young (Thompson and Young, 1980). During the first three to four months at each location members of the CEDAR team provided informal training, or ran the system for those staff not prepared to use it 'hands-on'. As staff accepted the system, visits by tutors became less frequent and users were encouraged to rely upon six levels of support as follows:

1. the on-line messages output by the system
2. diagnostic aids such as 'LIST' and 'HELP' facilities
3. user job aids such as an A4 sheet, folded to carry around easily in the pocket, summarizing the CEDAR command language
4. user documents, including reference manuals
5. help over the telephone from the CEDAR system adviser during normal office hours
6. personal advice given by a 'local expert' trained previously.

Training documents were produced and existing user documents improved using experience gained during this phase. The full CEDAR 3 documentation for users is summarized in Figure 2.

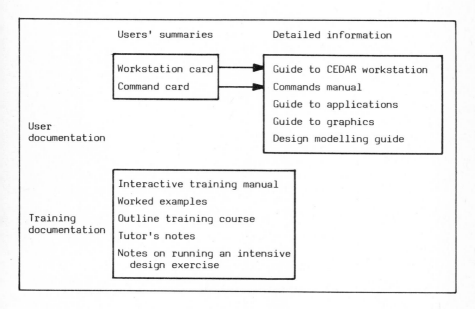

Figure 2 *CEDAR 3 documentation scheme*

The hard copy records of user interaction at the design work station already collected during the trials, were augmented by samples retained during production use and *ad hoc* training. The information obtained is incomplete because designers are not used to keeping exhaustive records of intermediate stages in their progress towards their eventual solution. However, as described later, a preliminary analysis of errors and command usage could be made with a view to establishing measures of learning performance and hence of the effectiveness of training.

Other approaches to training were considered:

(a) a five-day course followed by a two-day IDE using the trainees' own design problem to provide motivation for learning.

(b) an interactive training manual (ITM) using the method advocated by Grace (Grace, 1977) which can form the basis of a course or a self-teaching programme.

(c) an online 'System Tutor' using principles of computer-based adaptive teaching machines proposed from the MCI research work, (Cornforth and Mallen, 1981).*

The sequence in which material would be taught was suggested by the trial and error experience gained from this phase, paying particular attention to the user's own perceptions of difficulties in understanding the system.

The full investigation of alternative training methods

Unfortunately, the different methods could not be compared with the *ad hoc* training and learning as further support for the development of design systems was withdrawn, in October 1980, by the PSA. Development of CEDAR continues within the DHSS, however, and it is hoped that further investigation of training will be carried out there.

The remainder of this paper is limited therefore to consideration of the results of work completed in the second phase of the investigation.

The assessment of learning and training

Factors affecting learning

Learning takes place during initial training in the presence of a 'tutor' and continues with practice on real projects until skilled performance is achieved.

Learning depends principally upon:

1. the characteristics of the trainee, including professional discipline (architect, quantity surveyor, engineer), previous experience of design, ability to design and, of course, motivation. Group or individual use of the CAD facility may also affect performance.

2. the nature of the learning environment. For example, the physical, social and organizational characteristics of the design office, or the use of an uninterrupted IDE or training course elsewhere.

3. the characteristics of the teaching, including the users' perceptions of the 'personality' and ability of the 'tutor' (which may include software-based aids) and the order in which new material is presented. Total training time and the frequency of use of the system are also relevant.

4. the nature of the learning task. In this case to learn how to use a CAD system to help solve design problems. Hence the characteristics of the system itself and of design are important.

*The development and maintenance of this software, for complex systems such as CEDAR 3, may of course not be cost-effective in comparison with the alternatives. The shortage of good human tutors remains the strongest argument in favour of such an approach.

5. the stage at which learning has already reached.
6. During the practice after initial training, the new user will be affected by his frequency of use of the system and the nature and complexity of the projects he tackles using CAD.

Stages in the acquisition of skill and understanding

In Tables 2A and 2B the tasks outlined on p. 210 have been combined to form two groups concerned with building modelling and graphics (a + b + d) and with analysis (c). In each group, stages or levels of skill acquisition and comprehension can be broadly defined. These range from the use of separate commands or application programs, through the use of groups of commands to achieve subgoals such as the correct assembly of a design unit (BLOCK, STOREY, SPACE etc) to the understanding of the rules, conventions and relationships in CEDAR. The latter include the dependency graphs of design units and of applications where output from one analysis can be used as input to another (Thompson and Webster, 1978; Guide to Design Modelling, 1980; Introduction to CEDAR, 1980).

A skilled user should be able to evolve good strategies in the use of system facilities. These will minimise abortive command sequences and enable him to manage and manipulate large data sets whilst analysing his design for the interplay of different design variables (Thompson, Lera, Beeston and Coldwell, 1979; Guide to Design Modelling, 1980).

Table 2a *Learning tasks and stages for an integrated computer-aided building design system*

Stages of skill and knowledge acquisition	
1	Correct use of individual commands and operations.
2	Correct combination of commands in sequence to achieve sub-goals (eg blocks combined to give a valid building shell).
3	Understanding relationships and rules in system (eg hierarchy of design units).
4	Overall modelling strategies appear good in terms of: economy, relevance (eg to use of applications) and awareness of changes that are likely later when alternative design versions are explored.

Table 2b *Task and analysis of building performance*

Task: Use of building model and graphics to create and/or modify a design	
Stages of skill and knowledge acquisition	
1	Use of individual applications to calculate effect of design variables.
2a	Interaction between two or more applications.
2b	Interaction between two or more applications including when input of some calculations are derived directly from output of others.
3	Understanding interactions between applications from different design disciplines (architecture, quantity surveying, civil engineering and services engineering).
4	Effective strategies for searching problem space and managing large data sets during comparison of alternative designs.

Measures of performance during learning

Measures that were suggested by the trials are:

1. time and/or cost of design work to reach a solution
2. errors made by the user in the use of the CAD system
3. repeated errors before recovery from error conditions
4. confidence in the use of the systems facilities, shown by correct command sentences
5. frequency and nature of requests for assistance
6. opinions and suggestions about the system and the training made by the users themselves
7. judgements of skilled users on the basis of observation of use of the system during design.

Design is an 'open-ended' activity with exploration of possible solutions to a client's brief until an overall performance specification is satisfied.

In contrast, most published experimental studies of learning behaviour are concerned with relatively well-defined problems with a unique solution. Hence (1) is unsatisfactory in most real-life situations as it is more a measure of design skill than of use of CAD.

Measures (2), (3) and (4) can be monitored automatically using system software and *if sensitive enough* might be used in the System Tutor as well as by human tutors. However, (5), (6) and (7) were not recorded in detail, although they are essential to understand progress during the more advanced stages of learning.

Results of the preliminary investigation

The total number of mistakes made by users

Hard copy records from a sample of 54 sessions in design offices were analysed retrospectively for the total number of errors made. Figure 3 shows the variation of number of errors with the length of each session expressed in terms of the total number of commands.

Operators with greater than 100 hours experience and using CEDAR 3 at least once a week (usually members of the CEDAR team) showed a mean error rate (defined as the ratio of errors to commands per session) of around 10 per cent, whilst trainees with less than about 30 hours at the work station have an average of 26 per cent. In addition, architectural student trainees showed a similarly high error rate during their early contact with the system in laboratory experiments designed to guide the implementation of the prototype System Tutor (Cornforth and Mallen, 1981).

An individual learning history

It was possible to retain hard copy records of 14 of the first 40 sessions using the System by one volunteer trainee in a design office. The variation of error rate with his cumulative time at the terminal is shown in Figure 4.

For the first 30 hours (12 sessions) the trainee learned and practised under the direct supervision of one of two skilled users whose performance had been assessed from the results of Figure 2. Their *ad hoc* approach was typical of current industrial practice. Projects were not specially selected once basic concepts had been taught. Personal tuition was then withdrawn and the trainee forced to rely on the six levels of support above (documentation, system messages, etc).

Unfortunately, records for most of the tutorial sessions were lost. However, the few results available indicate that the mean error rate fell during training below the overall mean value for trainees. An abrupt increase occurred after support was removed and then decayed as experience was gained subsequently. As each new project was started,

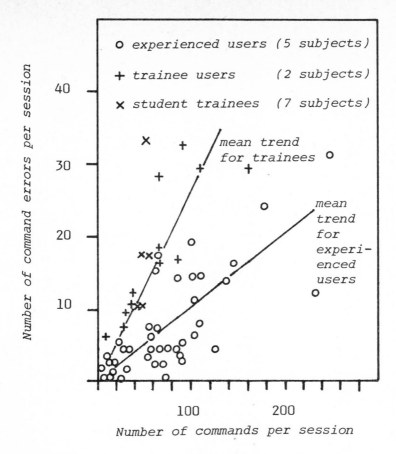

Figure 3 *User command mistakes found from hard copy analysis*

the mean error rate dropped from an initially higher value. There is a large scatter because of the many influences (project complexity, interruptions and external 'pressures' in the office etc) involved but the overall mean trend approaches closely the value for experienced users after about 80 hours at the work station.

Repeated mistakes and recovery by the user

Whatever training is adopted and however much practice a user has, it is undesirable to attempt to reduce the incidence of local mistakes (in momentarily forgotten command formats, mistyping etc) below which an individual works fluently. Natural conversation during problem-solving is far from correct grammatically, (Chapanis, 1976) and the important concern is that system feedback gives immediate recovery and that repetition of the same mistake is avoided.

Hence a proposed measure of recovery performance (really of mental 'set') is the relative frequency of occurrence of sequences of repeated identical errors before either the user requests assistance or resolves the problem himself. Figure 5 compares the cumulative frequency distribution for two trainees with that for two experienced users.

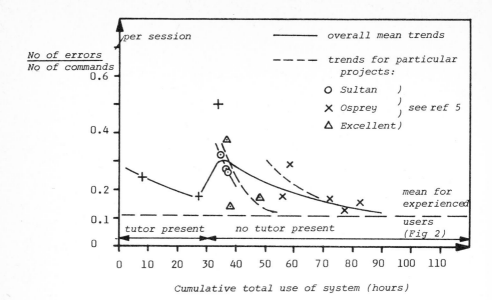

Figure 4 *Effect of training and experience on the performance of one trainee*

As expected, there is a noticeably larger proportion of longer error sequences by the trainees although surprisingly skilled users refused occasionally to believe they were wrong even after three attempts led to failure.

The trainee curves represent behaviour summed over a sample of 346 commands input during five sessions taken at random from their entire training period.

Curves for experienced users are aggregates over a sample of 1,115 commands input during eight sessions selected at random from several months of design work.

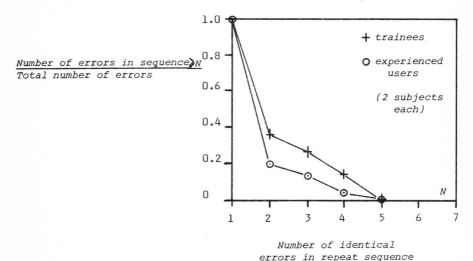

Figure 5 *Effect of experience on mental 'set' during use of the command language*

Grouping of instructions

Commands may be grouped together to form a 'sentence' on one line of input before the overall instruction is submitted to the computer. For instance, SWI AXES ON: SET SCA 500: DIS PLAN: DRA BLO ALL. The more confident a user is with the command language the more often he will be expected to use such sentences. This is confirmed by the cumulative frequency distributions of Figure 6, where experienced users are seen to use a higher proportion of longer command sentences than trainees. It should be noted that the 80 characters per line on the terminals used would normally limit sentences to six or seven commands (even with the maximum permitted level of abbreviation of command words) regardless of skill. This accounts for the cut-off in Figure 6. The sessions analysed are the same as on p. 210.

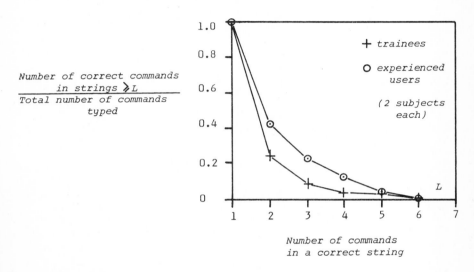

Figure 6 *Effect of experience on 'fluency of use' of the command language*

Categories of errors

In the first stage or level of understanding (see Table 2) the use of individual commands gives rise to different error conditions such as:

(a)   incorrect syntax
(b)   incorrect arguments
(c)   the use of a wholly inappropriate command altogether.

Mistyping can sometimes be clearly distinguished and the more obvious cases are omitted from the error count.

Examination of hard copy reveals errors associated with the second and third levels of building manipulation and graphics (Table 2A).

The effect of experience on the distribution between the different levels is shown in Table 3. The learning of rules and conventions caused trainees most difficulty and they made roughly five times as many mistakes in this level as experienced users. The design unit hierarchy caused most problems.

Table 3 *The effect of experience on errors per command for different levels of comprehension during use of the building design and graphics facilities*

Levels of Understanding (see table 2A)	Experienced Users (5 subjects)	Student Trainees (7 subjects)
Use of separate operations	0.04	0.084
Achieving sub-goals	0.027	0.024
Use of rules and conventions	0.028	0.145
Effective strategies	?	?

Summary

The limitations of the hard copy analysis

Table 4 summarizes the possible use of various measures considered earlier to assess achievement at different stages of learning.

In addition to 'local' errors, a user may execute a sequence of operations each individually correct but leading to a result that is valid but inappropriate for his design. This can lead to deletion of significant quantities of project information and the return to an earlier position in the design process with sometimes considerable wastage of input. A low error rate in the sense defined earlier will however be reported. Hence, even for sub-goal learning (level 2) the simple performance criteria proposed above may sometimes be of limited value.

Analysis of command sentences seems to indicate progress in achieving sub-goals (level 2).

It is important to record comments, suggestions and attitudes of trainees both during the tutored period and for at least three projects afterwards before the learning of rules, concepts and strategies can be understood and taught properly (levels 3 and 4).

Table 4 *The applicability of performance measures at different stages of learning*

Table 2A stage	Categorized errors (Table 3)	Repeated errors (Figure 4)	Command sentences (Figure 5)	Judgements of skilled users	Trainees' suggestions & attitudes
1	✓	✓		✓	✓
2	✓	repeated	✓	✓	✓
3	✓	sequences	?	✓	✓
4	?			✓	?
Table 2B stage					
1	✓	✓		✓	✓
2	?	repeated	?	✓	✓
3	?	sequences	?	✓	✓
4	?				?

The use of computer-aided learning

The comments above suggest that conventional computer-aided learning (CAL) techniques may be of limited value in the training and support of CAD users. Heuristics obtained from better observations of learning on CAD systems may be incorporated into artificial intelligence software to overcome this problem. The user's own view of his difficulties or uncertainties can be used to search a data-base holding explanatory text and graphics with references to detailed documentation (Cornforth and Mallen, 1981).

Lawson (Lawson, 1979) suggests differences in problem-solving strategies between architectural and science students from the results of short laboratory experiments. However, it seems that little has yet been done to characterize the patterns of use of man-computer systems in complex real-life situations of interest to practising designers.

The characteristics of users and human tutors

The present results are based upon very small samples of individual users and may not be truly representative. Motivation was provided in most cases by the personal enthusiasm of the members of the development team who naturally advocated use of CAD. The individual whose learning history was described on p. 214 was a mechanical and electrical engineer encouraged by the expectation of promotion as a result of experience gained using CAD. 'Cut-backs' in PSA work load and staff complement during 1979-80 reduced motivation of most design staff to learn new techniques, however.

Design teams used CEDAR through a skilled operator but no investigation of group training was attempted.

The characteristics of the training

Tasks were taught generally (see p. 210 in the order (c), (b), (a) and (d) over the training period with (e) and (f) reinforced in every session).

Sessions varied between a few minutes and about 2½ hours as it was believed to be essential to achieve tangible and relevant results of interest to the new user before leaving the work station.

The effects of changes in:

(i)   the order of introducing tasks
(ii)   the pattern of attendance at the work station, and
(iii)   the total training time

still need to be investigated.

Use of applications programs in CEDAR

The graphical interface ('LOOK AND CHANGE' templates, see Thompson and Webster, 1978; Introduction to CEDAR, 1980) made the applications easy to learn at level 1 (Table 2B). However, whilst error analysis led to inconclusive results, the subjective views of users showed that they found the interactions between data variables and between different analyses difficult to understand. The partially integrated mode where non-geometric data can be changed locally without affecting the building model was easily understood and used with great confidence. Automatic interaction between different analyses in the fully integrated mode required considerable time to accept, especially when it crossed traditional boundaries between different design disciplines (level 3).

Strategies, for instance in the choice of data sets used during a sequence of calculations, were not clear either from the hard copy or discussions. Incrementing values starting from one end of the likely range of interest seemed to be preferred to considering extremes followed by interpolation, however.

Conclusions and recommendations

(a) Tentatively, on the basis of the one learning history discussed here, it seems that about 30 hours of training with human tutors present, followed by a further 40 to 60 hours practice covering at least three real projects is required to achieve adequate skill in the use of an integrated CAD system.

(b) The analysis of hard copy from a small sample of users suggests that:

    (i)   26 per cent of commands typed by trainees had errors but only 10 per cent for experienced users.

    (ii)   system rules and conventions cause most learning problems initially but this is not true once users become experienced.

    (iii) measures of 'fluency' and of 'ability to recover from mistakes' in the use of the command language seem to discriminate between trainees and experienced users.

(c) A systematic comparison, in realistic design office environments, of alternative methods of training users of CAD systems should be made for each main area of application of CAD. Representative samples of staff, including members of all design disciplines, should be used. The effect of length of time away from the work station upon retention of learning and also the effect of differently structured groups using the work station, are among the factors that should be studied.

(d) The use of conventional CAL techniques is likely to be of limited value in CAD training but heuristics from the empirical studies could be incorporated by means of artificial intelligence software. The System Tutor concept should be developed further and included in the above comparisons.

(e) Academic rigour in experimental design should not be required when looking for the large effects of interest to industry. Nevertheless, monitoring and some repeated design work will require resources that firms and private practices are unlikely to supply without support from public funds. It is *recommended* that funding for (c) and (d) be made available from UK government or EEC sources to ensure the most effective use of CAD by industry.

(f) · The costs and resources required to develop and maintain CAL software and other advanced training and support aids should be carefully compared with those for more traditional methods. However, availability of good human tutors will probably be the factor deciding the use of CAL in the immediate future.

Acknowledgements

The author thanks the Property Services Agency for permission to use the preliminary results presented above. The views expressed are those of the author and do not constitute an official DoE statement or the policy of the PSA regarding the use of CAD systems.

References

Annett, J (September 1976) Problems of man-computer interaction in education and training. Proceedings NATO Advanced Study Institute on man-computer interaction; Mati, Greece: 273-86

Chapanis, A (September 1976) Interactive human communications: some lessons learned from laboratory experiments. Proceedings NATO Advanced study institute on man-computer interaction; Mati, Greece: 59-68

Cornforth, C and Mallen, G (Jan 1981) System Tutor – a study of learning processes involved in using CABD systems. Final report (unpublished) on DoE research contract no F3/3/429

Goumain, P G R (1979) A research programme in man-computer interaction aspects of CABD systems design. Proceedings PAr C'79 Conference; West Berlin: 663-76

Grace, B F (Winter 1977) Training users of a prototype DSS. 'Data base' Newsletter of SIGBDP **8**: 3

Guide to design modelling (1980) CEDAR 3 user documentation section 3 Ed A

Introduction to CEDAR (1980) CEDAR 3 user documentation section 1 Ed A

Lawson, B R (1979) Cognitive strategies in architectural design. *Ergonomics* **22** 1: 59-68

Thompson, B G J and Webster, G J (March 1978) Progress with CEDAR 3: a computer-aided building design system intended for the sketch plan stage. Science and Technology Press: Guildford, UK Proceedings CAD 78 IPC: 678-92

Thompson, B G J and Young, J S (May 1980) CEDAR 3 in practice: Using a large, integrated computer-aided building design system in Government design offices. *CAD Journal* **12** 3: 139-48

Thompson, B G J, Lera, S G, Beeston, D and Coldwell, R A (May 1979) Application of CEDAR 3: case studies from the pre-production trials of a large interactive computer-aided building design system in United Kingdom Central Government organisations. Proceedings PArC'79 Conference; West Berlin: 321-33

Discussion

J Gero, University of Sydney:

Your hard copy analysis struck me as being rather boring and difficult for someone to plough through. Was any thought given to getting the machine to do all that? The reason I ask is that we have been involved in a project not to do with CAD but to do with CAL in a rather different environment. There was no way we were going to go through all the output, so we put together a little routine that just simply tracked everything that was done and produced in essence something vaguely similar to what you have there.

B Thompson, Department of Environment:

Yes, we certainly had in mind building into the systems software a means of counting errors in various categories and counting commands in various categories. This would also have allowed us to have done more sophisticated things like calculating the frequency distribution of repeated error sequences, but we didn't have the funds to do that for the preliminary investigations so we had to flog through it the hard way just to prove the point that it could be of some value. The other long-term consideration was that these sorts of measures apply really to the lower levels of learning. At these earlier stages you are using single commands or single operations and just groups of them to achieve set goals. You can have problems as you move through the more complex situations where you are looking at the overall logic of what happened during the design process, for instance, when the user may correctly string together 20 or 30 commands but achieve the wrong result at the end. Do you regard that as something that should be trained out, or do you say, 'Well, we can't possibly monitor this automatically'? These are the sorts of things that we have felt were worth keeping in terms of annotated hard copy. It is rather labour-intensive, and extremely boring most of the time, but occasionally you spot some beautiful gems like the chap who typed nine times the same 'incorrect response' to a command when he knew perfectly well what he should have been doing but wouldn't ask. In fact he just logged out and gave up, but we would never have found this out if we hadn't some sort of record, but I agree that such repetition could have been monitored automatically.

24. Implications of CADCAM for training in the engineering industry

P Senker and E Arnold

Abstract: Results of a study carried out during 1981 and 1982 in the British engineering industry are presented. The paper is concerned mainly with the implications of CAD interactive graphics for training but CADCAM is also considered. Turnkey CAD interactive graphics systems started to diffuse rapidly in Britain in the late 1970s. The use made of CAD installations moved through three fairly distinct stages: installation and experimentation; use for production drawings, but inefficiently; efficient use on production drawings, involving operator productivity improvements of the order of 3:1 or more typically compared with manual operation. The time taken to move between these stages was extremely variable and was increased in some organizations for reasons such as: management failure to reorganize design; industrial relations problems; and supplier failure to deliver necessary software. The rate of operator learning was also highly variable, depending not only on operator aptitude and training, but also on management effectiveness in creating the right conditions. Increasing numbers of managers and design engineers will need to evaluate and plan CAD installations and there is an urgent need for training programmes to help them in this task. CAD has important implications for draftsman employment, training and skills – emphasizing the need for engineering rather than drawing skills. For the British engineering industry to prosper in the future, it will need to increase its expenditure on innovation, including design. It is essential to ensure that training programmes related to CADCAM are developed in the context of an appreciation of industry's overall skill needs.

Introduction

This paper presents some of the results of a study carried out for the Engineering Industry Training Board during 1981 and early 1982. The principal aim of the study was to identify the extent to which skill needs were likely to change as a result of the increasing use of CAD in the engineering industry.

Computers have, of course, been used in the design process for a long time to assist with design calculations. But we focused our study on interactive graphics, mainly because we thought that the increasing number of turnkey interactive graphics systems installed in the engineering industry was likely to have most important skill implications in the next several years.

During 1981, we conducted 36 in-depth interviews at UK engineering establishments which had either developed their own CAD systems or bought turnkey installations. Of these interviews 34 yielded results usable for analysis. We analysed our sample of establishments in terms of a broad industrial classification. The dates at which CAD interactive graphics had first been installed at these establishments are shown in Table 1.

Growth of CAD use in Britain

There has been interest in Britain in CAD since the 1960s. By the end of the 1960s, there were several CAD interactive graphics projects in Britain: Elliott Automation, Ferranti, ICL and Marconi were all offering systems based on their own hardware, and the CAD Centre at Cambridge was already in existence. Although the US was probably ahead in many areas, the number of CAD terminals even in the US was probably

Table 1 *Date of first CAD interactive graphics installation: by industry*

Date of first CAD installation	Vehicles and aerospace	Mechanical engineering	Electrical and electronic engineering	Total
Before 1970	1	Nil	2	3
1976	2	Nil	Nil	2
1977	Nil	1	Nil	1
1978	2	4	1	7
1979	2	2	3	7
1980	3	6	1	10
1981	Nil	3	1	4
Total	10	16	8	34

relatively small at this time. The growth of CAD use only really got under way in Britain after about 1977, when sales of turnkey interactive graphics systems took off. The vast majority of the 'general purpose' systems supplied to the British market so far have been marketed by US-based companies, although British suppliers are relatively strong in certain specialized areas – in PCB design applications, in solid modelling and in process plant design. Although some British software is very advanced, US-based turnkey system suppliers have larger and better sales and support organizations and it is difficult for British firms to compete with the established dominant firms.

The availability of turnkey CAD systems was important in stimulating the growth of demand for interactive graphics systems for several reasons. The most important of these is that the cost of a turnkey minicomputer system is much lower than that of previously available main frame based systems. The 'package' provided by the turnkey system supplier is much easier to use – to get it into productive use makes far less demands on users' computing and engineering resources. In addition, the selling and sales promotion efforts of turnkey suppliers have furnished potential users with economic arguments for convincing their managements that CAD systems would be beneficial.

The motivation for users' original interests in CAD were very variable. For example, large vehicles and aircraft firms sent engineers out to the US in the late 1960s who found out that firms in their businesses were experimenting with CAD. This provided the initial stimulus for a few UK firms to start their own experimental programmes. Particularly in these industries, firms were realizing increasingly that it was becoming important to reduce lead times in order to get new products to market ahead of competition – or at least, not too far behind them; and that CAD could help them to do this. At about the same time, UK firms in the semi-conductor industry faced the need to design large-scale MOS integrated circuits: this was only possible using CAD. Increasingly, the complexity of minicomputer and microcomputer circuitry necessitated CAD, or, at least, made it advantageous.

Such reasons were also of importance in the mechanical engineering industry as it became increasingly interested in CAD in the late 1970s. In addition, firms in all sectors of the engineering industry were suffering from shortages of draftsmen. They could seek to alleviate these shortages by employing contract draftsmen, but that was an expensive alternative. If CAD could deliver the draftsman productivity advantages which increasing numbers of turnkey system suppliers were claiming, then this had the potential to solve the draftsmen shortage and to eliminate the expense of employing contract draftsmen.

Managing CAD

In the late 1970s, increasing numbers of engineering companies were beginning to hear about the advantages of CAD – from conferences and exhibitions, from government sponsored bodies, from commercial companies selling CAD systems, and from the trade and technical press. Managers and engineers in the industry began to put together proposals to top management justifying the purchase of turnkey CAD installations – and justifying their selection of particular suppliers' systems. Managers had little or no experience of CAD at this stage. But they had to present formal proposals to justify the purchase of systems often costing £200,000 or more. In many cases, the real benefits offered by CAD were not quantifiable – certainly not to any degree of precision. For example, some firms needed to design new product ranges urgently so as to meet competition. Reduction of design lead-time by, say, a year, could sometimes mean getting a substantial slice of business. Failure to get a new product range to market in time could put a substantial amount of business in jeopardy. Realizing that, for various reasons, it was important first to gain experience with CAD, and then to benefit from its use, the majority of those who had to prepare formal investment appraisals justified the proposed purchase in terms of potential savings of the costs involved in employing draftsmen directly or on contract. Such appraisals often relied heavily on information provided by turnkey systems suppliers, generally envisaging draftsman productivity gains of the order of 3:1 compared with manual drafting.

These investment appraisals generally made no allowance for the time it might take for firms to learn to use CAD. Not surprisingly, we found that it generally took some time to learn to use CAD effectively.

When once a CAD system was ordered, establishments went through three fairly distinct stages in learning to use it. In stage 1, the CAD system was delivered, installed and used for experimental purposes. In stage 2, the system was used for production drawings, but not yet very efficiently. In stage 3, the system was used at today's state of the art – ie typically, operator productivity improvements of the order of 3:1 compared with manual methods were achieved on production drawings. On average, firms took about two years to reach stage 3 and all the firms which had had their first system delivered more than three years before our interview had reached stage 3 by the time of the interview in 1982.

However, if no undue delay was experienced, firms could often reach stage 3 in about a year. Significant delays were experienced in about half our sample, as can be seen from Table 2.

Table 2 *Total sample: reasons for any delays in reaching stage 3: sector*

Reasons for delay	Vehicles and aerospace	Mechanical	Electrical and electronic engineering	Total
Industrial relations problems	5	2	Nil	7
Supplier failure (in particular to deliver software)	1	3	1	5
Managerial inefficiency	Nil	6	Nil	6
No significant delay experienced	4	4	5	13
System delivered recently	Nil	3	2	5
Total	10	18*	8	36*

*Adds to more than number of firms in sample because both industrial relations problems and managerial inefficiency were significant in two establishments.

The table shows that principal reasons for delay were industrial relations problems, particularly in the vehicles and aerospace industries; supplier failures to deliver adequate software; and managerial inefficiency particularly in mechanical engineering firms.

The main union involved in negotiating the introduction of CAD was AUEW/TASS. Until late 1978 or early 1979, there appears to have been little attempt by TASS to, involve themselves in negotiations about CAD. But, in the second half of 1978, the union became concerned about CAD at an official level. The union circulated a statement by Ken Gill, its General Secretary, on 'Micro-electronics and Employment'. Later the same year, the Union published 'New Technology: a guide for negotiators'. Local negotiators began at about the same time to seek to achieve certain benefits and reassurances when managements proposed or installed CAD. The TASS negotiators often sought agreement that only their members should be allowed to operate the equipment; that all draftsmen, rather than just those who were going to operate the equipment should be trained; that there should be an increase in pay – sometimes just for those who were to operate the equipment, sometimes for all union members at the site; and they sought reassurances about safety and about job losses: they wanted to be reassured that no redundancies would occur as a result of the use of CAD. In many cases, negotiations were concluded without disruption. But, in other cases, there were strikes and delays. Where delays occurred in getting the system beyond experimentation (from stage 1 into stages 2 and 3) they varied from a few weeks to three years (so far) in the worst case.

It should be emphasized that these negotiations took place on a local basis. Sometimes failure to agree about pay was important. Sometimes – by management's own admission – delay or disruption was at least partly the result of management's industrial relations incompetence. The only areas in which TASS may well have gained significant advantages for its membership were in relation to pay. In relation to training, the only concessions we found that managements made – beyond training the operators they wished themselves to train – consisted of providing token, very short training periods for those not directly involved in using CAD. The reassurances given by management in relation to redundancies appeared to be virtually worthless. Such reassurances did, however, imply that, if management wished to make draftsmen redundant at some future date, they would be careful to attribute the need for redundancies to some other cause than the productivity increases achieved as a result of the use of CAD.

Managing a computer system was often a new experience for drawing office management. In addition to the need to deal with training and industrial relations, the work load, access to the machine and housekeeping all needed to be organized. In order to get good use out of the system, a data-base of standard components and standard procedures needs to be set up and the instruction 'menu' needs to be adapted to the firm's particular needs. Many users found it helpful to appoint one individual to be responsible for this. CAD often provided an impetus to reorganize drawing and related activities. At one firm, a new drawing numbering system was adopted for use on both the CAD system and the firm's main frame computer. This permitted parts list processing to be done on the main frame computer.

Several firms failed to get full benefits as quickly as they might have done because of slowness or failure in these respects. For example, one firm had installed a system a year prior to the interview but had not yet put standard parts into their data-base; it was also severely hampered in the first six months by its failure to appoint a CAD manager. Other firms may have limited the productivity improvements gained by not adopting a dedicated mode of operation.

In some other cases, systems suppliers had delivered systems with inadequate software and it had sometimes taken a year or more to get appropriate software from the supplier – a few establishments had not yet received adequate software at the time of our interview.

Operator learning

Of course, it is not only managements which have to learn to use CAD, but also operators. We found that there were considerable variations in operator learning. It could take between one and three months to learn to use the installation and achieve productivity equivalent to manual drawing. Typically, it took between nine and 12 months to reach productivity ratios of 3:1 compared to manual methods. But operator learning was not just a question of operator aptitude. High operator productivity depends on management setting appropriate standards, designing appropriate 'menus' of instructions and setting up appropriate data-bases. One user found that their success in setting up the right conditions for effective use in this way enabled them to reduce the operator learning time to achieve productivity equivalent to manual methods from two-three months to three-four weeks. Conversely, management failure to reorganize design activities often reduced substantially the rate at which operator productivity increased.

It is important to note, also, that high operator productivity can be achieved without necessarily implying that the CAD system as a whole is being used effectively or efficiently. For example, industrial relations problems sometimes delay CAD systems being used on production drawings; and management may fail to use the system effectively.

CADCAM

With the exception of the micro-electronics and PCB industries, CADCAM is not yet widespread. A definition of computer-aided manufacturing is a 'process employing computer technology to manage and control the operations of a manufacturing facility'. (Computervision definition (Machover and Blauth, 1980).

Electronics CADCAM applications – micro-electronics mask generation and PCB artwork – are inherently simpler than mechanical applications. The manufacturing elements of electronics CADCAM consist essentially of the manipulation of a light or laser beam under computer control. In effect, these technologies involve the superimposition of 2D drawings. But complexity of design and precision of mask manufacture in micro-electronics, and highly complex board design and very fine line widths in PCB artwork, both make CADCAM essential.

In contrast, in mechanical CADCAM, feeds and speeds have to be added to the design output of CAD terminals, as does data on tool offset etc. In vehicle manufacture, some progress is being made in the use of CAD to reduce the number of stages in the design process, but further heavy investment, for example in large and expensive computer-controlled die-sinking machines, will be necessary before the systems used can properly be described as CADCAM. The use of CAD is, undoubtedly, making an increasing contribution: for example, one firm now uses computer output to make templates, reducing the need for laborious manual processes. NC part programming is also relying increasingly on CAD output. Nevertheless, if the term CADCAM is used in the strict sense of using digital output from CAD systems as an input to computer-controlled manufacturing processes without additional manual input, then CADCAM activity in our sample of users was very small, outside electronics applications. In mechanical engineering, vehicles and aerospace, it seems likely that CADCAM is still at an early stage and may only really take off when CAD is well established.

But further rapid diffusion of CAD interactive graphics is likely for several reasons: several of the existing users in our sample have found that CAD offers significant benefits for their competitiveness, including substantially shortened lead time, increased draftsman productivity and better tender documentation prepared more quickly. Such users are likely to add substantially to their existing CAD installations, indeed several already have firm plans to do so. Knowledge of successful applications is spreading rapidly through the industry via personal contacts, journals, exhibitions and conferences and particularly through the marketing efforts of the turnkey systems suppliers. Over time, CAD is becoming increasingly good value for money. As more and better software is written, systems are becoming more versatile and efficient. Particularly as a result of continuing reductions in the cost of micro-electronic components, hardware costs are likely to increase more slowly than the rate of inflation, or even to decrease.

Implications for draftsman employment

In the period from the mid-1960s to the late 1970s, employment of draftsmen in the engineering industry fell by nearly a third – from over 90,000 in 1964 to under 64,000 in 1978. This was partly a reflection of decline in total engineering industry employment, but at the same time, the proportion of draftsmen in total industry employment also declined. This had very little to do with CAD – which has had a negligible effect on employment so far.

During the 1960s, the mechanical engineering industry employed about half the draftsmen employed in the engineering industry as a whole. In this industry – and in vehicles and aerospace industries too – between the mid-1960s and mid-1970s there were sharp declines in the employment of draftsmen – and also in employment in Research and Development. (See Table 3.)

Table 3 *Employment in research and development in UK private industry 1967–78, compared with draftsman employment: in thousands*

Engineering sectors	1967	1968	1969	1972	1975	1978	% increase/ decline 1967-78
Mechanical engineering							
Total R & D employees	19	18	15	10	11	11	– 42
Draftsmen	45*	45*	44*	37	33	29	– 30**
Instrument engineering							
Total R & D employees	5	5	6	5	3	NA	NA
Draftsmen	NA	NA	NA	2	2	2	NA
Electrical engineering (inc electronics)							
Total R & D employees	52	54	54	47	44	58	+ 12
Draftsmen	18	19	18	14	13	14	– 22
Motor vehicles							
Total R & D employees	14	14	14	13	14	12	– 14
Aerospace							
Total R & D employees	45	42	41	35	35	31	– 31
Vehicles (inc aerospace)							
Draftsmen	17	16	16	15	14	12	– 29

*Includes instrument engineering.
**Estimate – includes instrument engineering.
Sources: Draftsman employment – *A summary of information on the employment, training and supply of technician engineers and technicians in the engineering industry*, EITB RP/3/79, p. 40. R & D employment – Table 11, 'Occupational grouping of employment on research and development', *British Business*, 8 August 1980, 622; *Business Monitor* 1975, Industrial Research and Development Expenditure and Employment, Table 25, p 32.

It would seem that much of the decline in employment of draftsmen in the engineering industry as a whole, with the exception of electrical and electronic engineering, has been due to the increasing neglect of Research, Development and Design in the industry since the mid-1960s. The disastrous effect of this neglect of innovative activities on the fortunes of the mechanical engineering industry in particular has been documented extensively (Pavitt, 1980; Swords-Isherwood and Senker, 1980).

Electrical engineering (including electronics) is exceptional in that there was a recovery in R and D employment in the late 1970s (Table 3). But this was not accompanied by a recovery in draftsman employment. Possible reasons are that innovation in these industries increasingly involves electronics rather than electrical design. Electronic design tends to involve draftsmen less. Other categories of design occupations, in particular electronics engineers, gain the employment benefits arising from expansion in electronic design activities far more than do draftsmen.

Since 1978, increasing numbers of firms have succeeded in using CAD productivity. Establishments which had reached productivity levels of about 3:1 (stage 3 as described above) had generally made some progress in reducing their dependence on contract draftsmen. One or two had actually reduced the numbers of draftsmen they employed directly.

Management tend to explain the need for any proposed reductions in draftsman employment in terms of decline in business, rather than in terms of the effects of more widespread use of CAD, for fear of creating work force resistance to CAD. Nevertheless, the evidence available so far indicates that the more widespread use of CAD will tend to lead to the continuation of the long-established trend of decline in draftsman employment. Undoubtedly, the reduction in design cost afforded by the efficient use of CAD will lead to some increase in design effort – indeed, there is evidence that it has already done so. But it is rather doubtful that this will be sufficient to counteract entirely all the other factors which have diminished, and may well continue to diminish the demand for draftsmen.

Until recently, CAD was being installed in an environment of a continuing shortage of draftsmen. Recently, recession has reduced the demand for most skills – including those of draftsmen – and has resulted in the disappearance of all but isolated regional shortages of draftsmen. Given the trends outlined above, and in particular the productivity enhancing potential of CAD, it seems unlikely that general severe shortages of draftsmen will recur.

Conclusions – skill and training implications

The status of the drawing office and its role in engineering career progression have changed considerably over the past 30 years or so. In the past, the aspiring apprentice tried to get into the drawing office because it offered white collar status together with a good chance of promotion and entry into management. But, it is generally believed in the industry now that promotion out of the drawing office has become more difficult.

It is possible, however, that employment prospects for some senior engineering designers and design draftsmen could improve, as a result, for example, of the need to design increasingly complex products. But more widespread use of CAD is likely to reduce the need for more junior detail draftsmen. Increasingly, engineers and design draftsmen can themselves produce detail drawings as by-products by utilizing the strengths of CAD in routine and repetitive drawing.

As well as junior draftsmen, other lower level drawing office jobs are also threatened by CAD. During the last 20 years or so, technical changes in drawing materials, reprographics and micro-filming have gradually eroded tracing jobs, and most of the jobs of the tracers that remain are likely to be eliminated by CAD. The data processing capabilities of CAD (parts-listing, etc) threaten clerical jobs. If then, as we have suggested, CAD continues to diffuse rapidly, and if this results in continuing decline in the demand for drawing skills, there could be profound implications for training and retraining.

For the British engineering industry to prosper, it will need to devote far more resources to innovation, including design, than it has in the past.

It is important to ensure that the industry's needs for trained people are fully met and that training programmes meet the needs arising out of the more widespread use of CAD and, later, CADCAM.

In the next several years, increasing numbers of senior managers, designers, drawing office managers are going to be called on to evaluate and plan CAD installations. Few managers have experience on which to base cost-benefit analysis of CAD. Few design managers yet have much experience in tasks involved in computerizing design – in setting up computer data-bases and procedures, or in the industrial relations aspects of CAD. Yet, considerable experience now exists within the industry and could be drawn together as a basis for training programmes.

Design and design engineering training and retraining needs to be reconsidered in the light of the changes in design skill needs identified in this paper. As the jobs of draftsmen who are mainly skilled in drawing rather than engineering are under particular threat, it would appear that a review of training programmes for draftsmen is necessary to ensure that they are adequately trained to meet the needs of the design office of the future.

The actual importance of CAD and the potential importance of CADCAM cannot be denied. But it is essential to ensure that training programmes related to CADCAM are not developed in isolation, but are placed in the context of helping industry to cope with its future skill needs.

Acknowledgements

The research on which this paper is based was sponsored by the Engineering Industry Training Board. This research would not have been possible without the co-operation of the many managers, designers, trade union officials and others associated with the engineering industry who contributed extensive data and comments. The views expressed in this paper are those of the authors alone and do not necessarily reflect the opinions or policies of the EITB.

References

Machover, C and Blauth, R E (eds) (1980) *The CADCAM handbook*. Computervision Corporation, Bedford: Massachusetts
Pavitt, K (ed) (1980) *Technical innovation and British economic performance*. Macmillan: London
Swords-Isherwood, N and Senker, P (eds) (1980) *Microelectronics and the engineering industry: The need for skills*. Frances Pinter: London
E Arnold, University of Sussex, Science Policy Research Unit, Mantell Building, Falmer, Brighton BN1 9RF
P Senker, University of Sussex, Science Policy Research Unit, Mantell Building, Falmer, Brighton BN1 9RF

25. A practical approach to the training of engineers

D C Smith

Abstract: The paper describes the technique used at Davy Computing Limited for the training of drawing office personnel (draftsmen and designers) to operate a turnkey CAD system. It describes how this technique is based on Davy Computing's experiences gained in training engineers to use computers over a number of years, and specifically on the lessons learnt in training one particular company to use the CAD system.

The paper describes the training scheme inherited from the American turnkey system suppliers and the modifications made by Davy following the training exercise outlined. It emphasizes the importance of pre-installation education, understanding of the client's work and working methods, the formulating of good training objectives with the client and the role of system manager.

Training discussed in this paper concerns that given to engineers who have little or no experience of computer systems. It is anticipated that the problems inherent in this situation will gradually disappear, since the engineers now leaving the universities and colleges will have had experience of computers and will expect to use them to aid them in their future careers.

Introduction

Davy Computing Limited started in the early 1960s as a department at Davy Loewy Limited, a mechanical engineering company involved in designing plant for steel and non-ferrous industries. By the end of the sixties they were the computing services for the large Davy Corporation and are now operating as suppliers of bespoke software, a computer bureau and suppliers of CADCAM for the engineering, technical and scientific sectors of industry.

In 1977 following the acquisition of an American turnkey CAD system by Davy Loewy Limited, Davy Computing became the UK agency for the American company and undertook servicing and software support. At that time all operator training was carried out in America. However, the field service engineers undertook the training of system managers during the installation of the system.

In 1981 Davy Computing Limited undertook the task of operator training, and the first full training course closely followed the programme outlined by the American company. This programme is described later in the paper.

Training

Davy Computing have been training engineers to use computing systems since they developed their first software package, COMPAID, in the early 1960s. This package, as with other Davy packages is supported by a team of engineers, in this case since COMPAID is a software package for the piping industry, it is supported by a team of piping engineers. Part of their duty, if not their most important function, is training the clients to use the package. They have the advantage that they are able to converse with other piping engineers in their own 'language', and they have the experience of working on the type of project the client is involved with.

There is some scepticism and worry amongst engineers coming to use computing techniques for the first time. This is especially true when the package is a large one involving a number of different departments, and amongst draftsmen and designers who are generally the first to be involved with the introduction of CADCAM systems.

It is generally realized that the use of computing techniques to carry out tasks means that traditional methods of working will have to be changed. This could result in the loss of some jobs that are no longer necessary when the computer takes over, giving rise to a worry about job security.

Tasks which traditionally required some skill may be reduced to form-filling or key-punching tasks and so some protected positions and experience developed skills may be vulnerable. The computer works so quickly that fears about the ability of the engineer to control the situation might arise.

These worries are real and can develop into a strong resistance against the computer system. The way to minimize the effect of this resistance is to make sure that everyone likely to come into contact with the system is informed as early as possible about it. A comparison of the new ways of working and the traditional methods should be made, emphasizing the benefits to be gained from using the system and the new job opportunities that arise. I know of instances where terminals have stood idle because of this lack of communication between management and work force. This is very important when installing a large system like COMPAID or a CADCAM system, the influence of which is felt across departmental boundaries. The fact that we use engineers to impart information to fellow engineers enables them to build up a trusting relationship more quickly.

The applications engineers at Davy Computing Ltd are trained to work with the computer. They become the interface between the systems analysts/programmers and the eventual users of the system. Engineers, however, do not automatically make good trainers. The function of the training department is to support them in this task. One of the key areas of influence is in setting up good training objectives with criteria for success such as a list of the activities expected to be completed by the trainee at the conclusion of the course.

Probably one of the most decisive factors in the success of a large computer system is the appointment of a good system manager who is prepared to back the system and support it through any early teething problems. This appointment is crucial to a successful CADCAM implementation.

The Turnkey CAD system

To understand the training of users of the CAD system it is necessary to know something about it. The components of the system are:

1. A work station consisting of keyboards, menu pad, cursor controls, text screen and graphics screen. This is the most important part of the system as far as the draftsman/designer is concerned. At this work station designs are created and stored and plotters can be activated.
2. A graphics processor which houses the CPU, disc and tape drives.
3. A message centre for communicating with the CPU.
4. A plotter for hard copy output.
5. In some cases a digitizing device.

The system is operated either by the use of a 'simple' command structure or by activating instructions on a menu pad. Use of the menu eliminates the need to learn what is in effect a simple programming language and so is the most efficient way for designers to use the system.

How the system is sold

How the system is sold affects the way companies will view training needs and pretraining preparation. Turnkey CAD systems are sold on the following premises:

– They are installed 'ready to use'. That the system can do what a customer requires has been previously demonstrated and the need for a staff of computer software and hardware specialists is obviated.
– The system is easy to use. Draftsmen and designers with no previous computer experience can easily operate the system.
– The system is menu-operated and the menus can be changed or updated by the designers at any time.
– A series of commands can be combined so that they are initiated by one command caled a macro.

The system relieves the designer of the more mundane tasks of drawing, for example, cross hatching, dimensioning, updating drawings and writing bills of material. This leaves him or her with more time to devote to solving engineering and design problems.

The American training plan

Training provided free with purchase of the turnkey system is given in two parts; two weeks basic training and one week advanced training. The two weeks basic training originally took place in America prior to the shipping of the system.

The basic training consists of a series of lectures, demonstrations and hands-on time in which trainees are presented with information to enable them to create basic drawings and initiate basic system commands. Trainees are taught the concept of the command structure, the purpose and function of a variety of individual graphic and system commands and command parameters, system-operating procedures, use of hardware, creation of minor macros and symbols.

The aim of this basic two weeks training is two fold. First, the trainee is given a broad exposure to the basic commands needed to create drawings and second, he or she is taught to operate the system so that they can store drawings on disc and tape and use the facilities for digitizing and plotting.

At the end of this course trainees should be in a position to

– create drawings using the basic graphics commands.
– perform basic system operational procedures.
– construct simple macros.
– use the system menu.

Six to eight weeks after the system has been in operation, one week of advanced training is carried out on the client's premises. Emphasis is placed on the text file facilities, text merging, bills of material and optional packages purchased by the client.

The aim of this course is to increase the efficiency of the draftsman interacting with the system. At the end of the course the draftsman will be able to use the commands associated with any optional packages mounted on the system, write more complex macros and make full use of the text file facility.

In addition to these courses, a pre-installation seminar is given to the management of the company. The purpose of this seminar is to prepare management for the effective implementation of the system. Topics covered include

a demonstration session with some hands-on time.
a description of the support available in terms of application, software, hardware and training.
a discussion of field engineering support by the field engineering department.
a discussion of the facilities required for installation.
a discussion of organizational schemes, possible impact on the department and operator selection.

A case study

The first full training course involved a company that had taken a long time deliberating about CAD systems, and after finally reaching a decision to buy this particular system wanted to move very quickly. The management seminar was organized and implemented, and it seemed clear that the company were quite sure about the steps they were taking and the way the CAD system was to be integrated into their office structure. Unfortunately it was not until this time that the company selected the person who was to be their system manager. Problems were anticipated as the person in question was new to the office, had not been involved with the selection of the CAD system and had not any experience of working with computers or computer systems.

The duties to be expected of the system manager are summarized below.

- He controls and plans the work through the CAD system and ensures quality is maintained.
- He makes recommendations for long-range planning and budgeting.
- He keeps up to date with all new CAD developments and 'state of the art'.
- He develops new methods of CAD design and ensures optimum utilization of personnel and equipment.
- He develops and maintains estimating guidelines for CAD work and assists in the preparation of estimates for new products.
- He organizes the recruitment and training programme of 'operators'.
- He devises and controls the management of the CAD system with respect to back-up procedures, system security, keeping of records, tape/disc management and maintenance procedures.

The job is by no means easy, requiring a dynamic and forceful personality to carry it out. It requires a thorough understanding of the work of the designer and draftsman and also a strong empathy for this modern technology. The eventual success of the system depends a great deal on the skill and enthusiasm of this person. Of all the staff connected with the system, this is the man who will have most to do with the computer itself, the way it stores information, maintenance programs and system messages.

In this particular instance it rapidly became clear that rather than having an empathy with modern technology, the man was afraid of it. He was completely overawed by the system and was not entirely sure what was expected of him.

To help him overcome this fear, he was placed on a week-long training course prior to the start of the official course. During this week he was introduced to the command structure of the system. The purpose of the course was to give him a good grounding in the basic command structure, enabling him to see how the commands were constructed and so give him an advantage over the trainees who would take part on the first two-week training course with him. There was some concern that he would be 'overtaken' by some of the other trainees, resulting in a leadership problem. This in fact did happen. A much longer period of time, starting with more basic computer appreciation courses would have been required to bring the system manager to a suitable starting point.

Davy Computing were informed that the draftsmen who would attend the basic training course would be volunteers. In fact the company found volunteers difficult to come by and four trainees were eventually 'picked' and sent on the course. They arrived with absolutely no idea of what CAD was all about, what it looked like and what they were supposed to do with it.

This underlines the importance of educating the work force. They should attend seminars to show them what CADCAM is, what it looks like, what it does, how it is operated and how it will affect the way they work both in constructing drawings and in relation to other departments. Davy Computing now see it as an important function to provide these educational seminars.

In this case the trainees were dazed and unsure of what was expected from them. The first day of the course was therefore spent putting their minds at rest and explaining what CAD was all about.

The training course closely followed the pattern described earlier. This meant that the trainees were rapidly taken to the keyboards as the means for entering commands. There is no doubt that keyboard use is alien to the majority of designers and draftsmen. A certain amount of apprehension is detectable, letters seem to disappear and there seems to be a nagging fear that pressing certain keys will cause irreparable damage to occur.

This keyboard phobia is a real problem, though only transitory since most students entering the engineering profession will have spent some time working at computer keyboards. In fact all secondary schools are now equipped with microcomputers and some middle and primary schools also use them.

Problems therefore occurred through mistyping and even holding down keys too long. Error messages were generated which gave rise in the first instance to obvious feelings of panic.

It must be restated that the command language of the system is easy to learn and use. However, to the totally uninitiated with little computer experience, the importance of commas and colons can be dangerously underestimated. Errors occur and it can take a long time for the trainee to discover why they have occurred.

Most of the trainees rapidly overcame these problems, success fuelled motivation and by the beginning of the second week they enjoyed operating the system and looked forward to the challenges it offered them.

Here though arose a dilemma. Having overcome some substantial difficulties in using the keyboard and learning the command language, they now enjoyed working in this way. They had become interested in 'programming' the system and were gaining some job satisfaction from doing just that. Operating the system via the menu was too 'easy' and did not hold their interest, which had swung away from their engineering tasks. In effect the training had developed some bad working habits which seemed to be self-reinforcing.

Also system operators can work much more quickly using the menu and this worried a system manager who lacked confidence in his ability to control the CAD department.

The problems can be attributed to the lack of pre-installation education and to the structure of the training course. Rather than giving the draftsmen and designers a tool to aid them in their engineering tasks, there was a danger of training them to become high-level computer operators and programmers.

These problems have now been largely resolved with this particular company. The aim of Davy Computing, however, is to enable companies to make the most effective use of computer systems as quickly as possible and to this end the structure of the CAD training course was reviewed.

The revised training plan

Pre-installation education is now a definite part of the CADCAM training scheme. There are two basic areas, management education and work force education. The management seminar has already been described.

Work force seminars are now a necessary part of the training and should take place as early in the company's quest for CADCAM as is possible. They are aimed at all the people likely to be involved with the installation. Informed people are more likely to accept the system, and appreciate any reorganization that might take place. They should understand their roles in the total structure of the organisation.

The basic training course is aimed at operating the system via the menu. This involves some substantial preparation work on the part of the team of application engineers. Prior consultation takes place aimed at discovering the type of work the system is to be involved in so that relevant menus can be produced prior to the course.

Trainees are initially taught to operate the system via menu input and with a minimum of keyboard activity. They become accustomed to working at the work station, using the system in its most productive mode.

In effect the basic training course has been turned around. It could be argued that the initial scheme was aimed at understanding the way that the system worked, so leading to more meaningful operation of the system. The new scheme gets the operators working as quickly as possible. The command structure is taught during the second week of the course, as a means of enabling the draftsman/designer to create or modify his own menus. It is not seen as a method of drawing construction.

The future

Advances in CADCAM technology are rapid, more rapid than in the engineering industry at large. Methods of working will change, as will organizational structures. Education will become a more dominant factor than training as such. This is because systems will become less dependent on command structures and more totally menu- or prompt-driven. Training periods to use such systems will be shorter as the systems become more 'user-friendly'.

Computer-based training techniques are becoming more sophisticated and could be built into engineering computer systems. This will enable new operators to be trained 'on the job'.

Modern computer technology is redefining traditional engineering roles. It will also redefine traditional training roles.

D Smith, Education Officer, Davy Computing Ltd, Moorfoot House, 2 Clarence Lane, Sheffield S3 7UZ

26. Educating engineering designers: the introduction of desktop computers and software to the design environment

C M Jakeman

Abstract: Today every professional design engineer entering industry has experience of writing programs and using computers. However, the use of desktop computers is not widespread in the design office environment and there is great interest in the introduction of such machines.

This paper examines the advantages and pitfalls associated with the adoption of desktop machines with especial reference to software.

Preamble

The result of an international survey in 1981 (ESDU (M) Ltd, 1981) clearly indicated a trend by the engineering profession towards the widespread use of small computers: the desktop and microcomputer. In many ways the United Kingdom could be seen to lead this trend and, in the 12 months since then, the attendance at conferences and exhibitions together with enquiries coming to ESDU's Design Engineering Software Centre from industry and academic establishments have amply confirmed this.

The growing interest in the power and flexibility afforded by desktop computers reflects the increasing pressure on engineers to produce effective designs at an early stage and thus to avoid expensive development after the product is installed. The enthusiasm of younger engineers and their readiness to apply computer power whenever the problem warrants it has been a notable feature of design in recent years. Indeed access to a desktop computer is now a significant attraction when recruiting graduate engineers.

Psychology

It is unfortunate that the very eagerness of the younger members of staff combined with the easy-to-use features of modern desktop computers can paradoxically increase rather than reduce the strain on the design office.

The introduction of convenient interpretative languages, such as BASIC, has encouraged casual 'at-the-keyboard' development of programs. The running and de-bugging of these programs can account for a great deal of an engineer's time. In addition much of the challenge posed by a technical problem evaporates once a solution has been programmed successfully and so there remains little incentive for the author to document his work.

But without further time spent on preparing adequate documentation, neither he nor his colleagues are likely to be able to use the software successfully in the weeks to come. Hence without a suitable discipline, desktop computers can soak up, rather than save, valuable resources.

Use in the design office

Desktop machines are tools to assist engineers and can be used in a number of different ways.

For the test engineer conducting experiments or quality checks, the computer can store, manipulate and analyse the results. As well as these functions it is also possible to use the computer to control experiments and take readings. Some desktop computers are especially designed with this in mind and have facilities to make this easier. Commercial software is currently available to simplify the task of editing data and performing calculations on it and indeed this has been one of the first areas to be served by the market.

Another role for the desktop computer is in simulation. The ease with which the new interpretative languages can be acquired has enabled engineers, who have no training in computer programming, to explore complex design problems on a computer with a mathematical model of their own making. An example of the ability of a desktop computer to simulate engineering problems is a study of the arrangement of rolls in high-speed printing presses.

A third use for the machine is design analysis and performance estimation. The designer benefits from the speed of calculation and the elimination of errors in arithmetic and additionally the need for an intimate knowledge of the mathematics is obviated. The speed of the computer allows the engineer to investigate several solutions where previously there was time for only one hand calculation. The freedom from arithmetical chores and their attendant errors enable him to concentrate on the problem itself rather than the method of its solution. Lastly, the use of a program brings a standardization to the design. Only when similar problems are solved by the same method can anything worthwhile be learned from experience with the product in service.

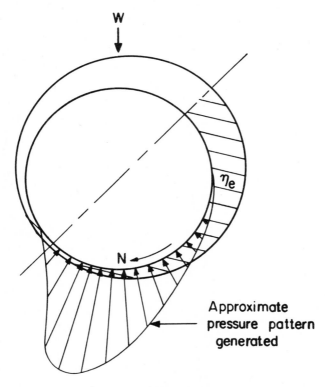

Figure 1 *JOURNAL bearing*

These benefits can only be realized when the software is easy to use and is of proven reliability. Otherwise the computer becomes merely another obstacle between the engineer and his problem. Analytical software is being made available by a number of organizations and it is becoming increasingly important that the packages and their contents be carefully scrutinized for origin, authorship and integrity. Incorrect data in printed form are hazardous enough but when translated into computer-readable form are virtually inaccessible and unrecognizable by even the most experienced of engineers.

An example of component design using desktop computers is provided by the JOURNAL program recently developed at ESDU. Although journal bearings are simple components, comprising a plain shaft rotating in a plain bore and lubricated by fluid under pressure as shown in Figure 1, their design is not so simple. ESDU produces, along with nearly 800 other data Items, a data Item or design guide for the design of journal bearings (ESDU Ltd, 1966), upon which the program is based. Although the Item is as concise as possible it still runs to 70 pages. Two additional Items (ESDU Ltd, 1977: 1977), extend the design to accommodate conditions of turbulent flow and lubricants other than oil. These account for a further 54 pages.

Naturally a work of this size requires some study before it is thoroughly assimilated and although it has been carefully arranged to be as easy to use as possible, establishing the performance of a bearing takes a little time. This is partly because there are one or sometimes two iterative calculations to be made where an initial calculation based on an estimate provides a result that can be improved by further calculations until the result is acceptable.

When using the JOURNAL program however, the initial calculation can be done with very little effort merely by choosing from the options offered. Once this stage is complete, the user may opt, for example, for a series of 20 calculations at different speeds of rotation. This requires no further work on his part and results in a set of graphs showing the performance at 20 different speeds, see Figure 2. In this way, the effort of calculation, with its attendant errors, is avoided and the engineer is free to study the behaviour of his design in depth and very quickly. The JOURNAL program comprises some 300,000 keystrokes but despite its size has been designed to run on desktop computers.

The ability to study the performance of a design, such as provided by the graphical presentations employed by the JOURNAL program, is one of the key features of CAE. With powerful software to help him, the designer can view every aspect of his design and investigate its strengths and weaknesses. Seen in this light, such software has a valuable role as an educational tool in addition to its use in production.

Cost reduction with CAE

Some of the benefits of designing components with the aid of the desktop computer were discussed in the previous section. When employed by experienced engineers these CAE techniques enable the performance of designs to be calculated more precisely and in greater detail than was possible before.

There are then opportunities for reducing the costs of manufacture as margins for error in design are reduced. Components become smaller and lighter, easier to handle and quicker to make.

However, although the performance of, for example, a gear or a journal bearing is known in greater detail and with greater precision, little advantage can be taken of this unless the operating conditions are known in equal detail. If an engineer is uncertain of the loads his machine might put onto a proposed bearing, then extra information about his bearing will be of only marginal assistance. It is necessary for the scope of research and development to complement the new CAE techniques before costs of manufacture can be reduced.

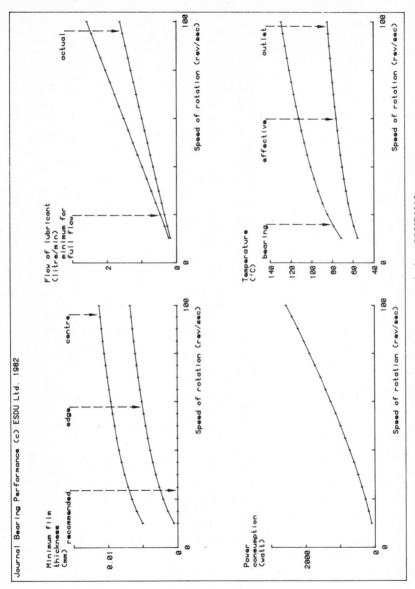

Figure 2 *Typical performance graphs from JOURNAL program*

Management

Managing the successful introduction of the first desktop computer into a design office probably requires more consideration than the selection of mere hardware. Not only does every office have its share of awkward and uncompromising staff, but, now that desktop hardware is providing benefits to organizations with no experience of machines for solving technical problems, they are appearing in offices where the older members of staff have probably never used a computer in their work. Consequently elements of ignorance, mystique and hostility must be overcome.

The first stumbling-block in managing the introduction of a computer system is unnecessary secrecy. It is essential to be open about the proposal, emphasizing that the computer should be seen as a tool to assist engineers rather than replace some of them. At this early stage a detailed idea of the benefits for the office is unlikely and the installation can be described as an evaluation exercise whose purpose is to assess the possible advantages.

If the intention is to use commercial software, that which is available should be examined carefully before deciding which computer to buy. It should be borne in mind that if the computer is not expandable, it may soon become inadequate and although the capital involved will not be great, other important factors come into play. For example the task of transferring software from one machine to another ostensibly similar machine is very time-consuming. Also, whilst graphics displays are a very valuable facility for technical work, the quality of graphics on desktop computers varies enormously and as the machines usually work with their own individual graphics commands translation of programs incorporating graphics is a major undertaking.

For those who plan to write their own programs, the purchase of some useful commercial packages is still worthwhile. These need not be technical. A word-processing package or even a suite of games programs will serve to introduce the machine to the staff whilst technical software is being developed.

Once purchased, it is vital that the computer be in an accessible position in the office rather than in a closed area; members of the office should be encouraged to use the system. It will be necessary to compile a brief user's guide for the installation and to arrange for the programs for general use to be readily available.

It will be advantageous to delegate a junior to supervise and maintain the machine, order spare paper and magnetic discs, arrange booking sheets and act as a librarian. A computer that is not fully operational will not be readily accepted.

Software pitfalls

The production of good software is a task far more demanding than the challenge of turning a calculation technique into a program. On first acquaintance with a desktop computer, the busy engineer can easily develop a host of bad programming habits to the despair of colleagues trying to use his software at a later date. Indeed the author himself often has difficulty running his own programs after a lapse of a few weeks.

When developing a new program, it is wise to keep duplicate or back-up copies. There will also be a printed copy or listing of the program. Without a rigorous dating and identification scheme, superficially similar version of code will proliferate to the confusion and frustration of the programmer.

It is essential to document programs in order to record not only what they do and what they do not do, but also how they work and how they were written. A program that proves useful will inevitably be worthy of extension or improvement at some later date. However it requires a prodigious effort to document the writing of a program so clearly that a newcomer can add to it successfully. The lack of adequate documentation is the largest contributor to the unproductive use of desktop computers.

Software standards are an essential tool for writing a library of programs. The use of standard routines for displaying and printing messages is necessary if all the programs are to show a consistent style to an unfamiliar user of the computer. In addition, use of a standard format for storing information such as user's values and calculation results is very productive as it enables the output from one program to form the input for another. An example might be gearbox design in which the results from the design of a bearing can be used as data for a shaft deflection program.

The writing of software that is easy to use, reliable and well documented is not therefore a task to be undertaken lightly.

Software – how to get it right

The features discussed here are given as means by which the quality of software can be assessed and should be considered by all those involved with buying or designing technical programs.

High-quality programs for desktop computers are truly 'interactive'. That is, the user is led step by step to the final result with frequent opportunities to step backwards through the procedure. No knowledge of computing is needed and detailed guidance and menus are given from which to select the next step. Programs including lengthy calculation do not simply 'go to sleep' but keep the user informed of progress and also allow him to interrupt and return to an earlier stage in the program.

Such programs will be predominantly self-contained with few, if any references to printed sources of information. All the data needed for the calculations are included within the program.

The user's values are recorded and offered as default values at the beginning of the next run. The range of values that the program will accept excludes impractical conditions and the subsequent calculations are guarded to prevent potentially dangerous extrapolations. The user is offered a choice of units and may view a picture of his outline design – a rapid way of checking for errors. Once an initial calculation has been successful, he may see the results of varying one chosen parameter as a set of graphs and he may instruct the program to read precise numerical values from those graphs.

Where the problem allows, an express option may be offered for experienced users and further use of graphics may be made to illustrate the solution (such as an exaggerated perspective view showing the mode of failure for a plate buckling under load).

Technical support is an essential service to users of design engineering programs. Naturally good software is based on the best available data, use the most appropriate calculation techniques and provides results that can be relied upon. It is also vital that users of the software have access to technically qualified staff. Since it is hardly possible for an engineer to become expert in both preparing software and the latest developments in engineering techniques, this support must come from both programming and technical staff.

Finally, it is probable that revisions will be necessary to accommodate improvements in engineering techniques, changes in standards and to overcome the errors that are to be found in even the most rigorously tested program.

Conclusions

Engineers should not be intimidated by the new and powerful technology of desktop computers. Advances in this field are being made rapidly and the only way to appreciate how small computers can assist in the design office is to become involved. The technology is young, and engineer who cultivate experience now will be preparing themselves for the developments still to come.

Engineers will be able to optimise design and manufacturing techniques and will also develop a deeper understanding of technical processes. Repetitive and error-prone manual calculation may be automated to free engineers' time and allow them additional flexibility for concentrating on the engineering issues in hand. All this is possible only if quality software is employed and until a broad range of evaluated and supported packages become available commercially, considerable time and effort may have to be expended in preparing programs.

References

Calculation methods for steadily loaded pressure fed hydrodynamic journal bearings (1966) Engineering sciences data unit, ESDU Ltd; London

Engineers and computers analysis of response to international questionnaire (July 1981) ESDU(M) Ltd

Journal bearing performance in superlaminar operation (1977) Engineering sciences data unit, ESDU Ltd; London

Low viscosity process fluid lubricated journal bearings (1977) Engineering sciences data unit, ESDU Ltd; London

C Jakeman, ESDU Ltd, 251-259 Regent Street, London W1R 7AD

Discussion

S Bennett, Ford Motor Company

First, do you believe that the spread of desktop computers should be controlled in some way, that the systems people should have regular checks on what people are doing with their microcomputers or should they just let them run wild and let them find their own way? Second, do you believe there is a case for a large company to specify a standard configuration so that people can always swap programmes and discs with each other? And third, if you believe that, do you believe that having standardized, one should then close one's eyes to the machines that come subsequently?

C Jakeman, ESDU Ltd:

I speak from experience of Baker Perkins, where we did introduce desktop computers in the same sort of manner that I have described. It will be very expensive in engineers' time and company resources, to allow piecemeal introduction of desktop computers. There has to be a learning curve, there has to be experimentation, but it must not be done in isolation, nor should each design office be allowed to go its own way irrespective of the others, otherwise you are going to waste an awful lot of time in translating programs. The time spent in trying to make programs compatible to one another between desktop computers cannot be underestimated, especially in respect to graphics. There are one or two things coming on the market now which do allow programs to be translated, the de-code system and various techniques of doing graphics which can be transported, but these are still very new.

Unidentified contributor:

We have a problem with the control of users. The desktops are proliferating and nobody seems able to stop it and there is a suspicion that many people are duplicating programs, and we suspect some people are probably sitting playing Space Invaders as well. Do you think that the systems people should make a conscious effort to make them fill in returns or say exactly what they are doing, or would this have the effect of people saying 'Well, if they are going to look over my shoulder, I am not going to try to do anything myself'?

C Jakeman:

I think that engineers should have free access to their desktop computers and it's up to their managers to make sure that they are not wasting company time, though quite a lot of experimentation isn't really wasting. There are two types of program. There is one that you develop just for yourself which is probably quite a quickie, and there is the other sort that is intended for everybody in the company to use. An example might be the configuration of timing belts and pulleys, which is a common problem in mechanical engineering design offices, or it might be the bending of shafts under load. I don't know how many different programs I have seen for the bending of beams and shafts under loads, but it would make sense to have one program that has been written for anyone to use, which would be kept just as one copy with everybody having access to it via a network. The other thing is standards in the presentation of programmes. If you are using a menu format then use the same way of presenting menus on every program, so you want a set of company sub-routines which everybody uses.

Contributor:

If you have a network like that, should it be controlled by somebody with an official hat, or would it be counter-productive?

C Jakeman:

I think you have to work through the man's manager. The person supervising the use of programs throughout the company is not likely to be senior enough to tell everybody to switch off and go and do something else. We heard about computer-aided design cutting across organization barriers and this is one of those cases.

A R Guy, Hatfield Polytechnic:

My experience when watching people introduced to desktop computers is that they want to expand and link their desktop computers with other equipment, such as different types of plotters and so on. This causes quite a problem, because the manufacturer's documentation in general is very poor. They also want to link into a network with a larger type of computer so that they can store their files on the main machine. The other thing which worries me is that most of these machines use BASIC and I am told on authority that the language on the new 16-bit micros is going to be Pascal, which indicates that we are possibly going to see a change which might cause a few problems.

C Jakeman:

When using peripherals and working with main frame computers, you must consider these things before you buy the first micro. Example of high-quality desktop computers are Tektronix and Hewlett-Packard, which provide an IEEE bus which they call the GPIB or the HPIB, depending on which company you come from. Their plotters connect into this with a standard plug and socket that has a standard number of wires to connect. Unfortunately the plotting commands are quite different so that Tektronix plotting commands must be in a little sub-routine at the bottom of the program so that you can cut those out and put in Hewlett-Packard ones when you run it on a Hewlett-Packard plotter.

On your second point, there are a number of programs which allow desktop computers to talk to the main frame and it's very wise to find out all about it in advance. Tektronix for example provide a back-pack which allows you to use the desktop computer as a terminal, and Hewlett-Packard do the same. We have just bought one of their interfaces and we are rather embarrassed to discover that we had to pay far more than the interface cost for a little ROM Chip before we could use the interface, which put our program back a couple of months as a result, so it's wise to find out all about this in advance.

A L Johnson, Cambridge University:

The most enormous pitfall in the use of desktop minicomputers is the susceptibility of the floppy disc to corruption, with the consequential loss of huge amounts of information on them. I have never seen engineers given formal training in the running of a back-up system, and those who do spend an awful lot of time doing it because to do it just with one's own files is a waste of time until the minute one needs it and then one is very grateful one's done it. It does seem that the ability to link up a minicomputer to a main frame so that all such archiving and back-ups can be done centrally is an absolutely vital feature of desktop computers in an industrial organization.

C Jakeman:

You do have to have a particular discipline if you are going to back up on floppy discs or data cartridges. It's essential to keep track of which program is which, because it's so easy to make changes when you are using interpretative languages like BASIC that you will have slightly damaged a number of different programs. This requires a definite discipline, and if you work with a main computer or even a large hard disc unit this will make all the difference, otherwise you will lose your programs from time to time, and there will be occasions when you haven't made a proper back-up. If you are using floppy discs you need to have three back-ups because if you are copying on to your back-up and the power goes down, you lose two copies so you must have a third. These are things that have to be considered as a company standard. To return to a previous question about Pascal on 16-bit machines. I mentioned that the P-code is a transportable system and an intermediate language. Pascal can be compiled into P-code as can a number of other languages including FORTRAN 77, APL versions of BASIC, Lisp and one or two others so that if you use the P-code as your intermediate transportable language you can have Pascal, FORTRAN and you can even mix languages in one application.

27. Computer-aided design for chemical engineers

P P Stainthorp

Abstract: The areas in which chemical engineers are increasingly using computers as design aids are reviewed.

The importance of training chemical engineering students in computer-aided design methodology is stressed and the implications for the computing resource requirements are examined.

Experiences with an interactive multi-terminal system are described as is the development of a general design tutoring package for chemical engineers. Some of the problems of constructing, transferring, and maintaining the package are discussed.

Introduction

In 1979 the European Federation of Chemical Engineers organized a conference dealing with Computer Applications in Chemical Engineering. One of the sessions at this meeting was devoted to chemical engineering education and discussions took place on the need to introduce computer-aided design methods into chemical engineering curricula and on the extent to which computer-aided learning was already being introduced into the courses. Contributions from Europe (Rose and Rippen, 1974), Japan (Matsushita, O'Shima, Takanpatra and Umeda), UK (Motard and Himmelblau) and USA (Stainthorp) demonstrated that in every country a growing band of enthusiasts was wrestling with the problems of introducing these new methods into crowded syllabuses often in the face of diminishing funding for new equipment and maintenance.

These enthusiasts are gaining ground and are now looking to exploit the interactive computing techniques that have emerged in the past few years. This contribution to CADED 82 describes the developments taking place at UMIST but these should be seen as part (a significant part we believe) of the total development in this area.

The term 'design' as used by chemical engineers has a range of meanings so the first section is devoted to explaining the scope we attach to the word and the extent to which computer-aided methods are being introduced into chemical engineering design activities is discussed. This is followed by an examination of some of the problems of teaching these methods. The solutions to these problems that are emerging at UMIST are then described and we concluded with some guidelines that, it is hoped, will help other schools and disciplines to take advantage of our experience.

Chemical engineering design

Figure 1 is an attempt to show in block-flow diagrammatic form the major steps in the evolution of a plant design from the conception to the materialization. In reality the clear-cut stages as shown usually overlap and according to the precise nature of the project, the contribution made by the chemical engineer can vary quite significantly. So, although the different areas shown for each stage indicate the distribution of chemical engineering effort across the profession for those engaged in 'design' they should not be interpreted as indicating a requirement for any one project.

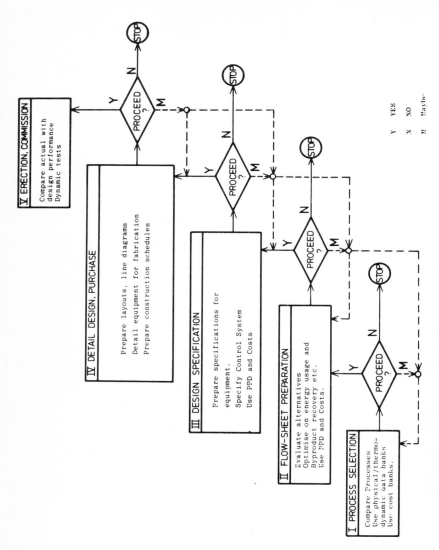

Figure 1 The major evolutionary steps in the design of a plant from conception to construction

These projects can arise in many ways. For example, a new product which will command a market in the area of agricultural, pharmaceutical or textile chemicals may have emerged from the research laboratories. Or a new route to a well-established product may be required because of a change in the availability of raw materials or the discovery of a new catalyst or a shift in the cost of energy. In all cases, however, the project will progress through a similar series of stages and a vital aspect of the scheme is the need to carry out some economic assessment at every stage in order to justify progression to the next.

In general, a team of people will be collaborating in any one stage of the project with, for example, a predominance of chemists in the early stages and a predominance of mechanical and civil engineers in the final stages but in all stages the chemical engineer will have some contribution to make and in the central design activities it will predominate.

In this article attention will be concentrated on the chemical engineering components of the various activities with particular reference to the growing use of computers in these activities.

Process selection

At this stage we are concerned to sift through the alternative process to narrow the choice for further, detailed study. The use of well-established short-cut procedures combined with an ability to recognize the potential for the various alternative ways of carrying out the reactions and/or the product separation and purification operations is required at this point. The engineer may well use simplified interactive flow-sheeting programs at this point and draw upon data banks containing costing information and the physical properties of the components used in the process.

Flow-sheet preparation

In Stage II the objective is to firm up on the process to be used for the project. This will involve a critical comparison of alternatives and this comparison is most conveniently made by preparing flow-sheets in which the possibilities of materials recycle and energy recovery by, for example, heat exchanger or coupled expander-compressors are explored in some depth. We refer to this activity as integrated or system design.

Over the last decade some very powerful computer programs have been written which enable chemical engineers to deal, accurately, with the immensely difficult calculations which arise when dealing with complex mixtures of components which exhibit 'non-ideal' behaviour. These same flow-sheeting programs also enable us to solve the problems that arise when streams of partially separated and purified reactants or solvents are recirculated. The earlier programs of this type were confined to fluid systems (vapour and liquid) and to specific components but we are now seeing the emergence of programs that will handle solids, quasi-components representing hydrocarbon-fractions as defined by their physical properties and even such difficult systems as coal, wood pulp etc.

Originally the major manufacturing companies and design houses developed their own flow-sheeting packages but there are now specialist companies offering packages which they will maintain on the users' computing systems. They also offer a service through bureaux to the smaller companies so that a simple console may give access to these very powerful design aids.

These packages are not small (a typical program runs to some 100,000 statements and requires nearly 256K of working space). They have to be used with skill and an appreciation that ill-prepared exercises and unnecessary runs including inessential changes to a basic flow sheet can soon run away with thousands of pounds worth of computation.

Design specification

The preparation of the design specifications requires the chemical engineer to determine all the principal features of equipment that is to be designed in detail or to be selected from manufacturers' ranges. In many instances the calculations required to arrive at an optimal design are quite horrendous and most companies have built up a library of programs that enable their engineers to prepare the design specifications for the equipment most commonly used. Many of the design methods require the engineer to make a compromise between alternatives and there is no doubt that interactive computing techniques are going to transform many of the traditional design methods.

One example would be in the design of one or more fractionating columns that are to separate multi-component mixtures. Another would be in the determination of the best arrangement and sizes of the effects in a multiple-effect evaporator system complete with the ancillary equipment required to achieve an overall economic optimal design.

The engineer would also be giving some considerations to the operability of the units under different loading conditions and hence to the structure and principal features of the control system that is to be detailed. This may well require an examination of the dynamic characteristics of the units to be controlled, both in their own right and in combination with all the other units in a particular section of the plant. Dynamic flow sheeting is still in its infancy but efficient simulation methods are now appearing and are being used by chemical engineers. Other programs that may be used at this stage of the design can be concerned with materials selection, corrosion prevention and hazards analysis.

Detail design

Once the design specifications have been prepared and it has been confirmed that the whole project is economically viable, the task of detailing the designs for workshop fabrication or for selection from standards can commence. The exact layout of the plant has to be settled and the utilities and process piping has to be detailed.

Programs exist that enable the engineer to carry out all these tasks interactively so that for example, the effect of changing the location of items for ease of maintenance may quickly be visualized on graphical displays. Line-diagrams and isometrics showing pipe runs can be drawn and purchasing and erection schedules prepared.

The control system will be designed in detail using programs to select valve sizes and characteristics.

As a part of all this detailed design activity there will be a final check on fire, toxic and pollution hazards and the means which have been planned to mitigate them.

In industrial practice a significant proportion of the computational effort will be directed to recording and monitoring procurement and installation of the equipment but the average chemical engineers are only involved with these activities to the extent that they must be reassured that the equipment is precisely what they require and will be installed on time.

A lot of CADCAM activities will be appropriate to this stage of the project but most of them fall more in the province of the mechanical or structural engineer and will not be further considered here.

Erection and commissioning

The chemical engineering component of this stage of the project really only commences when the performance testing and evaluation begins.

At this stage it may well be appropriate to use programs from the libraries referred to above so that efficiencies may be determined and there is nothing new in this. One area where additional computation may be required is when it is necessary to determine from perturbation tests the exact characteristics of the equipment as installed so that the best control behaviour may be obtained.

Teaching chemical engineering CAD

It will be apparent from the foregoing that computers are already having a significant impact on all aspects of chemical engineering design activities and we need to consider how best to introduce these new methods into undergraduate curricula.

In the author's opinion we have long passed the stage where chemical engineering CAD can be treated as an 'add-on' to traditional courses and we have to examine the resources that are needed to introduce these methods as an integral feature of the courses. Fortunately it transpires that the resources required for the design activities can also be exploited for other computer-aided learning activities and we have the prospect of enhancing our undergraduate teaching so that the students emerge with a much greater appreciation than hitherto of the relationships between chemical engineering fundamentals and the best design practices. Before we consider the interactive developments the present state of the flow sheeting work will be described.

Flow sheeting

In the early 1970s many chemical engineering departments developed flow sheeting programs appropriate to their computing resources and teaching objectives but it soon became apparent that the efforts required to expand, maintain and properly document these programs were beyond the capacity of the individual departments. Fortunately, in 1977, ICI donated a version of their company developed FLOWPACK (Programme Library, 1972) for use by the universities in this country provided that controls were incorporated to prevent unauthorized access. The program was installed at the regional computing centres of Manchester, London and Edinburgh and during the past five years has been used thousands of times by students in all the major departments with links to the regional centres.

In 1981 Simulation Sciences (SimSci) offered their 'PROCESS' under similar conditions and we at UMIST maintain on the UMRCC machines this very large design program for use by all the connected departments.

PROCESS is particularly attractive for our use because it contains a very extensive physical properties data bank, a library of thermodynamic sub-routines, a powerful and dependable sub-routine for the solution of multi-component multi-column fractionation problems and an ability to handle quasi-components as they arise from the assays of petroleum fractions. Two other features of this robust package are the helpful diagnostics that appear when incorrect or unfortunate data prevent the generation of a solution within a user specified number of iterations and the presentation at the end of every job of the royalty that would have been paid to SimSci had they not been so generous. We can use this latter feature to control access by giving each student an allocation of 'cash' to complete an exercise and thereby to inculcate a healthy respect for the need to think through a design before presenting it to the computer.

In one recent teaching exercise some 75 students used an average of 11 runs at an average cost of £48 per run to investigate a variety of parametric changes when applied to a basic design for a cyclohexane plant.

Currently, PROCESS jobs are submitted as card jobs via the remote job entry terminals but there could be clear advantages in submitting them via VDU consoles through a local pre-processor which checked and validated the data file before transmitting the job for batch processing in the large machine. A post-processor which lead to an examination of selected key results and thence to much reduced line-printer output is also required and we are engaged in the preparation of these processors. It is also clear that although the full resources of the package may well be required by final year students engaged in their design exercises (which can last for several weeks) it would not be practicable to use the full user's manual for introducing the student to computer-aided flow sheeting. For this purpose the author has prepared an introductory manual (Stainthorp, 1982) which presents the essential data requirements and a very restricted set of options. This seems to work well. In addition, we believe that there is a need to develop an interactive teaching program which will cause the student to compare hand solution methods with selected computer methods. How such a program would be used is indicated below, but an alternative should be mentioned first.

The availability of quite powerful micro-processors with disc back-up opens up the possibility of constructing a flow sheeting program that is suitable for first instruction and capable of solving small problems. Several departments are putting in effort in this direction. One successful micro flow sheeter that has been demonstrated is that of Hill at Heriot-Watt (Beveridge, Hill and Agilar, 1982).

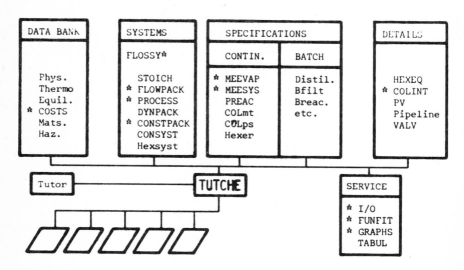

Figure 2 *Conceptual arrangement of TUTCHE package*

TUTCHE

The name for the package to be described derives from *TUT*oring programs for *CH*emical *E*ngineers. The conceptual arrangement of the package is shown in Figure 2 where it may be seen that a set of interactive terminals gain access, through the monitoring program TUTCHE, to five banks containing data files, program files or service files. TUTCHE is used to control access, record individual usage, maintain users' data and results files and generally service the data base. TUTOR is a program that enables the tutors to control access by setting allocations. They can also establish whether any particular student is having trouble with certain programs and, in general it provides a privilege access to TUTCHE. Some of the components of the banks are now examined in more detail. Particular reference will be made to those starred in Figure 2 since these are the ones which have been used extensively to build up the experience with this type of tutoring. Of the other titles those in capitals are at an advanced stage of development, the others are examples of those that are next in the queue for preparation.

System Bank

This contains the general packages which are required for consideration of overall design aspects and for performance evaluations. FLOWPACK and PROCESS have been described above. STOIC derives its name from stoichiometry and will form the basis of the teaching and assessment program required as an introduction to flow sheeting. It starts with an exercise in which the student has to match hand-generated solutions to computed ones for some progressively more difficult problems. It then leads on to an examination of the effects of changing, for example, tolerances and tear points in recycle systems. This work is confined to mass-balance studies but acts as a preparation for the larger flow sheeting packages which also deal with energy balances.

DYNPACK is under development to provide an interactive program related to learning about the dynamic behaviour of connected plants – especially those with recycles and incorporating significant line delays. It will also relate to the programs required for the computer-aided design of control systems.

Conversation with the above programs is currently via alphanumeric screens but there would be many advantages in using a symbolic flow diagram both for inputting topological information and for receiving results. To this end we have a program, FLOSSY, for use with a graphics screen which will serve as a common interface to any of the programs in the system bank that require a flow diagram.

COSTPACK was prepared to aid in the teaching of costing methods but it is now also used extensively by final year students when they are engaged on their design exercises. It contains three main sections, one dealing with approximate and exact methods of pricing items of equipment, a second one dealing with approximate and factorial methods of estimating the fixed capital required for complete plants and a third one concerned with product costing and project profitability assessment by various methods. This latter section also allows the student to examine the sensitivity of the costings and profitability to changes in the design, operating and economic variables.

Specifications Bank

The programs in this bank are intended both for tutoring in the techniques of preparing the design specifications for individual plant items and in the optimization of the design where this is appropriate. To illustrate the major features we will consider the evaporator design programme MEEVAP and MEESYS.

Multiple-effect evaporation as a concept is quite simple. We condense steam in a calandria in a vessel, called the effect, to boil off some of the water from a solution. This steam is in turn condensed in the calandria of the next effect operating at some lower pressure and evaporates more water from the solution. In principle one could conceive of a chain of those effects (up to seven are used) and if we ignore heat-losses and other factors we might expect to evaporate one kilogram of water from each of the effects for every kilogram of steam used in the first effect. In practice, as the solution concentrates, its boiling point at any pressure rises and the heat-losses – even with lagging – are significant. For any particular duty there is an economic design but to determine this by hand methods requires a tedious, error-prone, iterative procedure.

Nevertheless, it is one that chemical engineers need to know so MEEVAP was devised to help teach an understanding of evaporator design. For a particular problem set by the tutor the student is required first to generate a hand solution and then to put the same data into the program. The student's solution is compared with the unrevealed computer-solution and if it is outside a $\pm$ limit he or she has to repeat the hand calculation – after seeing the tutor if necessary. The filed data is recovered and the student is allowed two more failures before a lock is set to prevent access to the program until the tutor has discussed the problem again and unlocked the permit flag. Once the student has demonstrated competence a series of calculations are undertaken to reinforce an appreciation of the effects of changing design parameters and operating variables on the economics of evaporation. This type of extended exercise is one which could not have been undertaken in the past – there just was not enough time.

At a later stage in the students' training, when they are beginning to consider integrated design, they use MEESYS, the access to which may be conditional upon a recorded success with MEEVAP. With this program evaporator designs may be considered in which extra heat savings are achieved by recovering steam flashed from the condensates as their pressures are reduced and/or by preheating the feed-stream by heat-exchange with the vapour from one or more of the effects. There are efficient and inefficient ways of carrying out this type of optimization and the student has here an opportunity to discriminate between them. Again, this is a type of exercise which would have been difficult to accommodate in the pre-computer era.

A similar approach to the design of reactors, binary fractional distillation units, heat-exchangers and any other equipment for the standard continuous operations can be adopted. A related series of tutoring programs for batch operations is also required. At the moment these tend to be squeezed out of the courses although they can still be important in many sections of the chemical and process industries.

Detailer Bank

The programs in this bank are required to translate the design specification into detailed designs needed for the fabrication or selection of the equipment.

COLNT examines the effect of column internal details (tray type, tray spacing, tray details etc) on the efficiency of fractionating columns. HEXEQ has been programmed to enable heat-exchanger construction details (tube sizes, arrangement, spacing, baffle design, etc) to be considered. In using these and similar programs the student is expected to learn something about the economics of the detailed design and, in particular to establish that the equipment will still operate efficiently throughout the range of duties it may expect to experience in practice.

Computer hardware

We have some 75 to 80 students in each year of a three year course and in order to provide experience with the CADCAL work we clearly need some extensive and reliable facilities. What we have is in the nature of an experimental resource which, when it has been evaluated, may well form the basis for similar suites to be used elsewhere within UMIST and outside.

The basic concept is that all activities should be carried out at interactive terminals which have been installed according to the arrangement shown in Figure 3. The terminals are arranged in 'clusters' of five which share a matrix alpha/graphics printer. This, at the moment, is connected to any one of the five terminals by a simple multiple switch. All terminals can operate in upper/lower case alphanumeric mode and, of the five, two are being extended to work in graphics mode. The fifth terminal is intended to be a microprocessor which may work in stand-alone mode or as a console and will have alpha/graphics capability. The graphics is required to have a full Tektronix 4010 emulation with cross-hairs input and, at the time of writing, we are still experiencing some difficulty in acquiring terminals that will exactly match the specification at a price we can afford.

At the moment we have three clusters, ie 15 terminals in one CAD suite.

Any terminal may be connected via a switch to either a Prime 750 or a Cyber 172 both of which provide time-shared interactive operation at line speeds of 2.4 kB.

The advantages of the Prime are that we get a very good response rate up to the permitted loading and, especially, we are able to exploit a 'shared-code' facility that enables 15 students to work with just one copy of a program in the machine so that other users do not notice the extra load.

The potential advantage of the Cyber is that it is housed next to the CDC 7600 in which we run the large flow sheeting programs. It is our intention to develop the technique of preprocessing data in the interactive machine so that the large programs may run more certainly and productively in the big computer. Selected output from the results files will also be post-processed in the interactive machine so that a significant reduction in line-printer usage should ensue.

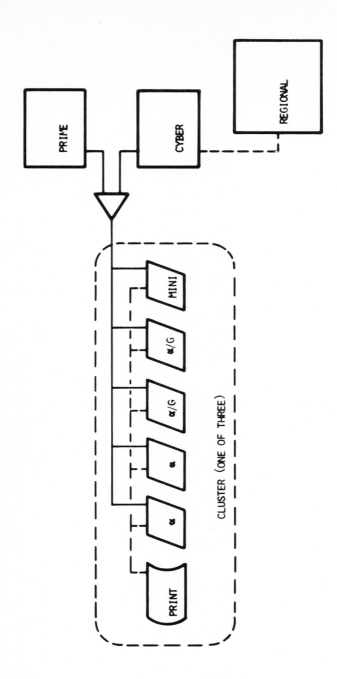

Figure 3

Experiences and observations

Whilst we have not yet accumulated sufficient experience with a wide enough range of CADCAL programs to classify ourselves as experts we have now run two successive years with the system and feel that we can record some of the experiences and make some helpful observations. Some of these latter may seem a little trite to the specialists but it has to be recognized that most of us, even those with years of experience at conventional lecturing, are having to come to terms with new requirements, new facilities and new opportunities. If this contribution saves some heartache and wasted effort then it will be worth making.

Programming

The effort required to produce a thoroughly reliable student-proof interactive program is many times greater than that required to generate a batch program for the exclusive use of the writer on a similar design task. However, it is all worthwhile if the resultant program can be used around the globe.

Since this is the objective then it has to be recognized that it pays to work to a strict coding standard. (We now use FORTRAN 77 but are still running into trouble on transportation because the compilers currently available are not as reliable as they should be.) Certain I/O, character handling and file handling procedures are still machine dependent and we have had to confine these features to our own 'standard' sub-routines which can easily be adapted to local conditions when we transport the package.

Once these standards are established and accepted it becomes feasible to consider setting up a co-operative exchange scheme and we are actively pursuing this possibility.

Before starting to program it is vital to have a clear idea of the way the tutoring task is to be carried out. We now would follow the dictum of Alcock and Shearing: 'Before any program is coded write the User's Manual in its entirety, including worked examples, mock-ups of output, and so on' (Alcock and Shearing, 1973).

The program should be written as an assembly of sub-routines none of which should be more than 250 statements long and most of which should be less than 100 statements. This approach makes it much easier to arrange for efficient over-laying and also eases the tasks of de-bugging and editing.

Write the dialogue at two levels so that it gives a full prompt for novices and a brief prompt for experienced users. (Provide a switch so that very experienced users may bypass the chat altogether.) Pay particular attention to clarity and brevity. There is nothing more off-putting than to be faced with WHAT NEXT?

Provide multi-level HELP at every point in the dialogue, first interpret the prompt then amplify, give validity limits etc.

Check all responses for relevance and validity and provide for the user to quit safely from any point and then to return reliably to the same point.

Structure the displays of results so that they are arranged on a page or half page that can be copied to the local printer.

Provide for the user to be able to select the values of variables from several runs and collate them so that key effects can be quickly identified. If a graph would be even more revealing provide for this at a graphics terminal. (We have some service packages that are used for these tasks.)

Organization

There is no doubt about the enthusiasm with which the average student will tackle a well-prepared tutorial exercise. (Indeed many of them extend the exercises on their own initiative.) Conversely, there is no doubt that an ill-prepared exercise which is just thrown at the students is completely counter-productive.

It would seem that one hour sessions are convenient slots provided that the actual terminal occupation is about 45 minutes. Clearly, with 15 terminals for a class of 75 students some organization is required and we find it possible to schedule the slots along with other conventional problem-solving or laboratory classes.

It is vital to have a responsive system for this type of work and we find that when a machine is so loaded that it takes longer than two seconds to respond during a simple data imputting session it is better to stop and use the time more productively elsewhere.

Acknowledgements

The participation of some 200 students at UMIST and elsewhere is gratefully acknowledged. The exposure of ambiguous dialogue and the discovery of programming bugs is enormously accelerated when so many users are trying to follow undreamt-of paths with unanticipated data.

The contributions of David Lomas are also acknowledged. He is responsible for maintaining the packages on our existing machines and has played a significant role in ensuring that the transportation problems to other systems are minimized. Jane McLean is the programmer responsible for generating the standard I/O routines and the graphics support programs.

The provision of time on the SERC Interactive Computer Facility (with the aid of Dr G Barney and W Swindells) has significantly altered the pace of development and enabled us to demonstrate the potential for this approach to the teaching of engineering.

References

Alcock, D and Shearing, B (1973) *Software – Practice and Experience* 3: 255-301

Beveridge, G S G, Hill, R G and Agilar, M R (1982) Interactive computer flow-sheeting on a micro computer. I Chem E Jubilee Symposium, paper H

FLOWPACK – program library (1972) ICI corporate lab: the Heath, Runcorn, Cheshire

Matsushita, M, O'Shima, E, Takanpatra, T and Umeda, T Education for the use of computers in chemical engineering decisions in Japan. *Computers and Chemical Engineering*: 209-12

Motard, R L and Himmelblau, D M Current situations on the use of computers in the education of chemical engineers in the USA. *Computers and Chemical Engineering*: 213-16

PROCESS – Input Manual (1979) – Simulation Sciences Inc

Rose, L M and Rippen, D W T (1974) The current situation of the use of computers in chemical engineering education in Europe. *Computers and Chemical Engineering* 3 221-24

Stainthorp, F P (1982) Process student's manual – Department of chemical engineering UMIST

Stainthorp, F P The education of chemical engineers in Great Britain in the use of computers. *Computers and Chemical Engineering*: 217-20

F P Stainthorp, UMIST, Department of Civil and Structural Engineering, PO Box 88, Manchester M60 1QD

28. A practical approach to training in the use of an integrated plant modelling system

D J Appleton

Abstract: An integrated CADCAM system is used by many different disciplines each of which may approach and use it in a different way. Thus a supplier of such a system if he is to achieve maximum acceptance and utilization by each of its various type of user we must be aware of their different requirements and organize his training for each discipline in the most appropriate way. This paper takes the case of process plant design and construction and illustrates the variations in use and approaches to training taken by the system supplier.

Recent reports such as Finiston, conclude that if British industry is to remain profitable it must adopt a more flexible role in its efforts to satisfy the needs of a rapidly changing world market. Continual emphasis is placed on the need to use new – or improved – manufacturing techniques in order to produce products which are both innovative and competitive. Ultimately this means exploiting technology in order to capitalize on the opportunities it offers.

It is perhaps a paradox that as technology becomes more complex it appears relatively less complex to use. Indeed developing countries are finding it much easier to apply, and therefore, if existing industrial nations are to maintain a competitive edge, they must resort to new production techniques.

With world attention currently focused on the application of robots and the fully automated factory environment it is more important than ever to ensure that any new idea, which could potentially effect our performance in this – and any other – area, is fully evaluated.

Obviously great advances have been made recently in our ability to introduce new technology both in terms of government support and in changing the attitudes of industry itself. Perhaps the greatest obstacle now is that of financing the development of these new ideas. Traditionally the government has been a supplier of financial assistance through the Department of Industry, the Treasury, and the Department of Education and Science but I would like to add that many industrial organizations (including my own) also have the ability to offer advice and to finance commercially viable projects. Perhaps it is significant that the USA invested more than £30B in research last year, of which 60 per cent was conducted through industry itself, and 26 per cent through educational and research institutions.

Inevitably new production techniques will be centred around computer-aided design and I would like to illustrate how training can help to communicate the advantages of CAD in the construction of refineries, drilling platforms and other similar types of plant.

As my company is heavily involved in this particular area of computer-aided design, much of what I am about to say reflects our approach to training in this field.

In developing a practical training strategy it is important to understand the demands made by the plant design industry and to appreciate the competitiveness of the markets in which it operates.

In situations where teams of designers are recruited for the life-cycle of one particular project it is necessary for them to become productively involved, as soon as possible, with the design process. This means that in practice they will be required to work at a computer terminal after only one day's introduction.

Challenges of this nature are probably only to be found in the plant industry. In developing a training strategy I suggest that as a starting point the trainer should address three potential problem areas.

First, he should understand that any new concept will normally compete for acceptance with an existing procedure which is already familiar, and which has an established and proven track record. The tendency will therefore be to absorb the new concept into the existing framework, before the old is discarded through evolution.

Second, he should appreciate that during any formal presentation, most potential users will experience difficulty in comprehending any new system's total capability.

Third, due consideration must be given to the decision-making process which must inevitably be circumnavigated before the implementation of any new technique occurs. By recognizing this process indirect assistance can be given to potential clients to help them prepare their own internal product justifications, implementation guides and so on.

The underlying implication in all these points is that one must be able to identify clearly the benefits offered in terms of the current operational objectives of all the disciplines which are likely to benefit from the introduction of CAD techniques.

In essence I am speaking of the trainer's ability to perceive advantages in terms of the user – not the product. My comments will therefore be largely based on the experiences Compeda has obtained in implementing software systems on a variety of computer systems at customer's premises throughout the world.

How then, can education be effectively applied to aid the use and implementation of CAD techniques in the plant design industry?

I suggest that the trainer must possess the following:

- a comprehensive product knowledge
- an understanding of the organization in which the CAD system will be integrated
- an ability to identify its benefits in terms of the user's requirements.

What application does CAD have in the field of plant design?

To answer this question it is important to realize the complexity of plant design and the implications of not getting it right first time. Those who have not been directly involved in plant design should nevertheless be able to appreciate the level of planning necessary to ensure that delivery dates of any sub-contracted material has minimal effect on construction schedules, and that the completed site will meet all environmental and safety requirements. This is in addition to ensuring the plant functions as intended and is completed on time.

Obviously the planning stage is critical and during the earliest design phase – the conceptual plant layout – decisions are taken which will dramatically affect:

- total plant costs – involving all aspects of foundations, steelwork vessels, piping, mechanical and electrical equipment and instrumentation.
- time-scales – with special reference to component delivery times and construction sequences.
- safety and environmental interests – both for plant workers and the surrounding populations.

One method employed by many companies is to construct an accurate plastic model of the plant which is continually developed until the level of detail is sufficient to be used as the basis of design. However this process takes time to develop, requires continual updating, and as a result is time consuming and therefore expensive. However a plastic model does offer the ability to easily obtain an overall impression of what the plant will look like when it is built. Designers will therefore use the model to identify potential problem areas and then adjust their design drawings accordingly.

CAD on the other hand, allows an accurate model of the plant to be constructed inside a computer which can then be retrieved, modified and analysed at any time. The inherent accuracy of the computer, working on a centralized bank of information, offers tremendous benefits in terms of obtaining construction details such as component quantity lists, lengths of pipe used etc, and in identifying positioning errors and inconsistency of design. Stages in the design process may be frozen, pending their acceptance by the design authority and the customer. To prevent further modifications being made by the design team whilst this process occurs, copies of the existing design information can easily be made thus allowing the design process to continue uninterrupted. Diagrams and detailed drawings can also easily be made (using computer-controlled plotting machines) from any viewing point in the site.

By constructing the plant in a common data-base all designers can have access to the latest information, and drawings requested by draftsmen will always show the latest stage of development. Information held in this form is accurate and rapidly accessible but – as you can probably appreciate – these benefits are not at first easily recognized (or believed) by established company personnel who have little or no computer experience. They may even be approaching the use of CAD with some degree of apprehension, either from the recommendations they know they will shortly have to make or from some fear of job security.

They will need to understand that the CAD system – used in plant design – will address five main areas:

1. design data
2. component catalogues
3. piping specifications
4. 2D and 3D drawings
5. project control information

The design data should be easily sub-divided into areas in the data-base equivalent to constructional areas within the site. The CAD control program should then be capable of allowing teams of designers to work simultaneously in their own design areas.

The component catalogue must be capable of listing all items available to be used in the project including their dimensions, required operating areas, insulation requirements and connection types. This information will be required for drawing purposes and also by the computer when it checks the consistency of the design process for alignment, connection compatibility and interference between piping components and structures.

Piping specifications aid the designer in the selection of components which can safely be used in individual pipe runs. Special consideration is given to temperature, pressure and fluid contents thereby ensuring that when a designer selects (for example) a valve, it not only fits but is also suitably rated.

Drawing also plays a very important part in the design and construction of a plant, so the CAD system must offer an easy method for keeping all diagrams and specifications compatible with the progress in design. This facility is particularly important to ensure that all drawings reflect the plant 'as built' at the end of construction phase. As the programmes which produce graphical outputs obtain their information directly from the design data-base, they can only produce drawings which are compatible with the current stage of the design.

The types of drawings required may be roughly categorized into two-dimensional and three-dimensional representations.

An example of the use of 2D would be in the preparation of flow diagrams and electrical wiring diagrams. Constructional drawings will require 3D representation, perhaps with sectioning, dimensions and annotations. A modern CAD system is quite capable of automatically evaluating distances between components held in the data-base. If the component's position is later modified, its original dimensional data will be automatically updated by the computer.

This capability therefore means that accurate, detailed drawings are available from the moment design begins.

Project control ensures the integrity of the individual design areas is maintained by, for example, preventing designers working in one area inadvertently changing the design in another.

I have just attempted to summarize an application of CAD to illustrate some of its capabilities. If these are now to be incorporated into a training strategy the trainer must now appreciate the environment in which the CAD system will be used.

In fairly straightforward terms a company will receive an invitation to tender for the construction of a particular plant. Accompanying this invitation will be a specification of the site's location, its purpose and process flow information. Normally the potential plant owner will have previously obtained provisional site surveys, outline planning permission and authorization to use any patented process held by third parties.

In the case of a medium land-based site the provisional plant specification would have originated through a marketing proposal which – if agreed by the main board – would have resulted in a conceptual design of the plant.

Typically this would contain:

- the type of process involved.
  This information may be supplied on a need-to-know basis particularly if the process is secret.
- production figures.
- operating standards together with vessel, equipment and piping application sheets which will indicate fluid types, operating temperatures and pressures etc.
- site survey results and building specifications.
- basic flow diagrams and an outline construction schedule.

This information may itself have been produced using CAD techniques.

The actual design of a project within a typical company will be conducted on a task force principle. A team of designers and draftsmen will be responsible for preparing quotational data, and when the order is eventually received the team will be supplemented by the addition of more staff appropriate to the delivery date. This is a summary of the disciplines which may be required.

The process engineer is essentially concerned with ensuring the process works and therefore may not be directly concerned with the physical design aspects. He will however make use of the 2D drafting facilities to prepare detailed flow diagrams.

Piping engineers will widely use all design aspects of the CAD system as they will be responsible for all aspects of pipe routing, sizing and stress analysis. They may have reporting to them a number of design draftsmen.

Civil engineers will use CAD to construct accurate drawings of underground services, foundations, buildings, access ways (including roads) and site services such as gas, water and electricity.

The vessel designer will be responsible for detailing the design of vessels, heat exchangers etc and therefore would benefit from the use of the 3D drawing facilities in order to communicate his designs to the vessel makers.

Structural engineers will be responsible for the design and positioning of equipment supports, pipetracks and access ways. They will therefore use CAD techniques to design these items with due regard to pipe routes, vessel/equipment locations and plant safety.

Instrument and electrical engineers will require the use of 2D and 3D drawing facilities in order to construct piping and instrumentation diagrams and electrical wiring diagrams and schematics. In addition they may be involved in specifying component data for inclusion in the catalogue.

Mechanical engineers will determine the nature of all pumps, turbines, conveyors and will obviously be involved in the siting of it and other pieces of heavy equipment.

Planning and control will co-ordinate all activities and will use information from the data-base to plan the construction sequence and to monitor its progress particularly with respect to the supply of sub-contract material. Special considerations may have to be taken if heavy lifting gear is required, as construction schedules may have to be modified especially when the equipment does not arrive at the specified time.

An organization such as they will have developed with experience will have many past successes to its credit. But CAD can now offer an ability to:

- produce drawings or graphical displays of detailed plant layout accurate to a level far greater than that required by the construction industry.
- obtain three-dimensional views of the plant from any viewpoint scaled to the size of the complete plant or any component within it.
- display any item with (or without) hidden lines showing, with (or without) section planes.
- obtain a computer analysis of all potential positional clashes between any (or all) items of plant held in the data-base.
- obtain up to date lists of all components and quantities of materials used.
- perform safety checks and pipe stressing analysis.
- identify any item within the plant by name and position, etc.

The list continues but the point I am making is that if this were presented to engineers or draftsmen – who may have little or no appreciation of CAD capabilities – one could perhaps excuse a certain degree of scepticism.

Similarly, one could accept a certain degree of foreboding in managers who, already steeped in company tradition, will be required to justify the acquisition of a CAD system to other (perhaps less involved) members of the management team.

If a smooth integration of CAD is to occur, all potential users must be made aware of just how useful CAD is in helping them to achieve their own specific working objectives.

So, having described the capabilities of CAD, and an environment in which it could profitably be used, one needs to analyse the benefits of its use in terms of the people who will use it. If this is successful they will soon begin to identify with the CAD system and suggest more ways in which it can be used within their company.

The resulting training strategy must therefore reflect the needs of the people and disciplines mentioned earlier and it must also recognize the need to be economically efficient. The approach I have chosen begins with an introduction to computer-aided design. This will involve a brief analysis of the computer environment in order to dispel some of the mystique which so often surrounds it.

Particular emphasis will be placed on clarifying the method and level of operator interface and in the explanation of basic CAD terminology. A brief analysis of the advantages offered by the CAD system to a typical company on a departmental basis should also be given to enable individual students to relate their own activities in terms of the global benefits offered. Although this section is relatively short, it does provide a common foundation of knowledge which can then be assumed and developed in subsequent training courses. After this introduction a number of options became available depending on job requirements.

Project managers, engineering and design staff, together with team leaders will certainly need to understand the design process and the facilities available to them. Careful selection of student projects will ensure that all participants receive training appropriate to their particular discipline but particular emphasis will be placed on piping, foundations and structural design because of the enormous benefits offered by CAD in these areas.

Pipe stress analysis has been included for engineers working in this specialized area of design. Special attention will be given to analysing the effects of internal pressure loading, anchor point displacements, externally applied forces and so on.

The administration course will be of benefit to the project manager, as it will describe how CAD can be structured to meet the needs of his design team and what precautions may be taken in order to protect the integrity of the design data. It will also illustrate the project control facilities offered by such a system. Normally one person within the company will be given the responsibility of maintaining the catalogue list of components available to be used within a project. They may also be responsible for the contents of piping specifications – these ensure that the system chooses appropriately rated components from the catalogue to suit individual pipe runs. The catalogue and specifications course will therefore detail the process necessary to describe to the system the exact geometric shapes and characteristics of all components to be held in the catalogue together with their method of selection.

The engineering drawing production course will be of particular benefit to draftsmen. It will explain how to obtain views of the plant from any direction and how to define section planes and annotations. Dimensioning together with the ability to define drawing symbols, roads, terrains and so on will also be included.

Draftsmen who are required to perform pipe routing and structural design should also attend the plant design course and conversely any engineer who is responsible for a drafting team will also be advised to attend the drawing production techniques course.

For electrical draftsmen the electrical drawing production course should illustrate the advantages offered by CAD in the area of creating symbol libraries and in the production of wiring diagrams.

A further advantage offered by CAD is its ability to assist buyers and materials progress clerks to monitor the acquisition of equipment from sub-contractors. With the knowledge necessary to interrogate the computer design model, listings of total quantities and types of components used can be readily obtained, making initial ordering and the monitoring of delivery dates that much easier. Similarly, instrumentation and control engineers, together with civil engineers will also need to be informed of how they can make full use of the system in their specialized areas of work.

Having designed this program specifically to meet the needs of CAD users, the trainer is now capable of communicating the benefits to all disciplines likely to be concerned, but there is one further advantage to be gained. This lies in the area of management training.

Many senior managers are preoccupied, by necessity, with the financial and administrative sides of their company. They are not necessarily over-impressed with processing speeds, or with abstract items illustrating the capabilities of a CAD system. What they really want to know is 'what can CAD do for my organization?' With a detailed understanding of their company's business objectives and methods of working, they will expect to be informed (in the macro sense) of exactly what the advantages of CAD are to their company. Having constructed this training strategy, I would like to suggest that these advantages have already been identified in terms of the various departmental needs.

Therefore, by using sections of each course it is possible to construct a management appreciation which illustrates, by example, the advantages of CAD throughout the complete quotation, design and construction phases of a plant. By indicating the amount of information and the quality of drawings available from the very beginning of a project, the benefits are very easily understood and soon begin speaking for themselves.

The training department – like all others – is required to be as cost-effective as possible. Accordingly, the presentation methods involved in the training strategy just developed, will determine the overall cost of each training course and will therefore be a major factor in deciding how many people will attend. Due consideration must also be given to the costs involved in training overseas students.

One obvious method in reducing overall training costs would be to use self-training techniques based on audio or video packages.

I am sure you are all aware of the implications of using these techniques but I would like to say that, if used correctly, this type of course is capable of reaching much wider audiences at relatively low cost when compared with formal training. As both audio and video standards are now world-wide, there are also advantages to be gained in using these techniques when training is to be conducted overseas. From a marketing point of view however one must always be aware that self-teaching techniques tend to transfer the control of training effectiveness from the teacher to the student. Care must therefore be taken to ensure an adequate monitoring system is available to measure the overall effectiveness of such courses.

In this training strategy I would like to suggest that the introduction to CAD could be designed as an audio/video package. It could be equally well be used as a general introduction for people not actively involved with the CAD system, as well as for potential users. There are other advantages. CAD can be installed on a variety of computers and can be operated from an even wider range of terminals.

Having adopted a structured approach in the introduction, this technique can be extended to include sections which describe how to operate the CAD system from specific machines and specific terminal types. The student can then choose the appropriate section to suit his machine environment.

Similar techniques may also be applied to other sections of the training program once formal training experience has been obtained. In time a completely integrated program will occur which is not only transportable but also available to be used by other companies who wish to train their own people. It may even be used by organizations such as the Manpower Services Commission should they feel the need to develop courses in this area.

Having developed a training program which provides technical and managerial courses for potential and existing clients, attention should now be given to the development of ones own staff.

It is important to provide staff with a structured career path, and having developed a strategy for customer training, such a strategy should equally well apply to the development of internal skills. With formal training, and time to develop the necessary skills at each level, the personnel department could equally benefit from using such a structured approach in formalizing its staff development programs.

The overall training strategy therefore reflects a common approach to all training requirements and yet is sufficiently flexible to allow product enhancements to be included as, and when, they occur.

In this paper, I have been attempting to illustrate a practical training approach to CAD in the area of plant design. The emphasis has been on adapting the system to the user's needs and not adapting the user to the computer as so often happens. I hope I have demonstrated that the benefits of a clearly thought-out training strategy, based on a structured approach, offers advantages which extend far beyond the classroom walls.

D Appleton, Compeda Ltd, Walkern Road, Stevenage, Herts SG1 3QP

Discussion

B Thompson, Department of Environment:

I wonder if you could give us any details about how you actually accomplish or propose to accomplish the monitoring of an effective training programme?

D Appleton, Compeda Ltd:

We make sure that when a student is provided with a self-teach training package he is automatically assigned a tutor, that is someone who has been through the training course before and is experienced in that particular area. The student is given questions within the training programme that do not have answers provided in the text, forcing him to go to his tutor. In addition to that we would expect to have some feedback of the student's progress through questions and answers, where some of the answers are not provided, and we will hope to monitor the effectiveness in that way.

29. Computer-aided design and development planning

M Bouyat, B Chocat and S Thibault

Abstract: We shall define development here as the provision of servicing facilities for the whole or part of a site. The training which the student developer receives must therefore allow him to propose servicing solutions consistent with the technical, economic and social characteristics of the geographical area in which he has been called on to work. The developer must be trained to carry out prediction studies in town planning. The goal of such studies is multiple:

- predict, for a given programme, the effects on the town, or the sector under study, of the planned operation
- predict, for the same programme, the various development plans possible
- compare solutions obtained with different programmes.

In order to reach this goal the student – like the professional – must have at his disposal reliable tools which can be used at a stage in operation planning when detailed working data are not readily available. CAD systems defined as 'organized and organizable sets of simple data processing aids' can provide such tools.

In the development field two systems of this nature are at present in use: the CEDRE system and the VERDI system.

These two CAD systems allow for immediate integration of any changes in routing and for the study of any type of variant. They are at present being used for the initial training of development engineers and for the further training of professionals concerned.

The use of these two systems facilitates training both in the technical field concerned and in CAD.

Introduction: training development planners

The growth of conurbations, the changing role of old quarters in towns, the rapid extension of suburban areas are all part of a phenomenon which is to be found at the present time in a large number of countries. This pattern of evolution is creating increasingly complex and expensive problems in planning, programming and managing servicing facilities. Random city building can be brought under control by co-ordinating measures but the inadequacy of servicing facilities and the copy-book architecture of previous operations have led to stereotyped housing estates and to disastrous living conditions.

Whatever the scale of the operation in question a great many planners are today directly involved in the various stages of the development process. Development planning can then be defined in terms of organizing and providing servicing facilities for the whole or part of a site. This being the case, the prediction, design and management processes must take into account whole sets of socio-cultural considerations in addition to the indispensable technical and economic constraints. By virtue of their existence and their quality in use, the servicing networks (which are the backbone of development programmes) have social impact and an effect on the life and behaviour of every single person. And, because of the irreversible choices they entail, road and facility networks are a fundamental consideration in a development programme, determining the overall plan and the living conditions of the estate under construction. Whether they be specialists of a particular technique or responsible for overall co-ordination, development planners must therefore be fully integrated members of teams covering all the domains concerned. Their technical training must allow them, in more precise terms, to supply answers to the following points:

- predict, for a given operational programme, the consequences implied in the programme for the town or the surrounding built-up area;
- predict, for the same programme, the different development schemes possible;
- compare solutions corresponding to different programmes.

These last two points should, by their very nature, contribute largely to a qualitative approach to development planning in so far as they allow for different solutions to be looked for and compared.

CAD and development planning

During their initial training only the study by simulation of a real operation can allow students correctly to grasp not only existing constraints and interconnections between servicing facilities and options but also the technical and financial consequences of one development option or another.

Whether the operation involves a single dwelling or the provision of facilities for an entire quarter of a town, practice allows the student to recognize that infrastructures evolve during the whole of the planning process and cover all of its stages from the project draft to the final instructions for construction.

Such study of facility systems, exclusive of variants, involves making repetitive technical calculations and handling large quantities of data. These difficulties are made worse by the fact that the student must apply in a very short time the considerable knowledge which he has only recently acquired. This state of affairs has two consequences:

- the overall view of the project is obscured by the profusion of problems relating to details;
- opportunities for the reappraisal of hypotheses and technical choices are limited in number.

As a result the student cannot benefit fully from the potential advantages of simulation study.

Since it is not possible to increase the length of studies in order to obtain greater quality both in teaching and in the object produced by the student it is necessary to improve study aids. The problem is to decide how this can be done.

Development planning studies make up a system which, in a first approach, can be represented by three interconnected elements:

- the object under study;
- the planning team (students working on a simulated project or a team of professional planners);
- the aids and methods of work related to the characteristics of the first two elements.

If the origin and abilities of the members of the planning team influence the project, then the aids available condition the quality of the work of the designers and, consequently, the results.

Given the impossibility of defining, *a priori*, the ideal operation, or of having a planning team capable always of producing it, it follows that improvement of design can only be obtained by supplying planners with new aids which are adapted to the design process currently in use.

For every development object, planning is in fact based on a linear approach which distinguishes between various successive levels from the initial draft to the final instructions given to the builders. From one level to the next only the scale of definition and problem-solving varies. However, each planning level must be subjected to a global view given the powerful interactions which connect the various factors and data involved in creating the final product.

As an example, we see in Figure 1 the different actions involved in the first design level of a dwelling development operation.

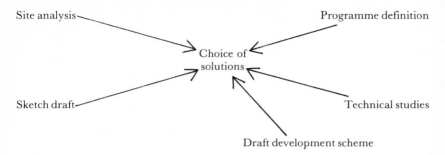

Figure 1 *The actions involved in first design level of a development operation*

1.   The **VERDI** system (road and other network planning).

The objective of the VERDI system is two-fold:

  – to allow rapid technical evaluation of a development proposal presented in the form of a draft masterplan;
  – to allow different variants of the proposal to be studied.

It is the combination of these five actions which will determine the choice of solutions (which will not then be subjected to further reappraisal).

It is then easy to understand that a satisfactory solution can only be found to the problem of planning if such a solution concerns the whole of a design level and allows the planner to grasp all the connections which exist between the different elements. The real solution consists in solving a problem of information exchange.

Traditional data processing – by definition the science of processing information – provides answers to precise technical questions but does not allow an overall view during the planning process. Since it is not possible, *a priori*, in a planning problem, to presume a determinist sequence in the planner's tasks, the objective at software level will be to provide the user with specialized modular aids which he can call on whenever he wishes, in order to solve one particular problem or another.

It is these requirements which impose certain qualities in a CAD system which we would then define as a set of software in conjunction with a specific planning task and possessing the following qualities (Botta, 1978):

  – internal qualities corresponding to what is expected of the system without necessitating user intervention (data management etc);
  – external qualities which are directly visible to the user (data clarity, right of error, language close to professional realities etc).

It can then be seen that this new data processing approach will allow systems to be designed which in practice will go well beyond the objective of initial training. Adapted to actual planning practice and composed of software of which some is the result of recent research, such systems deserve to be used solely for their operational qualities.

It is not our purpose in this paper to expose in detail the operational aspect of these systems since this has been done elsewhere (Bouyat, 1981: Chocat, Bouyat and

Thibault, 1981). We shall simply limit ourselves to stressing an important aspect concerning further training.

In this context the interest of such systems is two-fold:

– the demonstration to professional users of the immense contribution which CAD techniques can make to planning;
– the application in professional circles of new techniques resulting from development planning research.

These last two points are also introduced into professional circles by young development planners who have been trained in such techniques during their initial training.

Examples of CAD systems in development planning

Two systems are currently being perfected in the Methods Laboratory of the Civil Engineering and City Planning Department of INSA, Lyons: the VERDI system and the CEDRE system.

The system is designed to be used initially at the level of preliminary studies using a draft development scheme which represents concretely the basic choices of an operation. Such a system facilitates the examination of variants and simplifies the study of the interconnections between the master plan design and the network design, thus ensuring the transmission of information between the various planners involved in an operation. The object being designed can then be stored and a more precise description be produced later, using the same system, at the detailed draft stage.

The system consists of a set of data processing modules which allow a global approach of road and other networks and which can function independently.

The system treats the three dominant techniques:

– *Earthworks* (data acquisition, restitution, calculation of the volume of geometrically complex platforms and the cubing of road works);
– *Roadworks* (routing, restitution of the natural terrain cross-section; geometrical modules allowing not only elementary calculations but also the definition of longitudinal and cross sections, the inclusion of the data necessary for costing per meter, plan longitudinal and cross-section plotting);
– *Drainage* (plotting of data and of watersheds, rainfall calculations, plotting of networks, proportioning of networks, design of networks).

Certain modules have themselves been designed as real CAD sub-systems: a special program is then used for data management and digital processing.

A module which can be connected to the different aids mentioned above is used to construct a model of the natural terrain based on contour readings. This enables the planner to obtain automatic measurement of the natural terrain at any point.

Such a system – in addition to its conversational aspect – effectively allows the user rapidly to obtain technical assessment of a proposal. The user can use the data processing modules at different levels according to his own working practice or according to the different cases he may have to treat.

The type of user-machine communication which has been retained is that known as 'menu'. This technique proposes a control language which is easy to assimilate.

A 'menu' consists of:

(i)   a list of specialized actions in the domain given: ie road and other network infrastructure;
(ii) a list of objects to which the actions will be applied.

Examples of actions

Create, modify, print, destroy, trace, calculate cubing, calculate drainage flow, calculate surface area etc.

Examples of objects

Operation, natural terrain, master plan, network, axis, building, building plot, road, portion of road etc.

It is the general CAD system, PRAIRIAL (Foisseau, Jacquat and Vignat, 1981), which allows the corresponding programme to be implemented by decoding the action/object pair. It is obviously not possible to decide indiscriminately on any action whatsoever at any moment whatsoever. For example the user must stipulate the name of the operation on which he is working before he can modify an object. Or again he must know that it is meaningless to ask for the cubing of a building plot to be calculated.

The use of the 'menu' technique, as well as its implementation, allows for great adaptability. Adaptability concerning modifications or changes in existing technical programs as well as in increasing the number of actions or objects originally implemented.

The calculation aspect is therefore fluid from the outset and can be made more detailed in time without great modification of the system. Finally, the system can be made complete by adding a data bank which ensures the integration of existing connections between the following objects (Figure 2):

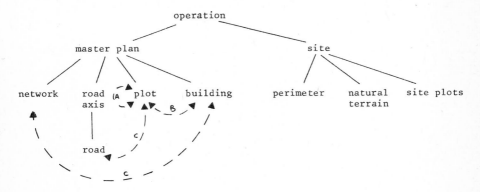

Figure 2 *The addition of a data bank to complete the system*

At the present time the VERDI system is based on the following equipment (Figure 3):

- a minicomputer 329K words of 16 bits (Mitra 125)
- a mobile disc of 50 MO
- a graphic tablet size A_o (Benson 6201)
- a graphic control screen (Tektronix 4014)
- a hard copy unit
- a drafting table size A_o (Benson)

2.   The CEDRE system

The CEDRE system aims to bring data processing assistance to the planning of rain and waste water drainage networks at the draft development plan stage.

It is based on a new method of calculating the flow rate of rainwater which is not subject to the constraints and limitations of the CAQUOT method – it takes into more detailed account the various hydrological phenomena which are involved in rainfall.

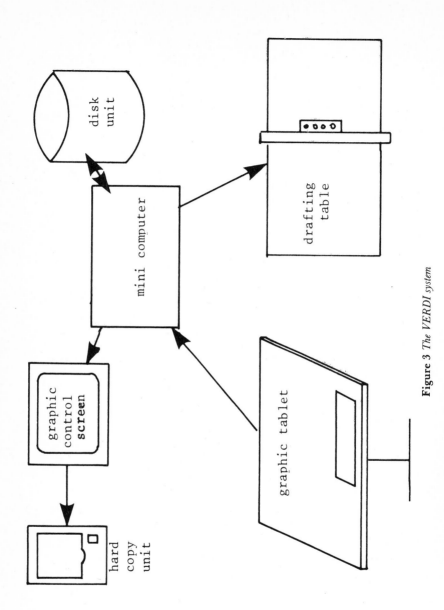

Figure 3 *The VERDI system*

It allows the integration of special works (storm drains, catchment basins etc) and the study of their influence on the working of the network.

It allows rapid analysis of variants.

It gives full importance to graphical display at the level of data acquisition as well as in presenting results.

The CEDRE system is made up of a set of interconnected programs with sequential or alternative linking:

- sequential, if several programs are to be used in succession in order to obtain the desired result;
- alternative, if the user can choose from several programs to obtain the same kind of result.

The system is organized around 5 main programs: rainfall simulation, surface run-off, sewer flow, sum of hydrographs, proportioning and secondary programs allowing study of special structures (Figure 4).

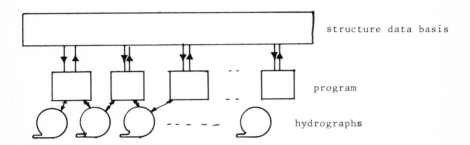

structure data basis

program

hydrographs

Figure 4 *The system is organized around 5 main programs: rainfall simulation, surface run off, sewer flow, sum of hydrographs, proportioning and secondary programs allowing study of special structures*

The equipment is composed of a microcomputer, a graphic tablet, a hard copy unit, a disc unit, and a drafting table if necessary. (Figure 5)

Each programme uses the graphic tablet for the graphic data and records its results on cassette or on floppy disc in identical form and in a place determined by the user.

It is thus possible to obtain at will and without data modification all the sequences required by the planner.

Training in CAD

The use of these two systems, which possess great clarity from the outset, is a demonstration of the importance of such aids in the simulation study of a project. Students – whether they be professionals or future engineers – then realize how necessary it is for them to receive training in the development of such systems so that they can take an active part in their evolution: this is what we mean by training in CAD.

Such training takes place in two phases:

1. Simple technique, applied on a microcomputer, through the description of the structure of CEDRE. This involves demonstrating how the basic notion can be implemented in data processing: a set of software which is organized and can be organized as the user wishes.

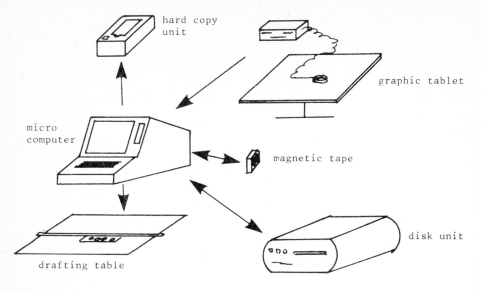

Figure 5 *The equipment is composed of a micro-computer, a graphic tablet, a hard copy unit, a disk unit, and a drafting table if necessary*

2.   More sophisticated techniques, both in hard and software, with the VERDI system which requires explicitations of data bank management systems and general CAD systems.

This double confrontation with development planning and CAD allows the professional to determine the specifications which will meet his requirements. It is then his task to develop the techniques and aids proper to development planning, and the computer specialist's task to perfect the data banks and general systems appropriate to the problem as it is posed.

Conclusion

We have presented two systems of CAD used in the framework of initial or further training for development engineers. The pedagogical value of these systems has been demonstrated at three levels:

Level 1   The study of the multiple connections between the various development objects: the possibility of very rapid simulation of the effect of one solution or another on the quality of the whole;

Level 2   The study of particular new techniques, especially rainwater drainage systems, which cannot be carried out by hand because of the complexity of the calculations;

Level 3   The study of data processing techniques used in the design and construction of such systems.

In the final analysis such systems lead us to new types of training. The planner will in fact have to master not only his own technical domain but also the data processing aids necessary for the use, the improvement or even the creation of CAD systems.

This dual competence can only be effectively acquired during initial or further training, if the students are familiarized early enough with the computer, and if computer specialists provide the development planners with aids which are both efficient and easy to use.

References

Botta, H (October 1978) CAD and civil engineering. Doctorate thesis: Insa Lyon

Bouyat, M (March 1981) A CAD system for development planning. Doctorate thesis: Insa Lyon

Chocat, B, Bouyat, M and Thibault, S (March 1981) Storm sewer design with micro-computer aid. Proceedings engineering software II; London: 914-25

Foisseau, J, Jacquat and Vignat, J C (July 1981) PRAIRIAL: General presentation. Centre d'étude et de recherche de Toulouse, file no 83121

M B Chocat, INSA, Dept Genie, Civil et Urbanisme, Laboratoire Methods, Batiment No 304, 20 Avenue Albert Einstein, 69621 Villeurbanne Cedex, France

M M Bouyat, INSA, Dept Genie, Civil et Urbanisme, Laboratoire Methods, Batiment No 304, 20 Ave Albert Einstein, 69621 Villeurbanne Cedex, France

S Thibault, INSA, Dept Genie, Civil et Urbanisme, Laboratoire Methods, Batiment No 304, 20 Ave Albert Einstein, 69621 Villeurbanne Cedex, France

Discussion

Unidentified contributor:

Could I ask whether your systems have been used in any real project work in France?

M Bouyat:

Yes, these systems are used. They have been used this year for the planning of first year students. We have also made some progress on a real project, and we have given approximately ten sessions to professionals using the two systems.

F Weeks, Newcastle Polytechnic:

I found this paper very interesting because I think it has some messages for us in relation to mechanical engineering, which is my own field. As you well know, there are many machines and systems in mechanical engineering that are either too complex, or too expensive for us to allow our students the opportunity of studying them. I pick up the point in this paper that the use of a large-scale simulation of the kind that's described here can enable students to gain an appreciation of the constraints and the interconnections involved. They can also get a view of the technical and financial consequences of changes which they make in the system, and this seems to me to be providing them with experience they they couldn't gain in any other way. Perhaps our speaker would just like to elaborate a little on what the students have gained from having been engaged in this work.

M Bouyat:

We use NUMIST to calculate run-off etc. For a student there are two phases. The first phase is to understand the method. If we don't have the system it's very difficult to understand why we have this method. The method exists because we can use such a system, and it's one way to make them understand this method. Using such a system is, I think, advantageous. The student can then understand the difficulties of a project with such a system. If traditional studies don't seem to see this point of view, students understand the difficulties of having a project. They see how difficult it is and they try to understand our work.

F Weeks:

Obviously there is a lot of work involved in producing simulations of this kind and I just wanted to say that reading this paper and hearing it presented has helped to convince me that in mechanical engineering it would be a worthwhile effort. I would thank the authors for doing that for me.

Chairman:

I felt the paper had some important messages in getting to a situation where we can put in front of students realistic models of the real world entities that they are concerned with in the design sense. Perhaps in civil engineering and architecture they are a bit closer to being able to put realistic models together than we are in some branches of engineering, where I think we have some way to go before we can truly model in a full sense the entities that we are dealing with, so that we can search for plausible solutions that meet all the criteria.

30. A case history of introducing CAD into a large aerospace company

C W Allen

Abstract: This paper considers the introduction of a large interactive graphics system into an existing organization with emphasis on the environmental, training, human relations and attitudes of staff in a high technology area and the methods used to overcome the ensuing problems.

Introduction

The Stevenage Site 'A' of British Aerospace Dynamics Group is a main contractor of guided weapons, which are principally for defence systems. These include Swingfire, Rapier, Seawolf and Sea Skua. These products require a high degree of innovation and technology across most disciplines and it was to assist in the increase of efficiency and productivity that computer aids in the design drafting, manufacturing and quality fields were introduced.

Justification and installation

During 1978 British Aerospace were suffering from:
1. lack of staff.
2. imbalance in level and type of staff.
3. rapidly changing technology.
4. high number of sub-contract staff.

Even today, regardless of any current problems of employment, managers suffer from the same problem of either not having exactly the right type of labour or not enough of the right type of staff to do the work at the time when the work occurs. This leaves the manager with the task of making the best use of the resources available, assuming of course, that the principle of hire and fire is not adopted.

Qualified designers are still a scarce commodity and if one is lucky enough to find some, then there is difficulty in retaining their services, especially if they are below middle age. It is therefore imperative that a designer, who is such a valuable asset, uses his skills efficiently. His greatest skills is, without doubt, his creativity. Here computers cannot help, but by using computer aids in the other areas of his tasks, more of his time can be concentrated on creative thought.

In the case where the right type of staff are not available, it is possible to gain a degree of help from computer aids. It certainly could not turn a mechanical type into an electrical type at a flick of a switch, but it can make say an electrical draftsman into a better electrical design draftsman by using the many prompts, design rules and checks which can be stored in its memory.

The advance in technology, especially in electronics, has made the introduction of computer aids an essential element of the design and manufacturing cycle, even four years ago, great difficulty was being experienced with the design and layout of complex printed circuit boards and today's integrated circuits would send the most tenacious designer screaming to the funny farm if he tried to draw the layout of multiple cells by the quill pen method.

It is so very easy to solve the difficulties of the moment by introducing short-term solutions. One of these is the use of sub-contract labour to supplement either the lack of staff or the wrong type of staff. This is an acceptable answer provided the number are kept to a low level. In our own case, the number of sub-contract staff was well beyond the 15 per cent figure that we believe is the maximum.

Because of the very simple financial case that can be guaranteed in introducing computer aids to the pure drafting field it was decided that the case for the purchase of our CAD system should be limited to the replacement of sub-contract labour by our own staff with the assistance of the Computervision equipment. It is now a well-established fact that improvement in efficiency of 3:1 ratio compared with manual method can be achieved by applying computer aids to pure drafting tasks. At the time the investigation was being made at BAe, these facts had not been verified. The method used to produce our results was to hire a system and an operator. Typical drawings which had been produced manually were selected to cover the major disciplines and then redraw on the system from the basic scheme by the new operator. Certain difficulties did occur, but these were through language problems (the trials were carried out in Holland) and the problem related to standards and procedures. In spite of these problems, an overall increase in output of 3.7:1 was achieved, averaged over all the drawings of the exercise.

The drawings chosen were at the lower end of the drawing complexity scale which gives the best efficiency ratio when using machine aids. Figure 1 shows the relationship between the machine time taken in producing the drawing and the complexity of the drawing. There is also an equation of machine time against thinking time, following a similar curve. It therefore follows that the lower order of detail drawings are more suitable for the machine if speed of through-put is the only criteria. At the other end of the scale are the most complex drawings. These are mainly schemes and layouts and usually classed as 'design' tasks and can often take up to three months to complete in the manual mode. It does not mean that there is a continuous drawing program during this period, probably about 15 per cent of this time is spent on the pure graphics task, the remainder being spent in thought, discussion, information retrieval and the like. On this basis, there appears to be very little help given to the designer by the use of the machine. However, as was later proved, once the original design data-base is established, the follow-up detail and assembly drawings are produced even faster.

By using the figures from our trial, a good case was made for the purchase of our first six terminal system. This was agreed by our board and the system was ordered.

Figure 1 *(see page 9)*

Environmental considerations

The main design drawing office is an area of approximately 1,250 square metres, housing 130 staff of mainly technical grades with the nominal clerical back-up. It has operated successfully over the previous ten years using a 'Product Structure' breakdown and split into four groups, each group responsible for a particular product with specialist staff attached to the groups as required. This open plan approach allowed the maximum cross-fertilization and flexibility.

It was essential to continue this philosophy with the introduction of the CAD technology. It is very easy for new thinking to become clouded in jargon and black magic and whilst not wishing to accuse the professional specialist of one-upmanship, it would help in the introduction of any new technology, if the layman user could understand the language enough to dispel his fears.

The method of introduction was to use the machines as a 'tool of the trade' to which all relevant staff could have access. To this end, the 'noisy' and 'hot' items were installed in an enclosed area and the terminals installed in an open area with free-standing screening to reduce noise and light reflections as shown in Figure 2. The

Figure 2

opportunity was also taken to refurbish the remainder of the office, with a view to placating the feelings of looking after machines at a higher standard than staff.

To make a tidy installation, it was decided to run permanent ducts in the floor. The floor is solid concrete. This presented some difficulty, but when completed, the system of cabling achieved the standard required. However, the error of our ways was soon shown when we extended the system for the first time. To extend the cabling requires major plant work with pneumatic drills, dirt, dust and vibrations, all of which would be very hazardous to computer systems. It was then decided to use ducting above the floor level for the extension of the system.

If one has the opportunity to start again, it is recommended that a raised floor system is used, giving approximately 300 mm clearance between the raised floor and the original floor for cables. This allows for the cables to be routed and rerouted without major plant work and extensions incorporated by the addition of further sections of the raised floor. The noise level of a terminal is zero but it is necessary to use an audio signal to indicate acceptance of a command. Trials were carried out by using visual-only commands but this is not successful. The noise, whilst not loud, does not affect the operators. However, the same cannot be said for other staff doing academic calculation tasks in the design office. Complaints have been received from the non-computer staff, hence high quality acoustic screens were used to reduce this nuisance to the minimum.

The other purpose of these screens is to reduce reflection from the lights in the remainder of the office. This can be a very difficult task and in an open plan office it is best achieved by siting the screens facing away from the main office lights. Local lighting should be used at the discretion of the operator. The arrangement at BAe is for the overhead fluorescent strip lights to be individually switched and local 'angle-poise' lights supplied to give the necessary illumination to the reference material.

   Temperature and humidity control is theoretically not necessary for the terminal, although we already had an office which was temperature controlled to approximately 21°C. However, to be on the safe side, the heart of the system consisting of the CPU and disc drives were placed in an air-conditioned room. The conditions of this room are not critical, but temperature and humidity are held to $20 \pm 1°C$ and $45 \pm 5°RH$ respectively. The photo plotter, which is used to produce printed circuit master artworks, also requires stable conditions. This is for reasons of accuracy of both machine and film. The accuracy of these machines is dictated by the technology and has no relationship to the computer installation.

Training

During all of the planning stage of this installation, a representative of the TASS union was kept well informed of the events and was also invited to send their nominated representative to the original trials which were held in Holland. It is recognized by all parties that the introduction of CAD is a major change from the long-established practices of design and drafting. This obviously caused concern to union members of staff and it was mainly to combat this fear of the unknown, that the members of staff, both union and non-union, were involved at this stage.

   It was during these trials that it was recognized that in introducing this new technology, the skills already held by the staff were not going to be changed, but that additional skills would be required. As previously mentioned, the computer system still requires a very skilled input and indeed it can be argued that because of the speed of response, the level of skill must be higher. On this basis, the management adopted the principle of recognizing a higher degree of skill, when training was completed. The additional payment as a reward for these increased skills was the subject of normal negotiation and a satisfactory result was obtained.

   One problem emerged in these negotiations. At what stage was an operator considered 'trained'? It is obvious that in adopting new technology the rate of change of technology itself would be fast. Therefore, it is impossible to define an achievement point based on the knowledge of the equipment. One can take the analogy of passing a driving test; it does not make the individual into a car mechanic, neither does it make him into an expert driver, it merely ensures that he is capable of using a machine to transport himself and others from one point to another in reasonably safe comfort. Similarly, it was decided the best criteria was to establish an achievement based on the existing manual method of performing the drafting task, this point being when the operator could draw with the aid of the machine, as fast as he could on the drawing board.

   The first group of trainees were monitored closely on this and the figure of five weeks from the start of training was established. This figure has been verified continuously since and whilst individuals will vary, the average remains the same. See Figures 3 and 4.

   Training is carried out in-house by our own staff subsequent to the initial training group. It consists of one week pure theory of operating systems and the remaining four weeks, examples to cover most aspects of mechanical drawing. An additional two weeks are allowed for specialized training on electrical/electronic layouts and structural analysis, where applicable. One major problem at commencement of training is the feeling of terror when sitting in front of the CAD screen for the first time. This is a problem which will probably only apply to the 'mature' generation of the moment. The younger generation have become accustomed to these devices in some form during their academic training. However, to get over this problem, it is always better to train at least two and preferably three members of staff together on one terminal, this eases the

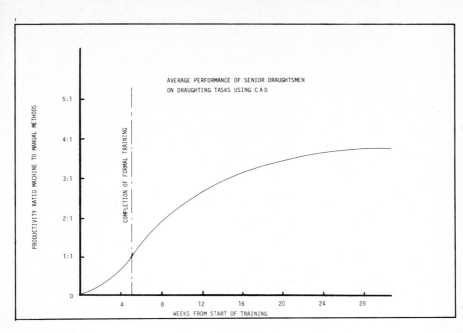

Figure 3

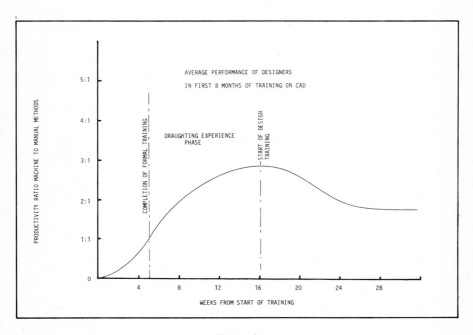

Figure 4

tension, allows prompting between the individuals and also some lighthearted repartee, which defuses many potentially explosive situations.

The training at BAe may not be typical in the way financial cover is obtained. This initial training is taken as an overhead payment but it is in the company's interest to reduce the overhead as quickly as possible, this being one of the many factors considered when establishing the training period. On completion of training, staff are on direct contract charge and as their contribution, by definition, must be at least equal to any other individual using manual methods, the contract costing is not affected. Indeed from this point in time, efficiency improves quite rapidly and the average time taken to achieve ratios of 3:1 in increased efficiency is three months approximately.

Training is not a once and for all activity. There are four main elements of training.

1.   Basic operator training.
2.   Specialized training.
3.   Increasing skills.
4.   Software and operating system updates.

One and two have been previously mentioned, the second continuing as the cases arise. The third, increasing skills, is something which is often hidden and can either be a blessing or an absolute disaster.

The introduction of CAD at BAe was caused by a critical labour situation, therefore it was absolutely imperative that the system was on maximum production as soon as possible. To this end, we did not allow any experiments or 'playing' with the hardware or software. The standard software packages we used in their entirety to a very good advantage. We were able to achieve our initial aims within our programmed timescale. However, as operators gained experience and system managers learnt from some of their errors, it becomes obvious that better ways can be found for doing certain tasks, small execute files can be compiled, special disciplines can be imposed ensuring easy recall, and as the more senior staff on specialized activities start to venture into more complex design problems, one can easily get caught up in a continuous training programme and a zero output figure.

It is absolutely essential to control this quest for knowledge for the sake of knowledge in an industrial situation, but it is also essential to ensure that the maximum benefit can be gained from the investment by using its facilities to an effective economic level.

Attitudes

It would be very easy to define the attitude of people by arbitrary descriptions such as management and unions, us and them, the 'haves' and 'have nots'. It is not that type of division which is the main stumbling-block to progress. It is the fear of change. This affects staff and management alike. The following are my own personal views of the events in trying to install a utopian facility into my company and in no way reflects the management or union policy at BAe.

It is not practical to try to change the whole documentation system and communication system in one step. Problems can occur by changing fundamental steps in a complex system. It is therefore better to maintain interfaces between departments until the new systems are thoroughly proven within departments. In line with this approach, we have maintained the existing communication system, with drawings being the prime interface. Gradually we will introduce direct data-base transfers and hence it is expected to overcome the human problems of change.

Choosing those members of staff to be trained in the initial training course had its problems. The feeling of a second-class citizen was very quickly expressed by many who missed out on the training. This was also noticed when the policy was adopted to train all apprentices in the use of the system during their training period. Some more mature members of staff were convinced they had been 'written off'. Following this comes the real problem of explaining to very loyal, long-serving members of the department who ask the question 'Why not me?' Why in the eyes of the management, are other people more suitable and are therefore to be given the opportunity to work with new exciting equipment and incidentally earn more money, while staff who have lost out, are expected to carry on with the same dedication and efficiency? By explaining to the staff the benefits in fine detail of the new systems, and explaining the necessity to maintain continuity of the old systems, staff gradually accept the new situation. This problem can be solved by full and proper communication from the beginning.

Whilst the CAD side is advancing rapidly in the new technology, other areas are equally keen to benefit and improve their efficiency by the introduction of CNC, DNC machines and the many systems required to drive these machines, inspect the output and purchase the raw material and components. The moment the first system is purchased, the barriers against change are erected. The task of interfacing becomes harder and the 'organization' rallies its forces to protect itself. The skill comes in redefining the existing organization whilst retaining the goodwill of all concerned without disrupting the existing facilities and staff. To carry this out, strong senior management policy is required to instruct middle management in the major policy issues for the future. With this well-defined policy, management and staff will then communicate and co-operate to achieve the desired aim.

It is obvious that future organizational structures, recognized job titles and interfaces will change and many other tasks will merge. Hopefully this will be achieved without too many traumatic experiences, but it must be accepted, that to make the maximum use of these new tools and hence increase the efficiency of industry, re-education and training of existing staff is required, coupled with a change in the education format, thus allowing a merging of the tasks between design, production and quality control. Industry must rely heavily on educationists both in schools and universities to create an attitude of mind within the future technologists, that 'change to meet new technology' must be the accepted rule and that of the old ideas of security linked to conservatism is not the way to approach the future. In this way the false barriers of current interfaces will gradually disappear and the new technology of today will become the norm.

C Allen, British Aerospace, Dynamics Group, Stevenage Division, PO Box 19, Six Hills Way, Stevenage, Herts

Discussion

S R Bennett, Ford Motor Company:

Could the speaker please tell us who did the training, how it was funded, and was any consideration given to doing the training yourselves on tailor-made courses?

C W Allen, British Aerospace:

The first course was run by the vendor company which trained the system manager and six operators. That was the initial training up to what we call operator skills. From then on we took over the training ourselves and we have trained 45 people to date. The initial five weeks training is paid for out of our training budget. This is an overhead budget and whenever you are a company in trouble it is always very difficult to get money for these sort of activities. However training is allowed on new technology.

The five weeks initial training is not all the training we do. More goes on with the specialized training, for example, if you have taken the operator course there are another two weeks that you need if you are going to be dealing with electronics.

J Garfield Davies, TI Creda Ltd:

Having trained the people, how do they share the 15 terminals that are used? Is there a group of people who use them all the time? Do they share them? How is that allocation determined? I would also be very interested to know whether the additional rate that was paid, was paid to all the people who were trained, irrespective of the way in which they used the terminals?

C W Allen:

The additional rate is paid to all of the people that have been trained. It's a skill like driving a car; when you are not driving you still have that skill. With respect to the splitting of the terminals it is obviously necessary to get the maximum use of the capital investment. We run a double shift system, starting in the morning at 6 o'clock. The first shift comes off at 2 o'clock, the second shift has half-an-hour overlap, starts at 1.30 and goes through to 9.30 pm. We haven't persuaded them to work through the night yet but I will if I can. Of course, people are not on the screens 100 per cent of their time. It does work out at around about 50 per cent. It's worked out by our system manager.

Unidentified contributor:

BAe have been very lucky in their negotiations with respect to payment, hours of working, etc. They are totally unique to each company, there is no national policy laid down.

C W Allen:

I think that the union TASS, the main union involved in this particular aspect, are very forward-looking from the central point, but with the individuals at local areas and the social problems caused by the recession you will find a different response.

D G Smith, Loughborough University:

You did say that very good draftsmen are needed to train as good operators and it was impossible to say that you could do otherwise. How do you see the implications of that statement with regard to the training of undergraduates, because clearly one hasn't got experienced draftsmen? What would you, as a practitioner, like to see universities and polytechnics doing in this area?

C W Allen:

One of the big problems here is the inverse status symbol of the drawing board. The one thing that undergraduates won't want to do is to learn about graphics; it just happens to be a very short phase that they were passing through in their academic training. Unfortunately, this doesn't make for good designers. There is a great deal of knowledge to be gained in a deep exercise in the use of graphics and in the transferring of information from the designer to the workshop. So the undergraduates should be taught a lot about drafting. I am not saying they cannot operate a graphics system without it. They can, but not efficiently. And if you want to make an efficient designer, he has to be able to communicate and unfortunately at the moment, graphics, three-dimensional drawings and models are the only way to get this information over. We can always tell, in the design office, people who have come through the old school, if I can use terminology like that, and those that have come through modern methods.

D Easthope, Davy McKee:

Was the capital investment for the system introduced into the charging rate for the projects that you were working on?

C W Allen:

We had exercises before where computers were charged at the phenomenal rates of £60-80 per hour, and this still happens. The first thing this does is make people think twice before they use it. We were trying to encourage people to use it, so the way we introduced it was this: it was basically bought by company capital and, like our machines and workshops or any calculator, it is just there to be used as a tool of the trade and, therefore, anybody that can use that tool will get a better response from the product and will also get cheaper products. Draftsmen working on that machine will give you three times the output they would on a normal one, therefore we are encourged by our project engineers to make sure that we put everything on the computer that is possible. There are snags, of course: you get a queue on the computer. That's good for investment, as far as I am concerned. It means that I get a more modern office much quicker.

K Ramsay, Paisley College:

I was very interested by your evaluation exercise which you carried out. But I wonder if you could clarify just what the background of the individuals were who went and did that exercise? Did they have prior training of any kind, or did they have a background in the technology before they carried out this evaluation programme?

C W Allen:

No, there was no prior training at all. These were four people that were chosen. The first was obviously going to be the system manager. The second was a senior draftsman, the third was a designer, and the fourth was nominated by the TASS union.

D Greatrex, EITB:

I was most impressed by your improved productivity factors, over a relatively short exposure time of five weeks. Could you explain what criteria you used to evaluate the improved performance both under the conventional method and the computerized method?

C W Allen:

There is no formula that I can give you. With out first two courses, we were monitoring these extremely closely and the exercises that they were given were all exercises that had been carried out before in the manual mode, so every time an individual used one of these exercises the time check was taken against what had been done before manually. Age didn't seem to make a lot of difference, the oldest person that was trained on the first batch was 57 years old and now obviously is about 60 plus and still working extremely happily at a senior draftsman level. Perhaps if you are talking about design and so on, then age may come into it a little bit.

Part 6:
Training course experience

31. Teaching CAD for electronics at the Norwegian Institute of Technology – present status and future ideas

E J Aas

Abstract: In this paper CADCAM courses for electronics given at the Norwegian Institute of Technology are described. In addition to regular student courses, the concept of continuing education courses to meet, for example, CAD training needs for designers in industry, is presented. These encompass both regular CADCAM courses for PCB and LSI design, and courses for LSI design with design method and computer aids emphasis.

The possibility of technology transfer between university and industry is demonstrated through the collaboration triangle university/research institute/industry.

Finally, ideas for future CAD education are given in lieu of current trends in data technology, networking, and CAI (Computer-Aided Instruction).

Introduction

The Norwegian Institute of Technology (NTH) at the University of Trondheim is educating engineers at the Master of Science level, but also offers post-graduate research leading to a degree equivalent to Doctor of Philosophy. In addition, intensive industrial training courses are given regularly. These are called 'Continuing Education Courses with options for formal examinations' (CE courses).

Close contact with industry is essential for a technological institute to ensure education of engineers suited for industrial needs. At NTH, this contact is established mainly through collaboration between NTH and the Foundation of Scientific and Industrial Research (SINTEF). Figure 1 depicts this community, which goes directly down to department/division level, eg the Electronics Research Laboratory (ELAB) affiliated with SINTEF co-operates with the Electrical Engineering (EE) Department of NTH. This environment is beneficial in many ways:

- Faculty staff work part-time on ELAB projects, and thus gain insight into industrial needs and problems.
- ELAB scientists teach specialized courses at the university. In this way, novel technological areas may be introduced in a way not hampered by traditional university staff turn-around, but more suited to the changing needs of industry and society.
- Access to common labs, equipment and commercial CAD programs is available at a total value far beyond the normal university budget.
- Subjects for term projects and dissertations may be chosen to fit either current research activities or industrial needs. Of particular importance for CADCAM is the newly established VLSI Design Centre at ELAB (VLSI – Very Large Scale Integration). One of its ambitions is to acquire a full spectrum of high-quality CAD tools for LSI design.

The above description is intended to give an overall picture of the environment the EE department is working in. In the following, the specific courses offered and planned for 1983 at the EE department, Division of Physical Electronics, are described.

Long-range ideas for student and industrial education and training will be presented at the end of this paper.

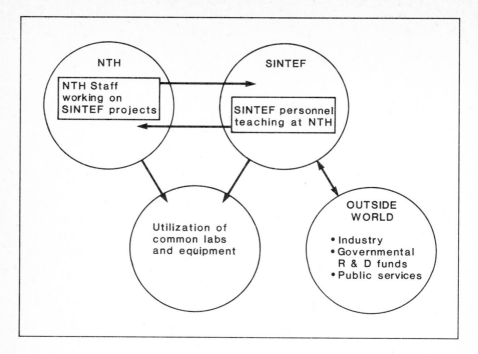

Figure 1: *The NTH-SINTEF community*

Current offers of CADCAM

Two courses where CADCAM holds a dominant position are described below. The first one is a typical CAD course where CAD tools of different types are emphasised more than particular design methods and styles. The second one puts emphasis more on the design process and trade-offs, and introduces CAD more integrated in the total design process.

CADCAM for electronics

This course is offered both as a normal one semester student's course in their last year of education, but also as a CE course for two full (non-consecutive) weeks.

Objective

To give the student a working knowledge of the utilization of CADCAM in the design and manufacturing of electronic equipment through practical experience. Underlying principles are exposed to comprehend usefulness, capabilities and limitations of CADCAM.

Syllabus

Man-machine interface.

Principles and programs for schematic drawing on graphics terminal, analysis and optimization of analog circuits, logic simulation on register transfer level, functional level and gate level, test generation, fault simulation, layout of PCBs (Printed Circuit Boards) documentation and manufacturing control. Integrated CADCAM systems encompassing specification, design, analysis and verification, testing, layout and post-processing are discussed. Different data-base techniques for CAD are touched on briefly.

Comments

The course has been given for three years now, last year to 43 students. The equivalent training course for engineers in industry attracts 12-15 participants each time. Emphasis is on own use of the different programs, and hands-on experience takes a substantial part of the course. Principles and methodology, as well as design examples, are the subject of lectures in the classroom. Due to shortage of graphics terminals, students are working odd hours at the terminals.

Till now, CAD for analog circuits has been included. These are now relegated to a lower level course, and will be deleted from the present course. CAD for VLSI design will be given more consideration this Autumn.

Exercises have been given on stand-alone programs up to this time, but starting this term an integrated system EPOKE (see Figure 2) will be utilized to demonstrate CAD flow from schematics input to manufacturing. EPOKE was developed during 1977-81 as a joint project between three Norwegian electronics companies and two research institutes, and partly funded through The Royal Norwegian Council for Scientific and Industrial Research (Bayegan and Aas, 1978; Tysso and Aas, 1981). The system is illustrated in Figure 2.

LSI-design and computer aids

This course was scheduled for the 1983 spring term, and was given only as a CE course at the first presentation.

Objective

To give the participant knowledge of how to design a digital LSI circuit, and experience in the utilization of computer aids during the design process. Understanding of how to evaluate and chose optimal technology, method, and design will be given.

Syllabus

Alternate ways to realize custom design LSI: full custom design, standard cell method, gate arrays. Technological overview: the design process, interface designer/manufacturer, design rules. Computer aids: schematic drawing, symbolic layout, electrical and logical simulation, layout software and hardware, test generation. Cost and design time estimates, design methods through case studies. Exercises and demonstrations.

Comments

Industry has expressed needs for a course as outlined above, and it is anticipated that 20 participants per year will attend the course. In conjunction with the course, designers from industry are invited to bring a particular design problem, work on it during the course, and perhaps stay for a while at the Design Centre to complete the design in collaboration with Design Centre engineers.

The participants of the course will evaluate the course, and revision will take place before the next presentation is given. This course will also become part of a postgraduate course for PhD students, where the Mead and Conway design style will be partially adopted (Mead and Conway, 1980), but symbolic layout programs will be utilized (STICKS, see Table 1).

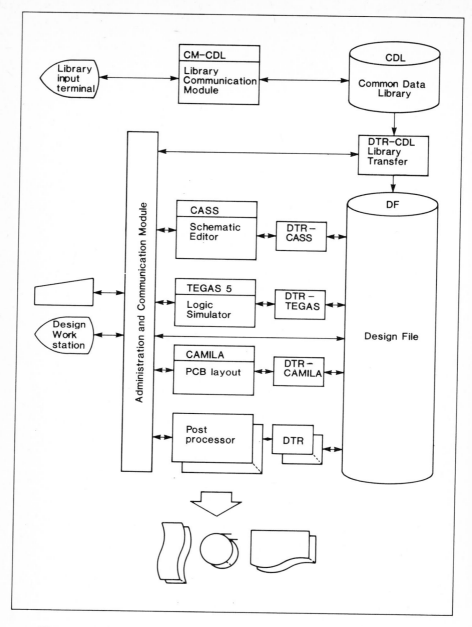

Figure 2 *The EPOKE system as utilized at ELAB/NTH. CDL is a common library residing on a commercial database system, while DF is based on a file system developed for this project. Each design is kept on a separate DF*

CADCAM software available at ELAB/NTH

The suite of CADCAM software available at ELAB/NTH for design of digital systems is presented in Table 2. This software is acquired in quite different ways. Some of it is purchased as commercial software, some is acquired as public domain software, and some is developed in domestic projects. It will be recognized that this environment has access to a fairly comprehensive suite of CAD programs useful at different phases of the design process.

Short-range plans for course enhancements

Short-range plans for course enhancements include:

- Acquisition of several cheap alphanumeric terminals with retrofit graphics capability, and one high-quality, high-speed colour plotter.
- Evaluation of new CAD software offered on the commerical market.
- Purchase or development of new CAD software as needed.
- Utilization of learning software packages for CAD, eg for the CALMA GDSII system.
- Acquisition of video tape courses for VLSI design.
- Exploitation of a general CAI (Computer-Aided Instruction) package for CAD learning purposes.

Ideas of future CADCAM training principles

When long-range training of personnel for utilization of CADCAM is discussed, current trends in computer technology, network development, and teaching methods should be considered.

I would like to focus on some trends:

- The rapid development of raster-scan terminals with local computing power, taking advantage of cheaper, faster and more compact memory modules.
- Increased utilization of new minicomputers, the 32-bit Virtual Memory machines appearing to become a *de facto* industrial standard for CADCAM.
- Standardization of operating systems.
- The evolution of data networks. Local, high-speed networks (eg the Ethernet concept) will connect different terminals and work stations to more powerful computers, mass storage and a diversity of output media, from cheap matrix printer/plotters to high-speed colour hard copy units (see Figure 3). The public data network can be used to reach remote main frames, but also to allow remote access to local networks.
- Finally, the maturing of CAI (Computer-Aided Instruction) is expected to impact many areas of education.

How will these trends affect CADCAM training? First of all, cheaper hardware will enhance availability of terminals and computers for instruction. Also, standard operating systems, or work stations with capability to communicate with different computers through one standard Man Machine Interface will lower the threshold for initiation to CAD and allow more time for the real work.

Of particular importance will be successful exploitation of CAI. I think CAD lends itself naturally to be taught through CAI, because they are both computerized. Some ideas of how CAI should be used for CAD training are expressed below:

- The CAI package should be interfaced, or preferably integrated to the CAD software itself.
- The user's manual should be a part of the CAI software, with different help options for the novice and the experienced user.

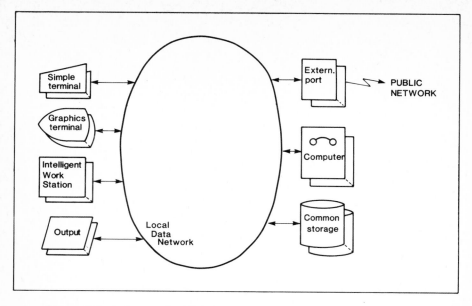

Figure 3 *In the future, a local data network will be installed at ELAB/NTH, with possibilities to share facilities across computer families. Ports to public network will facilitate industrial training in between course periods at NTH*

- The CAI part should contain a description of modelling aids and techniques, offer many worked-out examples, and give a guided tour through major application areas.
- Unsolved problems for the student should be offered, and a monitoring capability for the teacher to trace the activity may be also to examine the solutions.
- Within the framework of NTH's Continuing Education Courses, it will be possible for an industrial engineer to make connection to the training computers at NTH through the public network. The distribution of intelligent work stations could also allow down-loading of course material from the teaching institution to the student's terminal in the future. The network possibilities will diminish the need for on-site university instruction.

Conclusion

Current CAD courses for electronics have been presented, as well as short-range and long-range plans for amendments and new courses.

It has been shown that the fruitful co-operation between the university and a research institute yields an instrument to cope with changing needs in society and industry regarding CADCAM education.

Current trends in data technology and teaching methods have been discussed, and ideas for exploitation of new developments in these areas have been offered.

References

Bayegan, H M and Aas, E J (1978) An integrated system for interactive editing of schematics, logic simulation and PCB layout design. Proceedings 15th design automation conference; Las Vegas, USA. IEEE Computer Society and ACM IEEE catalogue no 78 CH 1363-1C: 1-8

Mead, C and Conway, L (1980) *Introduction to VLSI systems*. Addison-Wesley: London

Tysso, V and Aas, E J (January 1981) A common data library in an integrated CAD system for electronics. ACM SIGDA *newsletter* **11** 1: 30-42

Table 1 *CADCAM systems available at ELAB/NTH for the design of digital systems*

Program name	Function	Acquired from
CASS	Schematic editing	Central Institute of Industrial Research, Oslo, Norway
SPICE 2	Nonlinear simulation of electronic circuits	University of California, Berkeley, California
RTSIa	Register Transfer Level simulation	Techn. Hochschule Darmstadt, Germany
TEGAS 5	Logic simulation and test generation	COMSAT General Integrated Systems, Austin, Texas
CAMILA	Layout of printed circuit boards (PCB)	ELAB, Trondheim, Norway
CALMP	Layout of standard integrated circuit cells	Silvar-Lisco, Leuven, Belgium
CALMA	Graphics work station for LSI design	CALMA, Sunnyvale, California
STICKS	Symbolic layout of integrated circuits	CALMA, Sunnyvale, California (delivery July '82)
EPOKE	Integrated CADCAM system for PCB design	Norwegian joint project

E J Aas, Division of Physical Electronics, Norwegian Institute of Technology, N-7034 Trondheim, Norway

Discussion

D J Williams, Plessey Office Systems Limited:

First, can I congratulate Professor Aas on arranging a course which is so near to what I believe the requirements of modern high technology industry are. Right at the beginning of the sequence of events, when the students are training, I am not quite sure where the conceptual phase is dealt with, how the economics of the design that is to be introduced is brought in; how the function is described, whether there are restrictions on the components that may be used and, finally, whether there is any interaction with the marketing organization computers that he would have proposed using?

E J Aas, Norwegian Institute of Technology:

Of course at the University it is rather hard to have contact with a marketing organization. Let me emphasize that the first course in CADCAM mainly takes the students through the different CADCAM tools and concepts. The conceptional phase is dealt with in terms of schematic systems and high-level design languages and simulation. The economics of design is discussed briefly in terms of CADCAM advantages/disadvantages for investment, training, establishment of procedures and libraries. Design ramifications in terms of testability, use of production and marketing interaction are introduced in general terms.

In the second course I mentioned, 'LSI – design and computer aids', we are putting more emphasis on integrating design principles and aids. The economics of the design comes into discussion when technology, design method and design aids are chosen. Experienced data from quite a few designs in different technologies like NMOS, CMOS, 12L for the industry are collected for the purpose of presenting them in the course.

32. Post-professional education in computers in architecture at the University of Sydney

J S Gero and A D Radford

Abstract: Post-professional education in computers in architecture is particularly important because of the combination of the large potential effects of computers on the way architecture is practised and the low level of knowledge about computing within the profession. This paper describes the approach to post-professional education in the Department of Architectural Science at Sydney University. Its aim is to offer a range of courses each aimed directly at the needs and desires of a participating group, in terms of awareness, knowledge, competence, specialization or expertise. The nature and scope of these short and extended courses and of the extensive Diploma in Architectural Computing are described. The resources developed to serve these aims are a combination of personnel, computing equipment, computing software, library facilities and research specializations.

Post-professional education

The rate of change in the theory, techniques and technology available to all the professions has increased rapidly over the second half of the twentieth century. It is no longer realistic to expect the knowledge base acquired in a period of vocational training soon after leaving school to serve a practitioner throughout a professional life extending over 40 years. The necessity and difficulty of keeping up to date has been a subject of continued debate, particularly in the medical profession, where there is a strong public awareness of changes in methods of diagnosis and treatment, but extending to the other professions including architecture. In the UK, for example, a committee of the Royal Institute of British Architects has examined and reported on the need for continuing professional development and the responsibility of both public and private offices to provide time and facilities for their staff to develop and extend their skills.

There are three significant reasons why computers and computing should occupy a place in the forefront of this process:

- It is a field which is going to have a major influence on the way the architecture of the future develops and is practised;
- The rate of change is particularly rapid;
- Most practitioners in decision-making positions had no contact with computers at all in their professional education.

Further, many of the early undergraduate courses in computers for architectural students were not very successful in putting across the intended knowledge. There were a number of reasons behind this lack of success: insufficient time and resources allocated to the subject, insufficient clarity about its aims and direction, and its teaching by non-architects who were expert in computer science but unable to demonstrate its relevance to architectural practice.

There is, then, a demonstrable need for post-professional education in this field, a need perceived by many practitioners themselves. This paper describes the intentions and means behind the range of programmes developed by the Department of Architectural Science at Sydney University with the aim of serving this need.

Goals and means of the Sydney post-professional program

The overall goal of the post-professional courses offered at Sydney University is to equip the architectural profession with the knowledge necessary to exploit the opportunities offered by computers and computing. In this goal there are benefits for the general community in the improvement of the product and service of architecture and particular benefits for the profession and individual in maintaining the ability to compete and meet the demands of speed and efficiency in modern practice.

The specific intention of the program is to offer a range of courses each aimed directly at the needs and desires of a participating group. Courses are designed to fit under one of five broad headings: awareness, knowledge, competence, specialization and expertise. These headings may apply to computers in architecture generally or to one selected aspect chosen as the topic of a short course or seminar. Recent examples of the latter have been courses at the awareness and knowledge levels on productivity aids, layout planning, and computer-aided design systems.

(a) Awareness: about but not how. The aim is to impart an awareness of the state of the art in hardware and software, future developments and the economic feasibility of applications in practice. The medium is short, intensive one-day courses.

(b) Knowledge: about and how to use. The aim is to impart a basic knowledge about the different systems and options available and to provide some 'hands-on' experience of the use of commercially available computer hardware and software. The medium is short courses of one or two days, depending on the scope of topics covered.

(c) Competence: about, how to use and how to make. The aim is to impart a more extensive understanding of the implications and opportunities of computer technology for architecture and a working knowledge of how to program computers and use them in problem-solving, the latter within a fairly restricted field of applications. Two approaches have been tried: one-week intensive courses of 40 hours duration, and a 27 part series of evening lectures and workshops totalling 54 hours of course work and available for the first time in 1982.

A list of courses held in the period 1979-82 under categories (1), (2) and (3) is contained in Appendix 1.

(d) Specialization: in-depth knowledge. The aim is to impart a general knowledge and professional competence in a broad interpretation of architectural computing which includes computer hardware and software but also extends to computation methods, design methodology and operations research. The medium is a three-year part-time (two-year full-time) series of lectures, seminars and projects plus a substantial practical project, leading to the Diploma in Architectural Computing. The course is intended for architects and related professionals who seek to develop architectural computing as a significant (but not necessarily exclusive) specialization. It is described in the next section.

(e) Expertise: 'leading edge' knowledge. The acquisition of knowledge to the point of expertise is highly dependent on the experience and motivation of the individual and goes beyond any formal course teaching. Support and facilities are available for the investigation and exploration of diverse facets of architectural computing. Programs of research can lead to the degrees of Master of Science (Architecture), Master of Architecture and Doctor of Philosophy.

The diploma in architectural computing

The Diploma in Architectural Computing, introduced in 1978, represents the department's principal commitment to post-professional education in computing (Gero,

1980; Sydney University, 1982). It was not and is not conceived as a diploma in computer-aided design directly, but to fill the perceived requirement for architects with a specialization in computing who are able to work with and complement computer scientists. It also stems from a wider premise that many disparate parts of architecture and building can be better understood with a knowledge of the methodological background of systems analysis and computing. Admission to the diploma is not limited to applicants with a prior degree but may also be granted to registered architects or to others with recognized design qualifications.

The courses making up the diploma are interlinked, relying for their successful completion on knowledge gained in other courses undertaken earlier or concurrently. Computing is the unifying scheme. Thus, practical assignments in courses such as operations research and economic feasibility studies require the writing and implementation of computer programs to achieve the required results. There is a strong element of learning by experience, with workshop sessions occupying an important place in the timetable and assessment by assignments and projects rather than examination. A single substantial project undertaken towards the end of the course accounts for a quarter of the assessment for the diploma.

A list of courses and their objectives is contained in Appendix 2. Students can set their own priorities under a unit/option scheme which enables them to choose amongst the available courses; although not all courses are offered in any one year. Moreover, some courses map on to the requirements for three parallel post-professional diplomas offered by the department, the general Diploma in Building Science and the specialized Diplomas in Illumination Design and Building Energy Conservation. This promotes a cross-fertilization of ideas and ensures that a basic knowledge in computing is available to all post-professional students passing through the department. Students in any of these diploma courses can progress to a Master of Building Science by submitting an extensive research-based thesis in addition to satisfying the requirements of the diploma.

Resources

Five categories of resources are necessary to serve this variety of courses: personnel, computing equipment, computing software, library facilities, and a background of research specializations. At Sydney these resources have been built up over 15 years of activity in computers in architecture.

(a) The Computer Applications Research Unit in the Department of Architectural Science includes two academic staff, a tutor and a programmer, combining backgrounds in architecture, engineering, mathematics and computer science. At the present time, there are also two research fellows, two research assistants and two research programmers engaged on specific research programmes and contributing to the pool of knowledge in the unit. Teaching resources are extended by research students in the unit acting as part-time tutors and by two other members of the department who offer specialist skills. Further, visitors from other educational and research institutions and from the commercial computer world are used to provide teaching and expertise in their particular fields. In 1980-1, for example, the research unit had visitors from the UK, USA, Israel, Poland and New Zealand contributing to both diploma and short public courses and had visitors from local Sydney architecture and engineering practices already using computers to talk about their experiences. A range of backgrounds, specializations and perspectives is important in maintaining balanced and effective teaching.

(b) *Computing Equipment*: The aim of the department is to be able to offer experience on a range of computer types, from time-sharing access to a large main frame through minis and desktops to small micros. Initial teaching of computer programming is centred around time-sharing terminals to the University's CDC

Cyber computer, for which two teaching laboratories each with 25 terminals (mixed video and hard copy devices) are available to supplement ten terminals within the department. The latter includes two storage graphics screens, two black and white raster graphics screens and a colour raster graphics screen. Graphics plotting is provided by a Versatec V80 1200 line per minute printer/Calcomp-compatible hard copy device, an HP 7580 A1 size plotter and a small HP 7202A A3 size flat bed plotter as well as a Videoprint colour hard copy unit. As stand-alone systems there are an HP system 45T graphics desktop computer with a graphics tablet and Winchester disc storage and an HP system 85 graphics microcomputer. These will shortly be joined by a small graphics-oriented personal microcomputer. At the other end of the scale of graphics hardware, the department has use of the University's Evans and Sutherland Picture System 2 driven by a PDP11/34 minicomputer and a Computervision CADDS3 system in the Faculty of Engineering.

(c) *Software*: Software falls into three categories: teaching software, high-level languages and applications programs. Teaching software extends from the familiar self-teaching programs for computer languages to specific illustrations of some theory or technique. For example, an interactive program to rotate, translate and project three-dimensional objects will display the transformation matrices applied and an explanation of their use as well as the resultant image. High level languages include the graphics languages of Calcomp, Tektronix TCS, and Tektronix IGL. Sub-programs developed within the department carry out recurring tasks using standard calling sequences. Applications software includes analysis programs for general architectural use (sun penetration, daylight factors, heating and cooling loads, etc) as well as a small drafting system for domestic scale buildings and several specialized optimization packages. Access to bigger systems, particularly large integrated drafting systems, is available through local suppliers and bureaux.

(d) *Library*: In a rapidly changing field an up-to-date and comprehensive library is an essential resource. The architecture library contains a wide range of books on design methods, applicable mathematics, computing, graphics and operations research, plus conference proceedings, journals and research monographs.

(e) *Specialization*: The philosophy of research within the Computer Applications Research Unit is to utilize resources for research in a specialized area to the highest possible standards and to avoid spreading those resources too thinly. The major areas of research are design using optimization concepts within the unit and design methodology and geometrical modelling in other parts of the University with links to the unit. Active research specializations are important to the quality and authority of the post-professional courses and the insight into future developments that they allow. By bringing in visiting personnel with other specializations from other institutions in Australia and overseas, very different perspectives on computing and computer-aided design can be presented.

Conclusion

The influence of these approaches can be measured by the fact that the majority of the computing activities in architectural practices in New South Wales (which has one-quarter of Australia's population) trace their origins to one of the courses described in this paper. In the immediate future there will be an expansion of both hardware and software resources to keep pace with the rapidly changing technology of computing in architecture. In the long term, post-professional courses will continue to be needed to bring practising architects up to date on developments and in specialized areas of computing in architecture. At the same time there is likely to be reduction in the need for introductory courses.

References

Gero, J S (1980) *The diploma in architectural computing at the University of Sydney, CAD 80*. IPC Science and Technology Press; Guildford; 293-6

Sydney University (1982) *Faculty of Architecture Handbook*. University of Sydney

Appendix 1: Recent short post-professional courses offered by the department

1982

1. **Geometric Modelling in CAD**

    Professor C M Eastman
    Carnegie-Mellon University
    One-day seminar and workshop

2. **Introduction to Computers in Architecture**

    Dr A D Radford
    Sydney University
    a 27-part series of evening lectures and tutorials

3. **Microcomputers in Architecture**

    Dr A D Radford and Associate Professor J S Gero
    Sydney University
    One-day seminar and demonstrations

1981

4. **Architecture, Architectural Practice and Computers**

    Associate Professor J S Gero and Dr A D Radford
    Sydney University
    One-day seminar and workshop

5. **Developments in Computers in Architecture**

    Professor W J Mitchell, University of California, Los Angeles
    Associate Professor J S Gero, University of Sydney
    Dr A D Radford, University of Sydney
    One-day seminar

6. **Computerized Productivity Aids**

    Professor W J Mitchell, University of California, Los Angeles
    Associate Professor J S Gero, University of Sydney
    Dr A D Radford, University of Sydney
    One-day seminar

7. **Implications of Computers in Architecture**

    Professor W J Mitchell, University of California, Los Angeles
    Associate Professor J S Gero, University of Sydney
    Two-day working seminar

1980

8. **Science of Decision-Making**

    Professor A Kaufman
    Universities of Louvain, Grenoble and Paris
    One-day seminar

9. **Fuzzy Sets and Systems**

    Professor A Kaufman
    Universities of Louvain, Grenoble and Paris
    Two-day seminar

10. **Development in Computer-Aided Design**

    Professor W S Elliott, Imperial College, London
    Associate Professor J S Gero, University of Sydney
    One-day seminar

1979

11. **Computerized Layout and Space Planning**

    Professor W J Mitchell
    University of California, Los Angeles
    One-day seminar

12. **Computer Application in Architecture**

    Associate Professor J S Gero and Mr B S Forwood
    University of Sydney
    Three-day seminar and workshop

Appendix 2: Constituent courses of the Diploma in Architectural Computing

Subject Areas:

Computation methods
Operations research
Computers
Methodology
Project

Computation methods

The following six subjects are currently available in this area:

Computation Methods in Architecture and Building

An introduction to the basic notions of numerical computations and their applicability to architectural and building problems.

Introduction to Mathematical Models

An introduction to the notions of mathematical modelling.

Statistical Methods in Architecture

A course which commences with descriptive statistics and proceeds to multiple regression analysis and the use of computer packages, in particular SPSS, in architectural analysis.

Numerical Methods in Architecture

A course which covers a variety of methods for obtaining numerical solutions to mathematical models used in architecture.

Matrix Methods in Architecture

A course which develops matrix methods as a means of representing and manipulating information and goes on to examine their application in architecture.

Graph Theory in Architecture

A course which develops graph theory as a means of representing and manipulating information and goes on to examine its application in architecture.

Operations research

The following five subjects are currently available in this area:

Operations Research in Architecture I

A beginning course which examines the basic notions of operations research and then develops the techniques of simulation and optimization as used in architecture.

Operations Research Applications in Architecture II

An advanced course which examines complex simulations and the application of optimization concepts as aids to decision-making in architecture, particularly concerned with dynamic programming and multi-criteria decision-making.

Layout Planning

A course which examines the various formulations and solution methodologies for the layout planning of buildings.

Building Economics

An introductory course which examines the bases and methods of both macro- and micro-economics in its application to buildings.

Economic Feasibility Studies in Architecture

A course which develops the notion of the use of economics as the prime initial determinant of the feasibility of building projects.

Computers

The following eight subjects are currently available in this area:

Computers in Architecture I

A course which commences with the fundamentals of computers, computer modelling of problems and programming and goes on to introduce and develop a programming language, and concludes with a brief foray into operating systems.

Computers in Architecture II

An advanced course which extends programming principles, develops the notions of building representations, data structures, data-bases and examines the ranges of applications in architecture.

APL Programming

A specialist course devoted exclusively to APL programming and APL interactive graphics.

Computers in Building Structures

An advanced course which examines and develops the methodologies for the analysis and design of building structures.

Computers in Building Services

An advanced course which examines and develops the methodologies for the analysis and design of building services and building environments.

Computer Graphics in Architecture I

An advanced course which examines and develops the notions of graphical representations, data structures and techniques for passive graphics, interactive graphics, colour graphics and graphical languages.

Computer Graphics in Architecture II

An advanced course which develops geometric modelling, graphics data structures, colour graphics and the design of drafting systems.

Computer Applications in Architecture

An advanced course which reviews and critically examines the panoply of computer applications in architecture.

Methodology

The following four courses are currently available in this area:

Systems and Models in Architectural Design

A course which presents the basis for the use of both models and systems in architectural design.

Design Methods

An advanced course which critically reviews algorithmic design methods and develops an approach to handle design formulated as a multi-attribute, multi-criteria optimization problem.

History of Design Methods

A course which presents the developments in design methods as well as reviewing a wide range of approaches to architectural design.

History and Philosophy of Science

A course which presents the developments in the philosophy of science pertinent to a systems approach to architecture.

Project

The project involves the development and implementation of a computer-aided approach to a particular design problem selected by the student. It is designed to involve the student in about 200-300 hours' work.

J S Gero, Department of Architectural Science, University of Sydney, Australia

A D Radford, Department of Architectural Science, University of Sydney, Australia

Discussion

D M Brotton, UMIST:

Could you comment on the comparison between crash courses and the drip feed type particularly for the sort of teaching that you have been dealing with?

J S Gero, Sydney University:

We find the crash courses (such an unfortunate term – computer crash courses), to be very useful at the entry level, but our experience has been that when you get to the other level, beyond the awareness and knowledge levels, they are really quite unsatisfactory. There is a time frame which is needed to absorb the information and to gain some experience.

33. CAD in the Cambridge Engineering Tripos 1977-82

A L Johnson and R B Morris

Abstract: In the undergraduate engineering course at the University of Cambridge six graphics terminals are being used to teach 500 undergraduates by packages and routines especially designed for unsupervised operation by men in their first and second years. Topics covered include closed-loop control systems, slider mechanisms, projection theory, intersection of surfaces, elastic and plastic collapse and vibrating systems. Ordinary display facilities are also available for use by undergraduates, research students and teaching staff. The terminals are installed in the students' drawing office directly alongside the normal drawing boards and are in full-time use. The choice of a flat bed plotter and teleprinter as a terminal has been proved wise for teaching and economy. Gradual expansion into other areas of teaching and research is taking place, and the introduction of CAD methods, not as a separate subject but as an integral part of the normal Cambridge Engineering degree course, is proceeding smoothly and with enthusiasm from teaching staff and undergraduates alike.

Introduction

At CAD ED in 1977 we reported (Johnson and Morris, 1977) the results of the first five years of CAD teaching in the Cambridge University Engineering Department. Having started in a very small way, we had made sufficient progress to permit us to enter with moderate confidence upon the task of mass education. In Cambridge all 300 first year engineering undergraduates spend about 50 hours in the drawing office learning the graphical techniques of mechanical drawing, kinematics, structural analysis, and both conceptual and detailed design. In the second year, one-third drop out of design, but the remaining 200 have a strongly intensified course in all the above subjects. In both years, an examination has to be passed, and our experience has been that while project work educates the few involved, it is inappropriate to the mass education of 500 undergraduates at a time.

We have based our activities on the premise that whatever the difficulties inherent in teaching CAD, it is essential that all engineers should have at least some awareness of the potential of the computer in research, design, and manufacture. Most engineering graduates will later have to make decisions based on their understanding of what a computer system can and cannot achieve, and such awareness cannot be generated simply by the teaching of elementary programming. In addition we have to accept that although the advent of the computer as a design tool has given new significance to many areas of mathematics hitherto considered to be of purely academic concern, CAD cannot claim to be anything more than a powerful technique.

The nature of the Cambridge Engineering Tripos intensifies this limitation. The degree course has always been a theoretical one, concentrating, rightly in our view, on the fundamental principles of engineering in preference to techniques. During the first two years of the course, undergraduates receive a general grounding in all disciplines, specialization taking place during the third year. We are proud of the fact that any Cambridge engineer worth his salt can specialize later in life from the broad base of his education here. Our aim therefore must be to include CAD as a normal part of the broad base, eschewing narrow applications and promoting its universal employment.

Hardware and software

A lecture course is the ideal medium through which to outline the broader aspects of CAD, but an understanding of how the computer is actually used in design is less easy to convey, for practical experience is an essential part of the learning process. It is here that the problems of teaching are at their most acute, since demands are made on expensive equipment and undergraduates' time in return for what may seem initially to be nebulous gains. A commonly accepted approach is to instruct undergraduates in the use of a commercial CAD package in the hope that a general appreciation of computer-aided design will be acquired. This approach consumes enormous quantities of an undergraduate's time which he could more profitably spend learning relevant theoretical principles. It is vital that we recognize our responsibility to equip engineering graduates not merely with competence to use packages available today, but with full qualifications to sponsor the next generation of CAD software.

To this end we shifted the emphasis of our attack from the individual project work of 20 undergraduates at a time to the mass education of 500. We gave our first formal course lecture on CAD in December 1979. It formed part of a two-lecture course intended to develop the undergraduates' knowledge of computing; to demonstrate how simple elements of computer technology could be assembled to produce an integrated computer-based manufacturing system which would embrace aspects of planning, design, and cost control; and to highlight the economic advantages and social consequences of such a comprehensive system. To complement these first lectures we installed two graphics terminals in the drawing office next to the conventional drawing boards, we provided software so that a couple of the standard problems set for pencil-and-paper solution could be done on these terminals if an undergraduate so wished, and we obtained the tacit agreement of the chief lecturer that such solutions would be acceptable to him.

We now have six of these terminals available for use by undergraduates. Each terminal comprises a type 4662 flat bed plotter coupled to a type 43 printer. This combination has several advantages over the more usual storage or refresh display screen: it gives the undergraduate a permanent record of his work; it is cheaper; and it can be sited in a well-lit drawing office alongside normal drawing boards (Figure 1). These benefits, together with a remarkable degree of robustness, far outweigh the slight time penalty occasionally incurred when producing a complex drawing.

Figure 1 *(see page 9)*

First year work

The essential factor in CAD practical work is that it should not require continuous supervision by teaching staff. This means that the exercises must be self-sufficient in guidance material. We therefore employed a bright undergraduate, who had just completed the drawing course, to write under our direction a package (Morris, Johnson and Ireland, 1979) which produces on the plotter the direct equivalent of projection theory as taught in the first year. With this package undergraduates are able not only to solve the standard problems as set, but also to exercise their understanding of this subject. The package simulates normal activity at the drawing board, allowing the user to project orthogonally in any direction, and reporting the orientations and lengths of lines as projected. Thus to find the true length of a line or the angle between two planes, the user must carry out the correct sequence of projections (Figure 2). It soon becomes maddeningly clear to him that the computer could provide the required answer immediately, if only the right question could be put to it, and while gaining practice in conventional methods of projection (with enhanced accuracy) he rapidly becomes aware of the assistance which the computer could give him if it were fully orientated towards solving his problem.

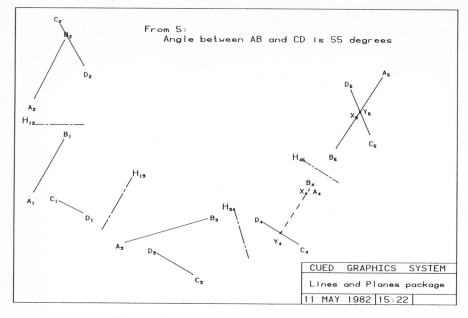

Figure 2 *First year projection exercise – two skew-lines*

From this simple exercise we learned much about the implementing of such packages. Their use on the central computer is logged automatically, and software development is aided by the ability to log any incorrect command typed in by a user. This facility has enabled us quickly to remove obscurities and ambiguities from the driving manuals. The manuals themselves are held on the computer, and any user can print a copy for himself without fuss.

The short lecture course in CAD is now an established part of the first year at Cambridge, though the undergraduates' knowledge of computing at this stage is negligible. Immediate practical experience of drawing by computer is made possible by use of a simply-controlled CUED Viewing Routine (Morris, 1979). This accepts as data a list of three-dimensional point co-ordinates and a series of instructions specifying the order in which the points are to be visited. The data thus describes a three-dimensional wire frame object, which the Routine projects onto the two dimensions of the plotter paper. The user is able interactively to alter the scale, orientation and position of his object; he can specify true views and take oblique sections; and he can then produce the type of drawing he wishes: orthogonal projections, first or third angle layout, perspective views or stereoscopic pairs to be examined fully in three dimensions.

The educational value of the Routine is considerable, for although it is powerful, it is easily handled by an inexperienced user. With the help of a carefully written manual and an example data file, the novice can acquire a passing familiarity after about ten minutes of use. The power and scope of interactive computer commands rapidly becomes apparent, as does the speed with which computers are able to produce high quality technical drawings.

At this initial stage data for elementary objects is supplied manually by the undergraduate, but as his first year progresses he learns elsewhere the mathematical principles involved and the role of sub-routines in computing, concepts which are the more readily assimilated as a result of his first-hand experience of their application. As a transitional stage we provide him with a number of ready-made sub-routines which link with the Viewing Routine to produce plots which would be laborious to draw manually, such as Whiteley and Nyquist diagrams as taught in the control-engineering course. He is able to produce his pictures without having to learn another set of graphical commands, and he finds himself longing to experiment with his own ideas. To help him we have written a library of sub-routines which he can call to manipulate vectors and to build a data file, effectively moving his drawing pencil in three dimensions from the results of calculations performed within his own sub-routine. He is thus conveniently placed between the Viewing Routine and the Library and has merely to write a little program of his own and to invoke a procedure by a single command. He is freed from any considerations of scaling and positioning of the plot, as these are handled automatically by the Routine, which sets up all the protocols of graphics on his behalf. The important result is that he concentrates on problem-solving, not computing, and he can look at his solutions in unconventional stereoscopic views, which give him not only a clear understanding of three-dimensional geometry but often provide his first experience of the practical use of vectors.

To say that the response from undergraduates has been encouraging would be wildly to understate the position. The terminals, installed in the students' drawing office directly alongside the conventional drawing boards, are in full-time use. The choice of plotter and teletype combination as terminal, costing less than £4,000 and giving hard copy with plots up to A3 size, has been proved wise for teaching and economy. We launched the system on the first year undergraduates and success was immediate. Indeed, we had congestion problems caused by second year men joining in the fun.

Second year work

Twelve months later we reaped the whirlwind as the first year became the current second year and pressed for more advanced packages and automatic graphics sub-routines. We had prepared for this expected response by building exercises firmly into

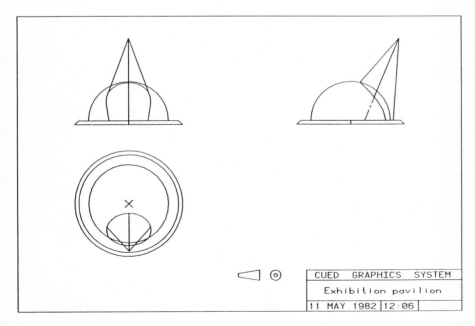

Figure 3 *Second year intersection exercise – exhibition pavilion*

the design course. As one exercise, undergraduates are now required to write a sub-
routine which calculates and draws the line of intersection between the conical front of
an exhibition pavilion and its hemispherical main block (Figure 3). The Library

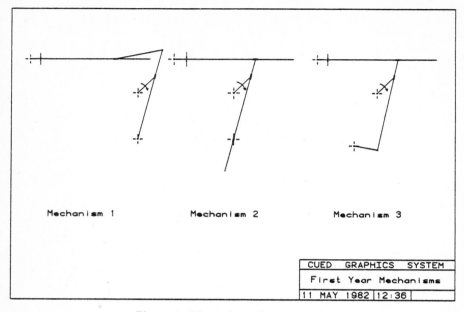

Figure 4a *First and second year mechanisms*

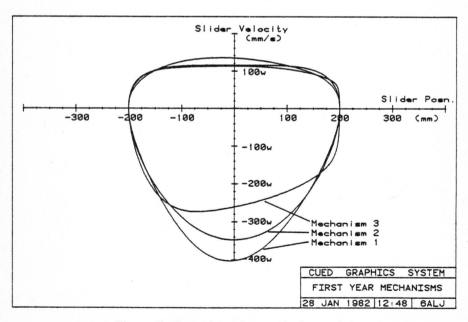

Figure 4b *First and second year mechanisms analysis*

provides the outline of the pavilion and vector manipulation sub-routines. Another exercise involves a package which calculates the velocity plot of a quick-return mechanism (Johnson and Morris, 1981) on receipt of the dimensions of the mechanism. The problem for the second year undergraduate is to determine the dimensions necessary to produce a particular shape of velocity plot: a first exercise in design synthesis rather than the simple analysis of the first year (Figure 4).

This we carry further. One of the new exercises in the second year course is to design in detail a diesel engine for a model aircraft. With library sub-routines to give combustion details, heat transfer, and mechanical analysis, the undergraduate can vary timing, compression, and other characteristics to produce simulated indicator diagrams for his design (Figure 5). The success of this particular exercise was again immediate and overwhelming, a major conclusion being that six graphics terminals cannot really cope with 200 users. Nonetheless, with tolerance and good humour they produced some excellent solutions.

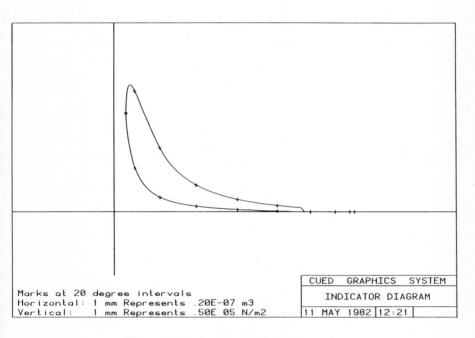

Marks at 20 degree intervals
Horizontal: 1 mm Represents .20E-07 m3
Vertical: 1 mm Represents .50E 05 N/m2

CUED   GRAPHICS   SYSTEM
INDICATOR DIAGRAM
11 MAY 1982 | 12:21

Figure 5 *Second year aero-engine design exercise*

For the ordinary undergraduate several other packages have been produced, including one for the elastic and plastic collapse of structures and one for the display of modes of vibration. For the computer enthusiast there is the option of choosing a computer-orientated project during his second or third year. Much valuable work has emerged from computing projects in the past (Morris, 1974) and the advent of CAD opens a new avenue, for the construction of even a simple package is a design activity in itself. As with any design, it requires a clear analysis of the customer's needs, and careful thinking to satisfy them in a simple and economical manner. Furthermore, in this environment the package can not only be built and tested, but also marketed to contemporaries, a design exercise far more realistic than is normally possible during a degree course. Subjects covered recently have included the synthesis of Fourier series and the comprehensible display of three-dimensional vector fields. Our experience is that it is not only those doing the project who benefit, but also their friends who are very free with advice.

Problems of mass education

Introducing CAD into our Cambridge syllabus has not been straightforward but has raised three major problems: time, cost, and motivation. It is difficult to incorporate even the most vital new topic into a degree course, as the insertion of a new subject into a congested syllabus necessitates the removal of some existing material. A university lecturer is well aware of the risk inherent in this process: some of the outgoing material includes concepts and exercises which have contributed subtly and significantly to his own comprehension, and removal could cut the rungs of the ladder up which he himself climbed.

Our introduction of CAD as a subject has reached a plateau. No existing material has yet been displaced, but this now appears to be a necessary step if further teaching time is to become available. Obvious candidates for pruning are some conventional graphical methods of analysis: displacement diagrams, velocity and acceleration diagrams, and the construction of surface intersections. These are all techniques whose use in industry has been abandoned in favour of quicker and more accurate computer methods. The strong arguments for their retention are that they provide valuable insight which would be lost to the undergraduate if the computer were to do all the work for him, and that they are useful techniques at the initial stages of design where the accuracy of expensive computing is unnecessary. These arguments are certainly valid in the case of displacement diagrams, whose intelligent use leads to comprehension of the way in which a structure deflects and the extent to which each member of the structure contributes to the overall deflexion. The equivalent numerical technique, virtual work, educates poorly by comparison.

The case for velocity diagrams is, however, weaker. They are limited to two-dimensional mechanisms, and only properly show the behaviour of a mechanism at one point in its cycle. This shortcoming is yet more apparent in the case of acceleration diagrams, where even the most enlightened engineer finds it tortuous to envisage how the diagram changes in shape as the mechanism proceeds on its cycle. Computer analysis is admittedly prone to withholding from the designer an intuitive feel for the behaviour of a mechanism. A computer-drawn animation can be of assistance, but the main CAD advantage lies in its ability to show the velocity and acceleration of parts of the mechanism throughout the whole of the cycle. Furthermore, the designer can alter dimensions and instantly obtain a new diagram to gain a different and more valuable insight into the contribution which each part of the mechanism makes towards the overall behaviour. This seems to be one area in which CAD might displace conventional techniques without loss to the educational process. The potential rewards are considerable, for we can expect that the vectorial techniques used by the computer can be extended to cover non-planar mechanisms in addition to the simpler planar ones which are within the scope of conventional techniques.

The second of our major problems is cost. Equipment is expensive, and although the cost of computers has fallen sharply in recent years, that of terminals has remained high. To replace every drawing board in the undergraduate drawing office by a computer-driven drafting station would entail enormous capital outlay and formidable maintenance charges, and is in our view both unnecessary and undesirable. The undergraduate needs hands-on experience no less of the drawing board than of the computer terminal. Our plans for expansion are modest, and we arrange schedules carefully to ensure that limited laboratory resources are used to the best effect.

Our third problem is motivation. The Cambridge engineering course is directed almost exclusively towards written examinations, with course work contributing in only a small way towards the class of the degree. While this method of assessment prevails it

could be difficult to stimulate an undergraduate's interest in CAD, for hard-pressed
students are understandably reluctant to pursue projects which ostensibly contribute
little to their final credit. We are, however, making progress. In May 1981 the second
year examination contained questions, one based on Figure 1 of our 1977 paper
(Johnson and Morris, 1977), which could be answered more satisfactorily by
undergraduates who were aware of CAD techniques than by those ignorant of them.
Past experience suggests that this practice will eliminate problems over motivation in
future, and we like to think it significant that one of us has been appointed examiner in
Drawing and Design for the second year of the Tripos, while the other has been
appointed to examine the first year.

Other work

We have dealt above with the mass education of our first and second year
undergraduates, and we have not here sufficient space to comment in detail on our post-
graduate activities. We hope to bring the commercial DUCT system and the BUILD
research package, both developed in Cambridge, into the third year, and a research
student has recently adapted a contouring package to solve first year mathematics
problems (Figure 6). Suffice it to say that we have mentioned above analysis of
mechanisms as one area in which computer-based techniques are poised to replace the
more conventional methods. Gradual expansion into other areas of teaching and
research is taking place with the help and encouragement of erstwhile sceptical
colleagues.

Conclusions

The introduction of CAD methods, not as a separate subject but as an integral part of
the normal Cambridge engineering degree course, is proceeding smoothly and with

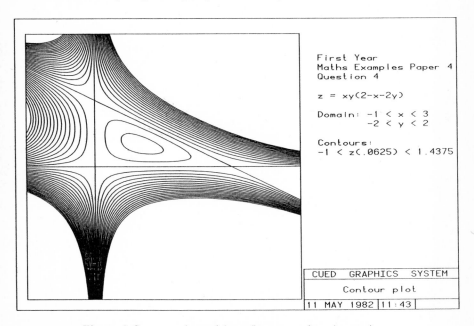

Figure 6 *Contour package solving a first year mathematics question*

enthusiasm from teaching staff and undergraduates alike. Many areas exist where an improvement in technique will not compromise the teaching of fundamental principles, and there is no university which can afford to ignore the challenge of finding these areas and bringing them up to date.

References

Johnson, A L and Morris, R B (1977) CAD in the Cambridge engineering tripos. Proceedings CAD ED conference: Middlesbrough, UK

Johnson, A L and Morris, R B (1981) CUED mechanisms analysis package. Cambridge University engineering department; Cambridge, UK

Morris, R B (1974) Computer graphics and its influence on manufacturing techniques. Proceedings international conference on production engineering; Tokyo, Japan

Morris, R B (1979) CUED viewing routine. Cambridge University engineering department; Cambridge, UK

Morris, R B, Johnson, A L and Ireland, D A (1979) CUED lines and planes drawing package. Cambridge University engineering department; Cambridge, UK

Morris, R Cambridge University, Department of Engineering, Trumpington Street, Cambridge

Johnson, A L Cambridge University, Department of Engineering, Trumpington Street, Cambridge

34. CAD in structural engineering at UMIST

D M Brotton, M A Millar and A J Bell

Abstract: Following an introduction the paper will give a brief description of the computer facilities available to the Department of Civil and Structural Engineering at UMIST. Subsequent sections of the paper will deal with our experience in teaching at undergraduate and post-graduate levels including the development through projects, dissertations and by research of powerful applications programs. Special short courses for practising engineers which have been run by the department will then be discussed. Many of the departmental programs have significant practical applications, for instance, their use in the design and construction of suspension bridges (Humber, Westerschelde); some commercial uses will be presented. Finally, concluding comments and future departmental developments will be given.

Introduction

Computer-aided design is interpreted in many different ways by different people and in different situations. However, in so far as this paper is concerned it has been interpreted widely to include considerations varying from computer structural analysis on large main frame machines to simple interactive design programs operating on microcomputers. Structural analysis was carried out at UMIST as early as the mid-1950s using the programs first developed by Livesley for operation on the Ferranti Mark 1 computer. The then Department of Structural Engineering in the late 1950s and early 1960s ran the first structural analysis computer service for industry; the service ran successfully for about five years, terminating only when commercial services came into being. This experience provided a firm base for subsequent research into structural computation and its natural development into the educational activities which are of fundamental importance in an engineering department in a technological university.

Two major factors have contributed to the enormous and continuing growth of computation and its place in engineering education of all types; the improved communications between computers and their users through the development of high-level languages such as autocodes and problem-orientated languages, the use of the interactive mode of operation and computer graphics; also, the remarkable growth in capacity with corresponding reductions in size and in cost of computer hardware. All of these facets and indeed others are continuing to change very rapidly, computers and computer programs have become more readily available to large numbers of students and engineers as portability has improved.

In the remaining sections of this paper the UMIST computer hardware will be described and the place of computation in the undergraduate and post-graduate teaching will be presented. Courses in computation for professional engineers will be briefly described and some of the more important practical commercial applications will be mentioned. Finally, a tentative look into the immediate future of university computation will be given.

UMIST computing facilities

One of the largest university computer configurations is located in the University of Manchester, UMRCC (University of Manchester Regional Computing Centre). In addition to providing large-scale batch processing to many of the universities in the United Kingdom, including the University of Manchester and UMIST, it also fulfils the function of providing the local smaller-scale batch and interactive service for the Manchester campuses.

The large-scale batch service is presently operated on two CDC 7600 computers each front-ended by an ICL 1900 series machine, one of the latter being accessed via five Remote Job Entry terminals on the UMIST campus. Each terminal has an operator who provides a service for the associated university community, undergraduates, post-graduates and staff. Departments have an allocation of computer resource which can be used at one of five different priority levels; the highest level provides for a turnround of approximately 15 minutes and the lowest level may be of the order of one week.

Interactive computer facilities are operated by UMRCC for the University of Manchester and UMIST only on twin CYBER 170-730 computers which support currently over 140 terminals, a number which will be increased to approximately 180 terminals in the near future. These terminals comprise teletypewriters (which are gradually being replaced), VDUs and storage tube graphics terminals; they are located singly and in small groups and also in larger groups for teaching purposes.

Interactive computation for SERC grant holders and some UMIST teaching can be carried out on the SERC PRIME 750 which is housed in the Control Systems Centre and is operated under contract by UMIST for SERC. It supports a large number of storage tube graphics terminals and a teaching cluster of 15 terminals, a number of which provide graphics.

Some three years ago UMIST recognized the very large contribution that computers can make to engineering drawing office practice. It met the need by the provision of a 64k word PDP 11/34 computer with a Benson 1322, 93 cm wide, three pens, drum plotter, a Summagraphics 1D digitizer table, a Tektronix 4014 large graphical display terminal and twin RLO2 discs. The configuration has the Ferranti-Cetec, CDA 40 software and hence provides a very powerful drafting facility. Alongside the above configuration is another PDP 11/34 computer driving a Ramtek RM 9351 refresh type colour display. These two systems together comprise the UMIST Advanced Graphics Unit which is supported largely for research use by a computer officer. It is however used increasingly for student and special course demonstrations, though its capacity is clearly not adequate for larger-scale teaching purposes.

Many departments possess their own computing facilities which are used for undergraduate teaching and project work and for running specialized courses. These vary from single microcomputers with floppy discs and printers to relatively powerful minicomputers and even main frame machines. Some of these facilities have been provided primarily for on-line use with experiments in the laboratories and as such form an integral part in departmental teaching activities.

Undergraduate teaching

Introduction

The teaching of computation to the civil engineering undergraduates was initially undertaken by the UMIST Department of Computation. The course was extremely professional but the students tended to think of it as peripheral to their engineering studies. As within the Department of Civil and Structural Engineering there were a number of members of staff who had in-depth experience of computer programming and who were also chartered engineers, it was decided to run the undergraduate computation course in-house, hopefully to improve the motivation of the students in this increasingly important subject.

The teaching within the course can be loosely divided into two sections, namely the initial teaching which is carried out in the first year of the course and the discipline teaching which comes in the second and third years.

Initial teaching

The broad aim of the initial teaching in computation is to give each student a confidence and awareness of the potential application of digital computers to assist in the efficient solution of civil and structural engineering design, detailing and management problems.

The students are taught about the functions of the various units that comprise a modern digital computer system, ie input/output peripherals, various types of storage, the arithmetic unit and the control unit. They are also encouraged to construct flow diagrams in order that they may formalize the logical path through their own programs.

Since the inception of this initial teaching course the programming language used has been FORTRAN. The early decision was based on the fact that IBM at that time had approximately 70 per cent of the Western world's computer market and that there existed a standard version of the language. This decision has been reviewed annually but a language has yet to emerge which has sufficient advantages over FORTRAN for the department to give up such an internationally acclaimed language which is truly portable. At present, approximately 80 per cent of the jobs run on the UMRCC computers are in FORTRAN and 90 per cent of the computer time is taken up in processing FORTRAN. All of the well known finite element systems are written in FORTRAN. FORTRAN 77 removes some of the more irritating input/output restrictions of FORTRAN 66 and the only major remaining drawback appears to be the lack of dynamic storage, a feature that may be more significant for virtual storage machines.

An ICL 1900 series FORTRAN in-core batch compiler has been used by the department for the initial teaching course over the past decade. The installation of the CDC CYBER computer with its cluster of terminals designated to be used exclusively for teaching meant that students could obtain experience of interactive development of their programming skills. The interactive compilers available tend to be less rigorous than the batch compiler and therefore allow code to be processed which is non-standard. The fact that it is possible to obtain code that apparently works more rapidly using the interactive system than the batch system generally appeals to students. However, the quality of the code obtained from the batch compiler tends to be significantly better than that obtained from the interactive system. It is healthy that students should have time to think before they respond to an error message. Batch processing fosters this thought process but it must be acknowledged that development is significantly more rapid in the interactive mode and therefore it must ultimately be adopted exclusively for initial teaching.

The initial teaching course is assessed entirely by course work. Each set project tends to be open-ended so that the students who already have a significant experience of computation before coming up to university can further develop their knowledge and expertise. A typical course consists of a first project usually involving some simple data processing of a geometrical or statistical nature. The second project may be taken from the structural analysis course, a third from the surveying course, and a final project from numerical analysis. As a typical example, the error correction of a closed traverse in surveying allows the students to process their own acquired data. The problem is ideally suitable for students to differentiate between function sub-programs, for processing angles in radians from degrees, minutes and seconds, and sub-routines which carry out the reverse process. The more advanced students can plot the corrected traverse using the lineprinter as a pseudo graph plotter.

Discipline teaching

The second year practical surveying course gives scope for the use of the programming expertise acquired in the first year. A similar traverse program is written in BASIC for a microcomputer. Groups of students process their own data, adjusting errors at stations, reducing levels and designing horizontal and vertical curves. The finite element course gives an introduction to one of the major advances in structural and stress analysis which has been made possible by the rapid increase in the power and versatility of modern computers.

During the third year of the course students are encouraged to use the versatile and efficient structural analysis bar element program to check their hand calculations. A wide range of departmental programs is also used extensively in final year project work. Many students write their own *ad hoc* computer programs to assist them in processing their laboratory project results. Increasingly, a number of students choose to undertake projects comprising the development of appropriate programs in the field of computer-aided design of structural elements.

Post-graduate teaching

Introduction

No formal teaching of computation or computer-aided design is given to students registered with the department for higher degrees. Such students however do make considerable use of computers and undertake research and development work in the area of computer-aided design.

Higher degree course

In October 1981 the department introduced an MSc Examination and Dissertation course in Structural Engineering. The course is run on a modular basis so as to be available to either full- or part-time students. Since it was assumed that all students taking the course would be relatively recent graduates it was not considered necessary to make any provision for the formal teaching of computation. Rather it was intended to teach and illustrate the use of computers in design processes, particularly those involving finite element modelling and steel design. Programs written and developed in the department are used for this purpose. Where appropriate, students are encouraged to use computer models to explore changes in the design assumption and parameters and determine how structures which they have designed respond to such changes.

An optional module in project management is also included as part of the MSc course. Students taking this option are required to use computer models; in this case, of construction processes, to investigate the effects on the overall costs, timing and execution of particular projects, of alternative construction methods, sequencing of and delay in construction operations, and of other management decisions.

All the students who actually registered for the MSc course were in fact very recent graduates, in the main, of British universities and polytechnics and it therefore came as a surprise to the department that many claimed to have little or no experience of FORTRAN computation. It was found necessary to put on an informal course in FORTRAN programming, which was in fact a heavily condensed version of that described above, to give students sufficient background to use the department's own computer programs in a design environment. Whilst most students did eventually manage to use the programs satisfactorily, a small number remained apprehensive of the use of computers, even when there was no requirement to write software. This fear apparently emanated from unfortunate experiences in their early computer education, and emphasizes the importance of adopting the correct teaching and approach to computation at the initial stage.

In future years more thought will be given to the provision of an optional formal course in elementary FORTRAN computation as part of the MSc course, though it is surprising that this should be necessary.

Many projects offered by the department for study for the dissertation part of the MSc course, and indeed offered for study leading to the degree of MSc (by research) and PhD are in the field of computer-aided design. Of necessity dissertation projects are of short duration and may make use of microcomputer systems. Such projects frequently involve writing programs for the interactive design of structural elements and connections. Similar projects are also set to undergraduate students as mentioned above but because of the greater time, and hopefully experience available, considerably more is expected of MSc students in terms of completeness of the programme produced and its suitability for direct use in a design environment. Such projects have a double benefit. They train the student in writing software for computer-aided design and once written the software is available for use by other students as part of their design studies.

For study at MSc (by research) and PhD levels, projects are inevitably longer and require an original contribution. They tend to form parts of continuous programmes of research and have to date made use of the larger computers available to the department The major areas in which work is currently in hand are:

  (i)   computer-aided structural design of buildings
  (ii) analysis and design of cable bridges
  (iii) analysis and design of cable and fabric structures
  (iv) analysis and design of telecommunications towers and masts
  (v)   development of input and data generation facilities for structures
  (vi) development of equation storage and solution systems
  (vii) finite element analysis.

Courses for professional engineers

The department's long-established practice of giving short courses for professional engineers in the application of computers to solve structural problems has in recent years been extended by the introduction of a part-time evening course and associated three-day course in BASIC microcomputer programming. The education and background of many senior practising structural engineers have not included any experience in computation and these courses have therefore been aimed at such engineers who wish to make use of the current generation of microcomputers in a design or other commercial environment.

These are the only courses taught by the department in BASIC and as such are a departure from its normal practice . The language was chosen for the courses because it is the most important and widespread language for microcomputers and it is the easiest for those unfamiliar with programming to learn. (*Note*: Undergraduate students in the department are expected to learn BASIC for themselves following their course in FORTRAN.)

The first course comprises eight weekly evening sessions of three hours. One hour is devoted to a lecture and the remaining time is devoted to 'hands-on' experience including writing and running programs. The course does not make use of actual microcomputers but rather uses a cluster of terminals on the PRIME computer Each participant is told only how to access the BASIC compiler and hence each terminal effectively is used as a stand-alone micro. An advantage of using this system is that it is possible to arrange for the user to obtain a hard copy of *all* that has appeared on the terminal screen during the work session. This is of considerable value as it may be studied during the week before the next session. Numbers on the course are limited so that each individual has his/her own terminal and to ensure adequate tutorial supervision.

The lectures deal with all aspects of standard BASIC including the use of disc files but excluding the use of graphics. In addition to teaching the BASIC language an approach to computer programming in general is also taught. Each week short exercises are set for solution (each involving about ten lines of coding) which make use of the commands and concepts introduced during the preceding lecture. In addition, participants are invited to develop continuously a single program making use of the BASIC commands as they are introduced. The program set for development has been the determination of shear forces and bending moments throughout a simply supported beam for different forms of loading. Program development starts with a single point load and finishes with a beam loaded by any combination of point and uniformly distributed loads.

Having now run the course three times for an average of 15 participants of the background described earlier, the following points have become clear:

(i)   The extended period over which the course is run is essential as it does take time for the material to be absorbed.

(ii) It takes at least the first work session for participants to become familiar with the terminal keyboard. Lengthy exercises are not desirable as too much time becomes devoted to typing.

(iii) Participants are encouraged to do homework and those who do so benefit most from the course. Those who have access to a microcomputer at their place of work are at an advantage.

(iv) The on-going problem is too much for most participants. When the course was first run, a copy of a program to satisfy each week's requirements was made available, but it was found many spent too much time copying and running that program rather than writing their own. The practice was discontinued but the listings are made available at the end of the course.

(v)   A full record of a user session at a terminal is very useful to most participants.

(vi) Within the eight weeks of the course it has not been possible to introduce graphics adequately. This is, however, an important aspect of microcomputer programming for structural engineers and it is proposed to extend the course in some way to deal with this.

The second course is a three-day full-time course and follows on from the first. Its purpose is to allow participants to develop and write a single computer program to a stage at which it will be usable in a commercial environment. A knowledge of BASIC is assumed and as such there are no formal lectures on this subject.

Unlike the first course the PRIME computer is not used, but rather a number of different microcomputer systems are made available and participants can choose which system to use for their work. In the past Apple, Commodore and Hewlett Packard systems have been available, but it is hoped to extend this range in future.

Participants are invited to bring their own problems from their place of work, but for those who do not a number of problems are available for solution, including:

– analysis of continuous beams
– analysis of retaining walls
– design of concrete sections
– design of masonry elements

Listings of programs to solve these problems are all made available at the end of the course.

Whilst the majority of participants' time is devoted to actually developing and writing programs with assistance from support staff, there is time devoted to a lecture presentation by each of the manufacturers of the microcomputer systems present.

Experience of running the course has shown that:

(i)   The majority of those attending have undertaken the first course.

(ii) Participants look on the course as an opportunity to spend a concentrated period of time on computer programming away from the pressures of a commercial environment.

(iii) A number of participants view the course as an opportunity to try out different microcomputer systems over an extended period of time without any pressure from manufacturers' sales staff. This is particularly useful for individuals from organizations contemplating buying a microcomputer system.

(iv) Few participants actually complete a program but most of those who set out to do so do produce something useful.

Practical applications and research

This paper would not be complete without a brief mention of some of the applications of the computer programs developed in the department; the value of the education and training provided by both the development and use of the programs is immense.

The first applications came through the structural computing service which dealt predominantly with structural framework analysis but also included carrying out the deck erection calculations for the Forth Suspension Bridge. The efficiency of the frame analysis program which was written for the Ferranti Mercury computer was considerably enhanced by the flexible and comprehensive 'alphanumeric' input system; it allowed compact data which minimized errors and simplified checking. The general format of the data provided much of the basis of the Report by the Institution of Structural Engineers, 'Standardisation of Input Data for Structural Computer Programs'. Data generation and manipulation facilities have since been considerably extended but the philosophy remains the same.

The speed and efficiency of arithmetical operations by computers continues to improve and the cost to reduce, but since structural engineers require the solutions of larger and larger analyses, there is still a premium on economical equation solution. A great deal of work has been carried out in this area and is still continuing so that full advantage can be taken of vectorization facilities of modern larger main frame computers.

In dealing with specific structural applications as distinct from matters of computational efficiency one has to be selective but foremost among applications of the department's work must be those concerned with suspension bridges. Most applications have been in association with the contractors, those responsible for bridge building; they have included carrying out the deck erection calculations for the Severn Bridge, Bosphorus Bridge and the Humber Bridge. There has also been a continuous investigation of the dynamic behaviour of suspension bridges, culminating in the determination of the flutter speed of the Humber Bridge in its construction phase. This work has recently been extended to dealing with both the static and dynamic behaviour of cable-stayed bridges in the complete condition, and also in the construction stage. The programs are being used for the first time at the design stage in connection with the proposed Westerschelde Suspension Bridge in Holland. The work is being undertaken by Dutch government engineers who visit the department at regular intervals.

Suspension and cable-stayed bridges comprise one class of non-linear structures; non-linearities of other natures, both geometrical and material have been included in powerful analysis programs developed in the department. In addition to dealing with a variety of non-linearities in two and three dimensional structures, the programs also provide facilities for automatically controlling the changes in the load parameter, a feature essential in ensuring convergence particularly when marked discontinuities in the non-linear response occur.

Building on experience obtained with cable structures computer methods are being developed to determine the cutting patterns for the material of fabric roof structures. A preliminary design has been produced for one such fabric roof structure using the software developed to date.

All the applications which have been mentioned up to now have been analytical in nature. Some 12 years ago work began on the application of computers to the design of construction details and the preparation of working drawings. Steel construction was tackled first, followed by some work on beam-column connections in reinforced concrete and a project concerned with the integrated design of structure and services in buildings is currently being carried out. It is believed that the economic consequences of this work which is heavily dependent on computer graphics are of major importance.

Conclusions

It will be apparent from the paper that the department has from the beginnings of structural computing made significant contributions to its engineering applications and its development to the present situation when it forms an integral part of undergraduate and post-graduate education and a basis for growth in special course activity. As mentioned in the introduction, much of the work at UMIST has concerned structural analysis which is an important and essential component of structural design, particularly in the cases of large special structures such as suspension bridges.

The contribution of structural analysis has been great, and will continue, but it seems likely that the economic significance to the design process of the interactive use of computers in the design of connections and the preparation of construction details and drawings may be greater. The economic significance of the widespread use of microcomputers is just starting to be felt, but it will clearly grow rapidly in the next few years.

The educational needs of structural engineers have changed significantly and although as mentioned earlier the assumption of adequate experience of computation of graduates was not borne out, the consequences of computation education in the schools will have a marked influence on our practice in the future. The major technological change in the immediate future will result from the introduction and widespread use of computer networks both local and national. The effects will be far-reaching not only in education but possibly more importantly in structural engineering practice.

D M Brotton, UMIST, Department of Civil and Structural Engineering, PO Box 88, Manchester M60 1QD

M A Millar, UMIST, Department of Civil and Structural Engineering, PO Box 88, Manchester M60 1QD

A J Bell, UMIST, Department of Civil and Structural Engineering, PO Box 88, Manchester M60 1QD

Discussion

H G Allen, Southampton University:

I was interested to hear about the part-time evening course for engineers which deals with microcomputers and I was surprised that Professor Brotton was still able to find customers for this course. I would be interested to know what sort of people are coming and where he finds them? My second question is, in the undergraduate course, excellent though the facilities are that Professor Brotton described, no mention was made of the interactive aspect of design work. I would be interested to know how we might introduce this into a standard civil or structural engineering course, and I would like to know if Professor Brotton has any thoughts on the interactive aspect of design work in that context?

D M Brotton, UMIST:

Our experience is that the majority of the people on these evening courses are from local authority engineering offices. There seems to be a flood of interest in authorities in getting involved in microcomputers; local authorities are notoriously short of money and microcomputers are the scale at which they have got to enter the field of computing. Some are consultants, generally the smaller consultants who have similar financial problems. As far as numbers are concerned, we ran the courses last year for the first time and we had to double up on the evening course because the response was so great. This year the response wasn't quite so great but we didn't want too many because we

PART 6: TRAINING COURSE EXPERIENCE

feel that it is important to provide a very strong tutorial component to the course. This is essential to make it a success.

I am not sure that we are yet very good at the interactive aspect of design work. We are still learning, as many people still are. It depends largely on the staff who are responsible for taking the individual courses, and we have had applications in particular subjects, in structural design for instance, we have students tackling the design of plate girders using a computer interactively. I am personally very keen on interactive computation because it gives the students very rapid and effective design experience. If they have to do all the calculations by hand they cannot consider as many design alternatives as they can by using a computer program. We have not done as much work of this nature as I would like.

F Weeks, Newcastle Polytechnic:

It was interesting to see Professor Brotton's data on the use of the various different languages and he certainly made a very convincing case for using FORTRAN. Am I right in interpreting the data on Pascal as indicating that there is a lot of teaching done in that particular language but that it doesn't get used a great deal subsequently?

D M Brotton:

I think that would be true. There is some continued use of Pascal in that a student who learns a particular language in the early part of the course has a natural tendency to go on and use that language subsequently. But FORTRAN is still the international course language for engineers and since the major use of computing is by engineers, and scientists to some extent, I think we are still going to see FORTRAN as the most popular and common language for some time yet. I wouldn't suggest FORTRAN is necessarily the best computer language. I am sure that there are those in the room who would extol the virtues of other languages, but one can't go beyond the large availability of engineering software in FORTRAN, not yet anyway.

35. CAD—the first year

R Metcalfe and M E Preston

Abstract: The first part of this paper details the philosophy employed in the setting up of a six-station graphics studio. It shows the way this was achieved by building on existing equipment and expertise where available and the way in which the studio and CAD can be introduced into the existing course structure.

The second part of the paper deals more with the actual setting up and day-to-day operation of the graphics studio. It details the problems with both hard and software and the various solutions to these problems where known. The student reaction is also included together with that of the staff. The paper concludes with a list of points to consider when purchasing a CAD system for operation in an education environment.

1. The philosophy

Introduction

Some years of experience developing graphics using a Verian 620 in conjunction with a Tektronix 611 screen had lead us into computer-aided drafting using two-dimensional graphics, whilst three-dimensional work together with other interactive programs was regarded as computer-aided design. Final year students have been involved in these activities for some time.

Before setting up a facility that would be used by a broader spectrum of students it was essential to understand the real needs of the courses that it was intended to support.

Engineering drawing in the early stages of a course is regarded as a communication ability. Drafting skills, particularly the ability to sketch, although useful to the budding engineer, could be regarded as not necessarily essential. First year drawing time would continue to concentrate on the acquisition of skills and the exercising of judgement, both of which require time and practice. The area to concentrate on was therefore the impartation of knowledge and for the use of the facility as a more efficient way of carrying out existing tasks in subsequent years. It was decided therefore to introduce CAD in the second year of a four-year thin sandwich degree course.

The role of a CAD studio had to be two-fold.

1. To give all students the experience of using a 2D system as an extention to basic engineering drawing.

2. To give a facility to more advanced students to use a 2D system in creating detail drawings for design and make exercises, also to have a 3D capability for more serious design work aimed towards the setting up of a data base to give an appreciation of computer-aided engineering.

The needs of the user

A fundamental 'friendliness' was seen as essential in order to maximize the experience gained by students, and allow full utilization of the facility. Not only should the software be relatively easy to operate and speedy in use, the facility itself should be available for a long working day with open access to its use for much of the time. The ability to take a quick copy was regarded as occasionally useful, but it should be possible to obtain a copy from a plotter fairly readily.

The criteria for selection of equipment

In addition to the above user needs a facility must be sufficiently large to support worthwhile groups in a teaching situation. The equipment was to be housed in a studio type of accommodation, which made its use more easily planned and controllable at the same time giving a certain status and therefore credibility. It was therefore to have a minimum of six terminals to be used in conjunction with six free-standing drawing boards (Figure 1). Groups of 12 students could then be accommodated, the drawing boards being used for preparatory work in order that terminals could be used as efficiently as possible.

Figure 1 *The CAD studio (see page 9)*

Hardware

The terminals used are Tektronix 4010 supported by an enhanced PDP11/40 computer (Figure 2). There is a Tektronix copier for quick screen size prints and a Calcomp plotter for the production of the final drawings. A digitizer is also available and is permanently linked to a raster scan type terminal.

Figure 2 *The Trent Polytechnic PDP 11/40 central graphics facility (see page 9)*

Software

The software to support computer-aided drafting had to be reasonably comprehensive in order to be able to produce fully dimensioned drawings to an acceptable standard. Anything less than this results in partial drawings that give a poor or even negative experience of using a drafting system. It should not be necessary to work inside this software nor is it desirable or even possible with the level of student using the system for producing detail drawings.

For these reasons PAFEC DOGS was chosen to support the drafting activity. The level of software used for design activities is dependent upon particular areas of specialization. GINO-F is used on the PDP11/40 based system although much work is still carried out on the in house developed Varian 620 system.

2. The reality

The second part of this paper gives a candid description of the reality of operating a CAD system in an academic environment. In September 1981 the final installation and commissioning of the graphics facility at Trent Polytechnic was completed.

In the consideration of running and use of this facility it may conveniently be divided into four main areas which will be dealt with in turn starting with the graphic studio.

The graphics studio

In the case of Trent Polytechnic the location of the room to be used as a graphics studio was initially dictated by the maximum allowable length of line between the computer and the terminals. The room finally made available was one with a high ceiling and a glass wall. In addition it was situated some distance from the mechanical engineering department up four flights of stairs. Whilst this was by no means ideal the equipment was carefully laid out so as to make the maximum use of the available space and facilities. A diagram of the studio is shown in Figure 1. The inherently high level of illumination is reduced by the use of suitable blinds. The rational behind the layout has already been discussed at the beginning of the paper but some practical aspects are detailed here.

The idea that a group of 12 students could effectively use the studio has not worked out in practice, the students are reluctant to vacate the terminals once they have gained possession and are quite happy to work in pairs. Hence the drawing boards do not get used. In fact they have mainly been used to shade the terminals due to the non-delivery of the blinds.

It should be noted that the layout of a studio is of prime importance and should not be allowed to just happen by default.

The computer graphics system (hardware)

The choice of the equipment has already been discussed and it is at this time intended to point out some of the problems encountered during the installation and operation. The equipment may be conveniently divided into two headings: the hardware and the interfacing.

The hardware

The hardware has performed well during the first year. The problems that have been encountered could, in hindsight, have been avoided.

The computer. This machine is by today's standards old and does not have the capacity or speed to run the software which was purchased. This is the main problem and is responsible for most of the other problems encountered.

Terminals (Tektronix 4010). The terminals are of the type considered to be the industrial standard, they are of the storage tube type and have given good service. The main problem would as seems so often to be the case in this type of equipment the poor mechanical connections to the computer network.

The high definition of the screens is in some degree mitigated by the fact that the screen is of the storage tube type and hence any change in the picture of messages needs a complete redraw to allow the current state of the drawing to be seen. In the case of our system this takes a comparatively long time to achieve.

Terminals (raster-scan). The digitizer was found to be incompatible with the standard Tektronix interfaces purchased hence an additional terminal was required to allow the digitizer to be used with the system. In order to reduce the cost a Dacoll raster-scan type terminal was purchased (cost approx £2,000). This has provided a useful comparison of the two types of visual display units and has hence allowed evaluation of the differing displays. With the exception of the jagged line appearance of non-vertical lines and the appearance of circles no real disadvantages of this type of screen have been found.

Printers-Plotters. The plotter used with the graphics system also is used on the main frame system (DEC 20/60) hence the plot files have to be down loaded via magnetic tape. The tapes are collected three times a day and then plotted on the Calcomp plotter via the DEC 20. In order to allow a quicker copy a Tektronix electrostatic copier was also purchased and installed in the graphics studio. A word on the operation of this is called for. The copier can only be used from one terminal at a time and the quality of the copy is related to the settings on that terminal. Hence it was not uncommon for four or five copies to be taken before an acceptable one was obtained. The cost of the special paper is not low hence a days check on unacceptable prints indicated a restriction of the equipment to staff use only. In addition the activation of the copier before the terminal has finished its drawing routine will cause a catastrophic loss of the user's file.

Digitizer. No problems have been encountered at present but this may be due to its limited use at the moment.

The interfacing

It cannot be stressed too highly that the interfacing of equipment in the area of CAD is one of prime importance. In the case of our system the computer is not ideally suited to the operation of a graphics package, but this has been made more acute by the interfaces for the terminals making the screens run at an even lower speed than is really necessary. This has been highlighted by the increased running speed of the raster-scan terminal which emulates the Tektronix type but at a higher Baud rate. It is suggested that the interfacing of the system should be very closely reviewed and some guarantee of performance gained, in writing where possible, or better still go and see an actual system working in an environment as similar to yours as possible.

The computer operating system

As previously stated, the Trent Graphics computer is not suited to running the DOGS system. The main problem areas are noted below.

Capacity

The lack of computing capacity is one of prime importance. In real terms it means that the parts of the DOGS package have to be overlaid to allow it to run at all. This reduces the speed and effectiveness of the program.

File space

There is a limit to the number of people who can be accommodated on the system at any one time and hence the effective student numbers are limited. In addition if a user overruns his allocated disc space he cannot return to the system without consultation with the computing department – this can be a problem at times.

The non-conventional exit

If DOGS is exited by a non-conventional means (only too simple to do) the system will not allow a return without a great deal of file handling.

Remoteness of terminals

The computer is based remotely to the terminals and the nearest source of help is four flights of stairs away in the computer room. In fairness there is a telephone link but often a graphics adviser is not immediately available and hence in the case of a student, time can be lost in what is an already tight schedule.

User-friendly

The system is not sufficiently user-friendly. It does not seem that enough thought is put into making the whole systems user-friendly, particularly for the computer-naïve engineer.

The graphics package

The graphics system purchased was that of DOGS (PAFEC Ltd). It is felt that at this stage a simple explanation of the basic method of operation of the system would help to explain some of the problems and benefits encountered.

Operation of package

In order to draw say a simple straight line, a series of menu options must be selected. This is done by positioning the cursors present on the Tektronix 4010 screens over the required options on the menu. A sample print-out of the screen is shown in Figure 3.

Figure 3 *DOGS menu on VDU screen (see page 9)*

The menu is shown down the left-hand side of the screen. It should be noticed that these refer to the options detailed on Figure 4, the DOGS menu. In other words, the *LT* will refer to *L*ine *T*ype, and the following figure will stipulate the actual type of line required, ie 2 would give a dotted line. Having selected the Line Type, the form of the line must be selected – a circle going left-handed would be *LN* 4. The circle may now be drawn on the screen by entering defined points on the circle in the correct order. This may be either achieved by the use of the cursor controls or by using Typed Input (TI) where more precise dimensions are required.

The great advantage of this system is the ability of a student to build up from a simple Etch-a-Sketch mode, which may be learnt quickly, to the more complex options in a progressive and planned way.

Figure 4 *DOGS menu (see page 9)*

Purchase agreements

In the purchase of graphics packages it is often possible to negotiate special educational discounts. If you take advantages of any of these deals, care should be taken that the educational package contains certain features, these being:

Training courses

In the case of complex packages such as DOGS there is a need to be fully aware of the best way to operate the package and the only way to get this information is to have a degree of training. Care should be taken that this training is included in the package and that a large bill will not have to be paid to receive this training. The alternative is a teach-yourself approach. This can be very time-consuming, frustrating and not at all cost-effective. To be frank, it is not in the interest of the company selling the software not to give this training and hence present its products in the best light to a group of future purchasers.

Manuals

The manuals supplied with any package must be of a good standard and before a package is purchased the manuals should be studied and if possible the examples and explanations should be tried out on a system. This is particularly important in the case of an educational environment where good manuals will reduce the load on the teaching staff. This would of course be equally true in an industrial situation. With regard to the manuals it maybe important to have the right to duplicate them in large quantities on your own reprographic facilities. This keeps the cost down to a reasonable level to allow the purchase by students.

General

In general, when purchasing a graphics package be sure of the extent of the agreement in all the relevant areas and make sure that you and your computing department are of the same mind as to the equipment needed and as to its method of operation. If possible get all the requirements and agreements in writing – it saves problems afterwards.

Practical operation of the package

The package purchased has in general proved to provide a good teaching tool for computer-aided drafting. In the operation of the system the main problems would seem to emanate mainly from the use of the PDP11/40 more than from the actual package. In this respect the main problem is that the package exits to the computer-operating system at the simplest mistaken input and as noted at an earlier point in the paper it will then refuse to re-enter the DOGS package. This is due to DOGS trying to save as much as possible of the crashed file it was working on at the time of the wrong command. In theory this is a good idea, but as no file has yet recovered it would seem to warrant a long look at the use of this facility, both in theory and practice. Problems such as that noted above are of particular importance to the student learning to use the system, as some seem to have an inbuilt facility to create disasters.

The management of the graphics facility

In the setting-up of a graphics (CAD) facility it was realized that one of the main problems would be to maximize the utilization of the system. In order to achieve this aim it was considered that a flexible approach to the operation of the facility would be required. Initially the computer services tried to impose a requirement of 24 hours' notice booking system on the graphics studio. This system is acceptable if all bookings are to be long-term but is not flexible enough for the operation of this type of system.

At Trent the graphics facility is essentially a separate sub-group of the main computing department and hence a better system has been devised. At the start of each term the main block bookings are made by the staff concerned with the lecturing of CAD. In addition a certain time is set aside for the maintenance and updating of the system.

The remaining time is available for the general use of the students.

It can be seen from the plots of system usage that the utilization of the studio has increased steadily over the year (Figures 5 and 6).

One thing which has become clear from the first year's operation of the studio is that some form of direct contact is required between the computer department and the major users of the graphics system. This should prevent any problems in the development or usage of the system or facilities.

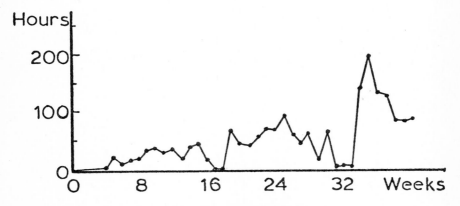

Figure 5 *Utilization of graphics studio by students*

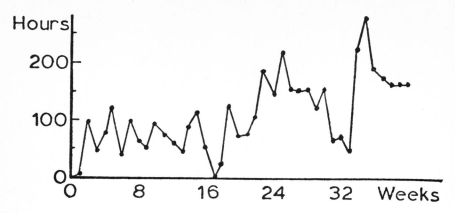

Figure 6 *Total connect hours—staff, students and operations*

Staff and student reactions

Having so far detailed the various parts of the system, both software and hardware, together with the management of the Trent system, there only remains the reactions of the staff and students to the teaching and users of CAD.

Staff reactions

The reactions of the staff of the design group at Trent to the introduction of the CAD system have been mixed. In some cases the support has been whole-hearted but it is fair to say that the degree of involvement has to some degree been dictated by the syllabus that each lecturer has to teach and its involvement in CAD.

From a personal point of view the authors have spent approximately 30 hours 'at the terminal' becoming familiar with the package and system. Additional time should be allowed for running up examples and 'on the terminal' learning – that is getting students out of undreamt of problems.

Student reactions

Now to move on to the most important area of this paper – how do the students react to the introduction of computer-aided drafting into the syllabus? As explained in the earlier part of this paper the initial exposure of the students to the CAD system is not carried out until the second year of the BSc courses. As this was in effect the first year of operation for all the groups, hence all will have been subjected to the same material, but in order for the reaction to be as accurate as possible the efforts and reactions of the second year have been given most prominence. This is because the third and fourth years will finally be built on these foundations.

In general the reactions given are those of the combined second year mechanical and production students, they are normally taught BASIC in the first year so all have some knowledge of computers but in general a wide spectrum of ability may be expected. The numbers in this years group are 86 students in all; these are broken down into seminar groups of approximately 15.

Teaching methods

The method of teaching adopted was as shown below.

(a) 2 × 1 hour lectures (all students)
(b) 2 × 1.5 hour seminars (15 per group)

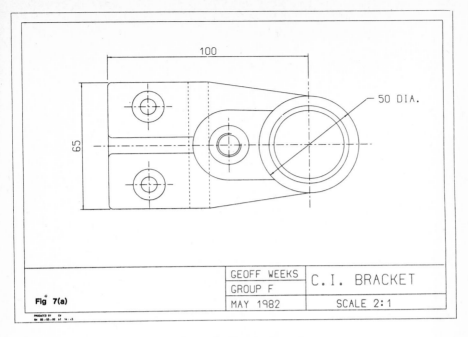

Figure 7a *Drawing example*

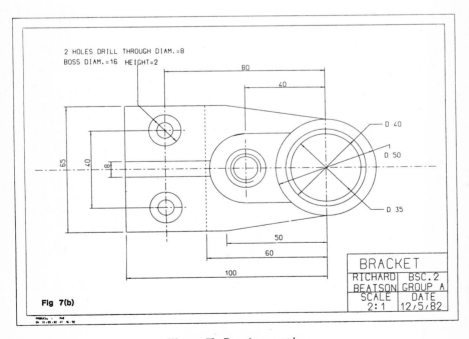

Figure 7b *Drawing example*

(c) Estimated 5-hour exercise to be carried out in their own time.

In the case of the lectures these consist of a step-by-step introduction to the graphics package and operating system working up to the steps required to draw a simple shape.

This is followed by the seminar periods which introduce the students to the terminals and initially allow them to play with the screens and system in the form of a very expensive Etch-a-Sketch. They soon want to try the other options available and are soon drawing shapes and using the cross-hatching functions etc.

Having progressed past the first hurdle, they will start to carry out the set exercise. This consists of a simple plan view of a well-known drawing example (Rhodes and Cook, 1975), see Figure 7. This allows them to practise the setting-out of a drawing and as a final exercise they have to produce an A3 size plot of their work. This is then included as a part of the second year course work.

Initial student reactions

The initial student reaction usually fall into one of the following areas.

- Not very keen, they have usually had bad experiences with computing in the earlier part of the course.
- They cannot be removed from the terminals. Computing is like a drug. (Can be a problem when other work starts to suffer.)
- Finding time to carry out the exercises. (Some students are always short of time.)

Continuing reactions

In general the degree of interest shown by the students in the subject was good. The main problems encountered were those involving the poor standard of manuals, both those supplied by the software company and initially those provided by the computer department. In addition the many unexplained exits from the DOGS package after which the system would simply hang until a large amount of file deletion was done, or in some cases the system would return to normal after a long wait, but in most cases the urge to press other keys in the meantime was irresistible.

An additional irritation was that when a group of students first entered the CAD studio and tried to log on to the system, it would seem that the computer could not cope with all the files being created at once and would simple hang for anything up to 15 minutes. This being the students' first acquaintance with the system it did not bode very well for the future.

Final reactions

The final reactions of the students were canvassed on their completion of the CAD exercise: this was done by means of a questionnaire. A brief resumé of the results obtained from this survey are given below.

Lectures and seminars

The first set of questions dealt with the lecture and seminars. The general reaction to these was that the lectures were acceptable but could have been improved by the inclusion of more material on the actual operating system; also the inclusion of a more detailed example would help. In addition, it was suggested that it would be a good idea to split the lecture periods so as to give a short time of hands-on experience between each formal lecture. This may be a good idea, but in the case of our students would cause very difficult timetabling problems.

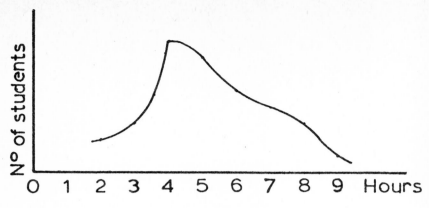

Figure 8 *Student hours required to complete first example*

Time to complete example

The time taken for each student to complete the example was requested, a plot of the times are shown in Figure 8. It should be noted that the difference in times taken graphically indicates the different learning curves for this type of system and so account of this should be taken when planning a system.

Terminals, package and system

In this area the student was asked to comment on the terminals, package and operating system. The results from these areas are listed below. (*Note:* the comment in brackets gives an indication of the proportion of the group)

- System crashed for no reason (many)
- Fillet option did not work (many)
- Redraw took too long (all)
- Problems with circle drawing (some)
- Better manuals (many)
- Dimensioning did not work (some)
- Lost files (many)
- Finger marks on screens (some)
- Lost part of drawing (few)

Suggested ways to improve the system

The students were asked for ways in which they thought the system as a whole could be improved. Their suggestions are listed below.

- Faster system (all)
- Terminals in mech/prod department (some)
- Better manuals (many)
- Staff nearer to the studio (many)
- Good novice system (two)

Staff conclusions

The conclusion drawn by the staff on the first year have as may be concluded from the remainder of the paper been mixed, but in general the main problems have come in the areas outlined in the replies to the student questionnaires. It is hoped in the future to develop the system to eradicate most of the problems. This could be achieved by the purchase of a more suitable computer to run the software at greater speed and to allow the package to be utilized as it was intended. In addition, a better standard of manual is being compiled more as a beginner's guide to both the system and the package. This again should remove at least some of the problem areas.

The second year will also be progressing on to the next year of CAD which will no doubt be another learning year for both staff and students.

Conclusions

In conclusion it is felt that a number of very relevant areas have been defined throughout the first year of operation of the CAD system at Trent Polytechnic. The most important have been listed below.

1. The initial reaction of the students to the system is all-important and may give them set views on the use of computers in the design task. In other words, it is important not to have the initial file creation routine hang just because all the terminals just happen to have tried to log on at the same time.
2. The system should be friendly and should be as foolproof as possible. It should not crash for the simplest of errors, and even if it does occasionally crash the situation should be easily redeemable.
3. The management of the system should also have a friendly face and should above all be flexible. This is particularly the case in the area of facility bookings and the method of logging the user into the system.
4. The facilities available within the package must be constant as it is difficult to keep up with constant changes and the examples always use the deleted or changed function.
5. When purchasing a package you must be sure that the correct type of instruction is available and is included in the cost of the package. If possible get it written into the contract.
6. When considering the purchase of a package make sure that the documentation is good and is up-to-date for your version of the system. It is also a good idea to come to some agreement with the software vendor as to the rights and costs likely to be incurred if you want to purchase or copy their manuals for sale to the students. Cost should be not more than £5 per copy.
7. Considerable attention should be given to the compatibility of the various makes of equipment purchased and make sure that some guarantee is given as to the performance of the equipment. There may be some benefit in purchasing all the equipment from one source: at least you have only one company to argue with.
8. The response time of the system should be as high as possible, as this will dictate the work rate of the students. In particular, the redraw time should be noted, as this may be the limiting factor in the system.
9. The members of the lecturing staff who operate the system should have enough knowledge of that system to allow them to carry out simple trouble-shooting exercises.
10. Know the shortcomings of your system and be able to explain how these are overcome in industry.
11. Ensure that all groups that are to use the graphics system are represented on a user group. It is important that good relations are formed between the users and systems operators. It makes the whole system much more manageable.
12. The selection of the type of terminal to be used needs a good deal of thought, should it be the raster-scan or storage tube type. It may be dictated by price but do try and see the system selected working on each type before you make a choice.
13. The layout of the graphics studio is of prime importance and should be planned with the aim of presenting the most credible image possible to the students and staff alike.
14. Possibly the most important of all, the overall system should present a 'user-friendly' face to its users.

The future at Trent

To conclude the paper, a few comments about the future of CAD at Trent.

Courses

In addition to the BSc courses detailed above, CAD will be introduced in some of the other courses run in the Mechanical and Production Engineering Departments. The involvement will vary from a simple introduction in the case of TEC students to a higher and more comprehensive involvement for those involved in our MSc in computer-aided engineering.

Equipment

It is hoped in the near future to be able to improve the computer facilities for CAD at Trent. The increasing use and high utilization of the system will, it is hoped, justify the increase in computing facilities. An overall development plan for the Polytechnic Computing facility is at present being considered and it is hoped that this will include an improved graphics computer.

The siting of the terminals remote to the Mechanical/Production departments does cause logistical problems and it is foreseen that there will be a need to provide a local studio for departmental use.

Staff

In the future, various members of staff will become more involved in the use of CAD. The use of pipework design for our Plant Degree course is under consideration as an addition to the physical modelling at present carried out.

Staff are at present registered to carry out research into the use of desktop computers in various areas of CADCAM.

Acknowledgement

PAFEC Ltd, Strelley Hall, Strelley, Nottingham

References

PAFEC Ltd. Drawing office graphics system (DOGS)
Rhodes, R S and Cook, L B (1976) Pitman. London
R Metcalfe, Trent Polytechnic, Burton Street, Nottingham NG1 4BU
M E Preston, Trent Polytechnic, Burton Street, Nottingham NG1 4BU

Discussion

Mr Youssefifar, Ward and Goldstone:

Have you tried larger Tektronix systems and, if you have, would they have been more descriptive? Have you tried packages using a raster-scan and what happens when you use it?

R Metcalfe:

We haven't tried the larger Tektronix. I think the main problem is cost again. We would like to try this system and it would be better if we had a larger screen, but the ones we have got now have done remarkably good service. We have a raster-scan terminal. We have six Tektronix ones, and we have one Dacoll terminal. That runs the digitizer. Certainly there doesn't seem to be any major disadvantage with the Dacoll terminal.

R Metcalfe:

A quick comment here. With the Dacoll we have it operating at a higher Baud rate, so one can in fact sit behind that terminal and work at a much faster speed. This was something we didn't anticipate, it just happened to turn out that way.

Unidentified contributor:

When you use a raster-scan, won't that eliminate the need to re-draw? I noticed that you pointed out that the re-draw takes a long time.

R Metcalfe:

Yes, the re-draw is still quicker because of the higher data transfer rate but I think it's still the same system. It still emulates a Tektronix screen, in other words the image is still consistent but what PAFEC are working on now apparently is a complete re-fresh system. I think maybe we would look at this if and when it comes out. A lot of our future efforts in this area will be dependent on what computer we get. This is one of our main problems at the moment and is very often the problem in education. Until you get something running and working and prove a need for it you won't get anything better. You have got to start at some sort of level.

I C Wright, Loughborough University:

How much lecture time do you give prior to letting the students loose on this test piece – the plan view that you ask them to produce?

R Metcalfe:

At the moment we have done two one-hour lectures. That includes a general introduction to CAD and then explains how things work. It's about two hours, depending on how the timetabling works. They have been on for an hour after the end of the first hour lecture and they seem to cope remarkably well really.

I C Wright:

And would you have expected them to have read the reduced manual before they get on to the system on top of that lecture?

M E Preston:

In the case of this year, they certainly didn't because the manuals have only just been produced so there were ringed copies of the manual in the graphics room and they could see those. What I in fact did was to go through most of the very basic commands and the possibility of covering menus up which you could do with a system which is bit 100. You can't in fact do that with this system. Also, we have since found out we don't think we can fit that sort of menu to this type of terminal, at least the way we have got it at the moment.

36. Six years of teaching computer-aided design at the University of Stellenbosch

H T Harris.

Abstract: This paper examines the course entitled 'Computer-Aided Design' given to fourth year mechanical engineering students.

For reasons which are explained, a broader view than normal is taken of what CAD is, and the course consists of the following:

(a) Optimization applied to design
(b) Interactive computing for design
(c) Introduction to computer graphics
(d) Computer-aided NC programming

A description of the computer facilities available is given, and some programming projects which have been used are described.

Introduction

A course entitled 'Computer-Aided Design' has been given to fourth year students in mechanical engineering at the University of Stellenbosch for six years. Initially this was based on *An Introduction to Computer Aided Design* (Mischke, 1968), but since 1968 the meaning of 'CAD' has changed considerably and this book is now insufficient. Nevertheless it is still a good text as an introduction to using a big computer for certain engineering applications – Mischke deals mainly with optimization in mechanical design, with some linkage design. The age of the book is reflected in the fact that only batch-mode computing is used, and graphics is not even mentioned. Thus the present understanding of what CAD is, is not dealt with at all.

Although we are now using interactive graphics, due to restricted interactive computing facilities we have retained some of the batch-mode work, using the Mischke approach. This, and the fact that we include a certain amount of programming for computer-aided numerically controlled machining, means that we take a broader view of CAD than is normal. It is gratifying to note the emergence of the term 'computer-aided engineering', which seems to cover what we are teaching!

Stellenbosch computing facilities

We have, of course, access to the University's central main frame, via an efficient self-help system of card readers and printers. The advantage of this facility is that it is geared to handle large numbers of students efficiently. Our students are taught FORTRAN 77 in their second year. BASIC is also available but at a fairly low level and not interactively.

The hardware in the Department of Mechanical Engineering is much better suited to CADCAM. At present the system is based on three Varian minicomputers, but in the near future we will be plugging in to a VAX in the Department of Civil Engineering, using our present peripheral equipment.

The second Varian 77 with its three terminals has not yet been made available, but has been used exclusively for laboratory work through a CAMAC system which can also be used with either of the other two Varians.

The Tektronix 4054 was donated to the University by industry, mainly for CAM development work, and is not used by students except for one or two final year projects. This looks as though we are very well equipped, which we are – until one wants to teach CAD to an undergraduate class! We will examine this problem later.

We have considered going to the low-cost minicomputers which are now available, since an Apple (or similar) plus floppy disc unit, printer and TV screen, costs less than a single Tektronix 4006 terminal and has a greater memory than the Varian, but decided to stay with the expensive equipment because of its high quality graphics. The graphics resolution of the low-cost minis is unsuitable for CAD, in our opinion.

Table 1 *Computer facilities, Department of Mechanical Engineering*

Processor	*Terminal*	*Other*
Varian 620f	Tektronix 4014	Printer Paper tape reader/punch Magnetic tape Disc Calcomp 32″ drum plotter A1 sized digitizer
Varian 77	3 Tektronix 4006	Printer Paper tape reader/punch Tektronix flat bed plotter
Varian 77	3 Tektronix 4006	
Tektronix 4054	Tektronix 4054	Printer Paper tape reader/punch Tektronix flat bed plotter

The Stellenbosch course

This course is given during the second semester between July and November. There are 14 weeks of classes and three of exams. The course has five hours per week contact time, officially designated as two 'practicals' (laboratory sessions in other subjects). Students would be expected to work for about three hours a week on their own in addition to this.

From the beginning this has been a 'hands-on' course and we usually give four projects. A certain amount of lecturing is also necessary but most learning takes place through the students writing and de-bugging their own programs.

For the first five years we had 20-25 students, then 35, and expect 40 when we run the course this year.

The course consists of the following elements:

Based on Mischke (Mischke, 1968)

Only a superficial knowledge of the optimization mathematics is required, the emphasis being on the application. Batch-mode computing in FORTRAN is used, the optimization being carried out using a suite of sub-routines taken from Mischke's book.

We have found the experience of using a package written by someone else very valuable to our students. They learn that each person will structure a program in a different way and that it is always very difficult to understand someone else's program!

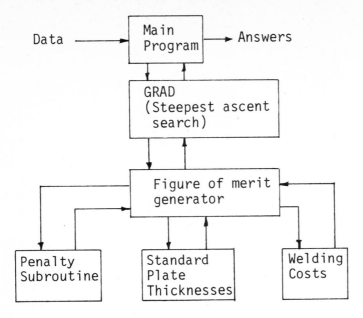

Figure 1 *Typical optimization program structure*

It also teaches the concept of modularity, of linking up a number of sub-routines each of which could be of more general use than their particular problem at the time. We are able to supplement Mischke's optimization sub-routines with design information sub-routines, like electric motor characteristics and standard steel plate thicknesses.

All Mischke's optimization sub-routines work on the basis that there is a main program, which controls the input and output and defines the problem for the user, and calls up an optimization sub-routine. This is totally independent of the problem, so a further sub-routine has to be provided which calculates the 'figure of merit' for the particular problem. It may be necessary to call on penalty functions, standard design sub-routines or whatever. An example is shown in Figure 1.

It is usually our students' first experience of writing a program for general use, with proper documentation. Mischke emphasizes that any CAD program must be suitable for use many times by many different people, and one must make sure they are able to understand it and evaluate its suitability for their particular application, and of course be able to use it. This implies many 'comment' cards, a well laid out printout of the input data as well as the answers, and especially a clear indication of the assumptions made to define 'optimality'.

A spin-off for our normal design courses is that students are forced to quantify optimality – far too often one can get away with a subjective evaluation of what is the 'best design', but the computer cannot. In deciding to make, say, minimum material using the optimality criterion one's eyes tend to be opened to the influence of other factors. Then students realize that in real life design a trade-off is usually required, that a simple optimality criterion such as 'light', 'cheap', 'small', 'strong' is seldom satisfactory.

Another thing they learn is that the computer must work systematically, and that systematic design procedure is possible. All the computer is doing is generating alternatives and evaluating them, by putting a number to optimality.

Interactive computing for design

This means writing a program for the design of some machine element, a spring for example, in as general a way as possible. The large number of different ways of specifying a spring make an interactive program invaluable. The language used is BASIC, on the Varian 71 minicomputer with Tektronix 4006 terminals.

We have discovered that mechanical design programs use plenty of computer memory – the spring design took 2K to 5K words depending on the level of sophistication, and sometimes the Varian 71 could not cope with all three terminals at once. Having the second Varian 71 has meant that all the CAMAC calls could be deleted from the first one, making more memory available for programming.

The small available memory limits the number of terminals which can be used to about four, and this has caused a critical bottleneck situation at times. This is a very strong reason for retaining the batch-mode optimization work in the course – the class can then be divided into two groups, only one of which works on the Varian system at a time. In the future if we can have more terminals available we would like to make more of this part of the course, as we believe that an engineer who is a user of CAD must have a feel for how his system works – and this he learns best by doing his own programming. In particular, it would be valuable for a CAD user to be able to add to his system things like proprietary company design methods or calculations which he performs regularly.

Introduction to computer graphics

Although it is recognized that the practising designer will probably be using a graphics package and thus will not need to develop his own software, we believe that it is

Figure 2 *Simulation of NC milling machine cutter path-square to round transition piece*

essential for the engineer to have a basic understanding of 'what goes on' in the software. Time limits how much one can do, however, as one does not want the course to simply be an extended computer programming course. The Tektronix 4054 computer with its very high-powered refresh graphics is ideal for demonstrations in the classroom, due to its portability, large size, and repertoire.

We try to link the computer graphics to the interactive design part – for example, if the design project is a shaft design program, then the students will be asked to draw the final shaft and plot out its deflected shape.

NC programming

The emphasis here is on the use of the computer for producing the NC program, a technique used for parts where a cutting path is repeated a number of times with varied parameters. A good example is a propeller blade – we have used NC very successfully to machine experimental propeller blades where basic profile is kept constant but chord, thickness, angle of attack and camber can all be varied as the cutter moves from the root to the tip. The computer is, of course, used for the propeller design so that one can produce the NC tape directly with the design. For undergraduates, simpler problems are given, for example a square-to-round transition piece shown in Figure 2. The program is written in BASIC and has as its output the NC tape. The program included one cycle of the cutter path with variables defining the co-ordinates, and it goes around a loop repeating this cycle, calculating the actual co-ordinates for the NC tape. In Figure 2 the square-to-round is shown with ten cuts, but it is no more effort (only more paper!) to have 50 cuts. All our NC work is for the department's three-axis milling machine. One should mention that they come to this course with a knowledge of NC programming, learnt in a course in manufacturing processes.

Use is made of a computer simulation of the milling machine to check programs by reading them in and plotting the cutter path on the screen, which is how Figure 2 was produced.

The simulation program reads the NC tape line by line, and draws the movement on the screen before reading the next line. This has saved countless hours compared with checking by running the tapes on the machine itself, because changes can be made so much more quickly. It also eliminates the risk of the machine destroying itself due to a wrong instruction which drives the cutter into the bed, and eliminates the need for workshop personnel to help de-bug the NC tape.

We cheat our undergraduates slightly, in that the NC tape they produce cannot be used on the milling machine. This is because it uses the EIA code whereas the computer produces according to the ASCII code. On our Varian 620f we have a translation facility so that we can produce usable NC tapes.

It is recognized that NC is not design, but we believe it is valuable to give our students some experience of NC and of ways that the computer can be used to speed up programming and program checking, and also to show the trend towards a direct link between design and manufacturing without the need for drawings.

Lectures on CAD

A little time is spent lecturing on what the term CAD (and CAM) means today – what is available in hardware and software, what can and is being done with it, what the market looks like, and what the future trends might be. Without having a system to show, though, there is very little one can really 'teach' on the subject.

Examples of programming projects

These are some of the projects which have been successful:

Optimization

(a) Tin can proportions, with material allowances for manufacturing. A good project to show the limitations of the method! (Because one can question the choice of minimum metal – what about the cost of manufacture, the convenience of packaging, human factors, if it is a beverage container which cannot have too large a diameter, etc?) This is a simple project with only one variable.

(b) Pressure vessel – a variation on (a) where it is possible to set up cost equations for the manufacture, for example weld cost per metre as a function of plate thickness. It can have one variable, when a certain volume has to be stored at a given pressure, or two if a certain mass has to be stored and the pressure may be varied.

(c) 1 beam – A steel 1 beam made up of three plates, thus having six dimensions, and constrained to conform to the Standard Building Regulations. The beam can be designed for minimum cost or minimum mass. Some of the constraints can be used to eliminate variable – the shear stress in the web for example – but others like buckling can only be tested for, after the dimensions have been chosen by the search method. This means many combinations of dimensions are unacceptable and a 'penalty' sub-routine is used to assign these beams a negative merit. Here one has to be very careful with the choice of optimization method. We basically use two, proposed by Mischke (Mischke, 1968), a gradient search and a grid search. To vary this problem, cast iron 1 beams or T bars can be designed.

(d) Acceleration of a metro train – For a given metro unit, with the electric motors already chosen, optimize the gear ratio to give a minimum time between stations. (Larger gear ratio gives greater acceleration but lower top speed.) We have found this a very enjoyable problem – it is realistic and sufficiently complex to be challenging, and appeals to the mechanically minded! It also gives the students an opportunity to use their knowledge of numerical methods to solve a differential equation with variable coefficients. For example:

$$\frac{dV}{dt} = \frac{F}{m}$$

F depends on the motor characteristics, air resistance and rolling resistance, all of which are speed-dependent.

(e) We have used, but now tend to avoid, various problems posed by Mischke which seem rather unrealistic. (Optimizing the ratios of a train of nine gears to give the minimum overall moment of inertia, and designing a cam of minimum size to generate a particular function, constrained by a maximum pressure angle, for example.) Another poor problem was optimizing an aircraft's descent – by the time it was simplified for mechanical engineering students to understand it had become unrealistic.

Interactive design

(a) Shaft design – a generalized program where the user can specify the number and position of supports and loads, and the program draws shear force, bending moment and deflection on the screen, and finds the diameters necessary. One of the advantages of this problem is that it can be posed simply, and some new requirement added each week, overcoming to a certain extent the problem of getting students to work a little every week on a project instead of leaving it to the last minute. This 'night-before panic' is particularly a problem when computer terminals are limited.

(b) Helical spring – perhaps the best machine component for this purpose, although it doesn't lend itself to computer graphics. There are so many ways a spring design problem can be posed and the program should foresee them all. Second, one cannot calculate directly the necessary dimensions, so the computer – which can exercise judgement – must work as a team. Until you try it, you don't realize how difficult it is to write a program which can think of all the alternatives, give guidance with decision-making, and allow the designer to change his mind.

(c) Pontoon bridge – a real problem for which one of our personnel wanted an answer. The bridge was made up of aluminium beams which were laid on pontoons which floated in the river. The beams were connected to make a continuous beam, but of course the supports were highly flexible! An interesting computer graphics problem was to draw the bridge with the deflection exaggerated, and then 'drive' a vehicle across.

Computer graphics

Here one can simply ask the students to draw a picture on the screen and make a copy of it on a plotter, but since this is not design as such we have tended to include graphics with our interactive design projects.

NC programming

(a) Square-to-round transition piece. A wooden model was required as a mould for a glass-fibre transition piece for one of the department's gas dynamics facilities. The simulated cutter path is shown in Figure 2.

(b) Delta wing – also required for gas dynamics. A small wing with circular-arc aerofoil section. The cutter follows a repeated path in the XZ plane, the size of the section being computed for each Y value.

Conclusions

To conclude one is tempted to simply ask a number of questions!

Should this type of course be included in an undergratuate curriculum? (We say yes.) How can one keep up with the very rapid growth of the technology, most of which is coming from industry and not from the universities? What computing facilities are best? How does one overcome the problem of ever more expensive hardware (the peripherals, not the computers!) and large student numbers? It is hoped that this paper may open up a discussion of our questions and that the conference will provide the answers.

Some of our own conclusions are the following:

(a) This course deviates quite widely from the currently accepted definition of CAD, but we accept this and we believe that the course is worth offering. We don't believe it is the University's place to train engineers to operate today's CAD systems, but rather to prepare them for the environment in which CAD will be used. They must be aware of what is currently available and what it can do, and have some idea of how it works.

(b) If nothing else the course gives the students an opportunity to explore some realistic applications of computers in mechanical engineering, and improves their programming knowledge markedly.

(c) Coping with a large number of students is a problem, because it is essential for them to spend a lot of time at the terminals. We have already divided the class in two and run double sessions, but still had about six students per terminal. With six terminals available this is only three per terminal, but even then one has to leave them to schedule most of their work outside the available time on the timetable.

(d) Only four projects can be run in the 14-week semester, so they need to be carefully selected.

(e) Evaluation is very difficult. If a program works, it automatically tends to get a high mark, whereas one which doesn't work may only have some minor error one can't find.
(f) It is easy for students to copy one another's programs and just change the 'comments'.
(g) We are situated far from the bigger industries who have installed CAD systems and could demonstrate them to us.

References

Mischke, C R (1968) *An introduction to computer aided design.* Prentice-Hall; Englewood Cliffs, New Jersey

H Harris, University of Stellenbosch, Department of Mechanical Engineering, Stellenbosch 76000, South Africa

37. CADCAM education at Cranfield Institute of Technology

M J Pratt

Abstract: Cranfield Institute of Technology is strongly involved in CADCAM education on three main fronts. Specialist MSc and PhD studies are available covering various facets of the subject. A broad range of short courses is offered to enable industry to update and expand its knowledge. Furthermore, CADCAM is now used as an everyday tool in the normal teaching of engineers at the Institute.

Introduction

Cranfield Institute of Technology is a university institution which is in many ways unique in Britain. More than 80 per cent of its students are working for post-graduate degrees in advanced technology, applied science or management. Cranfield is also the largest centre within the university sector for applied research and development in industrial technology. It is estimated that this activity will bring in £7.5 million in contract funds during 1982.

Cranfield currently has over 1,400 full-time students. Additionally, some 5,000 participants in various short courses will pass through the Institute this year. The academic staff numbers about 200, and the research staff 300. The main campus covers more than one square mile, though much of this is taken up by the Institute's own fully operational airfield. It is located in Bedfordshire, near the new city of Milton Keynes, and is conveniently situated roughly equidistant from London and Birmingham, Oxford and Cambridge, any of which can be reached in about one hour by road.

Overview of CADCAM at Cranfield

Cranfield first became active in the CADCAM area in the early 1970s. Initially, various elements of CAD and CAM were taught in diverse departments, but more recently the need for integration has been perceived and a committee has been formed to co-ordinate this work across departmental boundaries. The Computer Centre acts as a focus for these activities, and an increasingly wide range of CADCAM software is available for general use on the Institute's DEC VAX computers (see Appendix 1). The main emphasis is currently on applications in mechanical engineering.

In this paper CADCAM education at Cranfield is reviewed under three main headings. First, the routine teaching of CADCAM methods to engineering students will be covered, followed by the teaching of CADCAM specialist courses. Finally, the spectrum of CADCAM short courses for industry will be described.

CADCAM in the engineering curriculum

An overview of CADCAM from the point of view of the practical and economic benefits its use can bring is available to engineering design students throughout the Institute. This is an integral part of the Design Core Course, given by the Centre of Engineering Design, which was set up in order to co-ordinate the teaching of that discipline at Cranfield.

From October 1982 most of the engineering departments at Cranfield will be giving practical instruction on turnkey CADCAM systems to their MSc students. The systems at present available are CDM300 (Kongsberg), Medusa (Cambridge Interactive Systems) and Unigraphics (McDonnell Douglas Automation). These systems are all widely used in industry, and it is felt that students of engineering should have some exposure to their use so that they will know what they must increasingly expect when they leave Cranfield.

As regards the analytical aspect of the design process, the use of the finite element method is routinely taught to many of the engineering students. The packages currently available include PAFEC, a suite of programs from Structural Dynamics Research Corporation including SUPERB, SUPER and OUTPUT DISPLAY, and ABSEA, a system developed at Cranfield specifically for teaching purposes.

On the manufacturing side, NC and CNC techniques have been taught to MSc students in the School of Production Studies for some years. Lectures are now also given on the use of computers in process planning, production planning and control, and quality control. More will be said later concerning this school's activities in robotics and flexible manufacturing systems.

At present electrical design is a minority interest at Cranfield. However, the School of Electronic Systems Design teaches CAD of control systems using a suite of programs developed by themselves, and also uses the SPICE package for circuit analysis in both time and frequency domain. The only system currently available for PCB design is a Racal-Redac CADET stand-alone system based in the Cranfield Unit for Precision Engineering; more software is needed in this area, since the School of Production Studies has an interest in the computer-aided manufacture of PCBs.

To conclude this section, three examples will be given of the use of CADCAM in specific engineering courses.

(a) In the College of Aeronautics the students of aircraft design undertake each year as their group project the design of a complete aircraft. Students of this department are at present comparing the relative labour of preparing general arrangement drawings of an executive aircraft using manual methods and using CDM300. Next year the entire group project will use one of the turnkey CADCAM systems in conjunction with the APTNMG (Numerical Master Geometry/Consurf/APT140) system used by British Aerospace.

(b) The School of Mechanical Engineering runs an MSc course in Applied Mechanics. Students on this course collaborate in a group project which involves the design of a complete piece of machinery (for example an industrial gas turbine). In future they will make extensive use of CADCAM systems for the design of components and assemblies. Interfaces to finite element software and to other analytical programs will be written as required. Consideration will also be given to the manufacture of components. Optimization for forging, casting etc will be looked at, using finite element analysis for the plastic flow and heat transfer problems involved.

The Applied Mechanics group also offers a preliminary year's course for entrants who do not have the usual good honours degree required for immediate entry to an MSc course. These preliminary year students are given an overview course on CADCAM and are set drafting exercises to give them some ten hours experience of a CADCAM system.

(c) The Department for the Design of Machine Systems runs a unique two-year MSc course. The multi-disciplinary first year consists of lectures and practical design studies, while in the second year the students work as a group on an industrially sponsored project – designing, manufacturing and testing an advanced £50,000 machine to be used by the sponsoring company. Recent sponsors have included IBM, Metal Box and Molins. Next year's students are expected to make heavy use of the Medusa system for the design stage. This department intends to acquire its own numerically controlled machine tools, and the eventual aim is that each student will design and then actually make at least one component of the machine under construction, using the Institute's CADCAM facilities.

To summarize this section, although CADCAM systems have only recently been installed at Cranfield they are already widely used, and much thought has been given to efficient ways of using them in the engineering curriculum.

CADCAM specialist courses

Two specialist MSc programmes on different aspects of CADCAM have been running at Cranfield. A third course commenced in October 1982, and a fourth began in 1983.

MSc in Applicable Mathematics – specialist programme in computer-aided engineering techniques

The pace of adoption by British industry of computer aids to design and manufacture is rapidly accelerating. Many companies are purchasing ready-made turnkey CAD or CADCAM systems. Other companies are developing their own systems in-house, while still others are acquiring separate packages covering various aspects of the design and manufacturing processes and interfacing them to each other. No truly integrated system yet exists which will handle all the separate aspects of design, analysis and manufacture in a unified manner, but such systems are under active development. Eventually, all processes will operate with respect to a common data-base, information transfer between processes will be handled efficiently by the computer, and the system will present a unified interface to the user. The greater the degree of integration, the greater will be the potential increase in productive efficiency.

The current explosion of interest in computer aids to the industrial process has led to a grave shortage of specialists in the planning and management of CAE (computer-aided engineering) systems and their component sub-systems. This situation can only become worse as the current recession ends and there is more industrial investment in CAE. The Cranfield one-year course in CAE Techniques aims to alleviate this situation by providing training in the mathematical and computational techniques required for the planning of an integrated CAE system and the development of its actual software. Graduates of the course will be valuable not only to developers and established users of software aids, but also to companies who have newly invested in CAE technology, since their breadth of knowledge will enable them to advise as to the choice and best possible use of new equipment or software.

The CAE course is primarily concerned with applications in mechanical engineering. It adopts a bottom-up approach, being based on the viewpoint that product description methods are fundamental to any CAE system. If the product to be made cannot be represented within the system then no automated manufacturing process can make it. Furthermore, in an integrated system it must be possible to interrogate the product representation in various ways in order to perform analytical calculations and to determine appropriate manufacturing operations. In the case of 3D mechanical engineering components these operations are in general only feasible if the product representation is generated by a geometric modeller which is capable of providing a complete three-dimensional representation of the product. The techniques of geometric modelling are therefore covered in some detail in this course. The specialized methods used in the computer representation of complex curved surfaces are also covered. A range of geometric modellers and surface-defining systems is available at Cranfield so that demonstrations and hands-on experience can be given as required.

Other major topics covered in this programme include finite element analysis, data-bases, software engineering and interactive computer graphics. In order to put all this material into its proper technical and organisational context a series of Engineering Appreciation lectures is also provided, in which the organization of a typical engineering company is described, together with the roles of the various departments involved in the design and manufacturing processes. The impact on this organization of the introduction of CAE is discussed, both from the point of view of the problems which can arise and of the benefits which can result.

The Engineering Appreciation lectures are reinforced by weekly seminars and demonstrations, mostly given by speakers from industry or software companies, but also including some given by leading academic workers in the CAE area. These seminars are also open to visitors from outside Cranfield, who are always especially welcome, not least for the illuminating contributions they make to the ensuing discussion. A recent seminar programme is given in Appendix 2 to illustrate the range of topics covered.

Students are assessed on the formal course work by a combination of examinations and project work. The projects set are usually based on the use of one or more of the CADCAM software systems available at Cranfield (see Appendix 1). During their final six months the students work on individual projects which they write up in their dissertations. Some recent dissertation topics include:

- Use of rational parametric curves in CAD of aircraft fuselages
- Computer-controlled inspection of moulds and dies
- Least-square fitting of parametric surfaces
- Real-time generation of shaded pictures
- Optimal node numbering to minimize bandwidth in finite element analysis
- CAE applications of the 'Oslo' algorithm for recursive sub-division of B-spline curves and surfaces

The majority of graduates of the CAE course go on to work in industry or in software companies when they leave Cranfield. The skills and knowledge they have acquired enable them to play a valuable part not only in organizations wishing to expand their established CAE expertise but also in organizations which are newcomers or potential newcomers to CAE.

The course described here is run within the Cranfield Mathematics Department by the CADNAM Group (CADNAM stands for Computer-Aided Design and Numerical Analysis for Manufacture). This group has close links with CADCAM interests in the School of Production Studies. It has been active in this area for eight years, initially through the organization of short courses, later by engaging in contract research and development work, and now by running the MSc programme in CAE. Contract work for CAM-1 (Computer-Aided Manufacturing International, Inc.) in particular has enabled the group to make many valuable international contacts with companies who are actively using CAE in their design/production cycle. Information thus acquired is fed back into the course material, while other contacts with software houses and university research teams ensure that the group is aware of many forthcoming developments before they reach industry. The teaching therefore encompasses both the state-of-the-art and established technology in CAE.

MSc in manufacturing technology – specialist programme in industrial robotics

The rate of advancement in automated manufacturing technology continues to accelerate, and this progress brings with it a worldwide demand for experts in the new multi-disciplinary field of robotics and automation. If the UK is to improve its position in the league of industrial nations it needs an increasing supply of industrially oriented technologies with a wide knowledge of this fast developing field.

Cranfield's one-year course in Industrial Robotics – thought to be the first such course in the world – is run by members of the Cranfield Robotics and Automation Group (CRAG) within the School of Production Studies. The objective of this unique programme is to produce well-rounded specialists who, on leaving Cranfield, will have the perspective, motivation and training to enable them to seek out and fulfil the potential for cost-effective applications of robots and automation in British industry.

The course provides a balanced blend of group working, lectures, seminars, laboratory work, industrial visits and personal project work, allowing the development and assessment of many aspects of a student's ability.

Theory is backed up by practice. A wide variety of the latest commercially available robot equipment is available to provide a wealth of first-hand experience. Different elements of robot technology are taught during the year so that those who wish to go on to design special-purpose robot equipment will have the necessary skills. However, the current major requirement of British industry lies in the selection, systems engineering justification, and implementation of commercially available robot technology. For this reason the main orientation of the course is towards robot applications and the best means of achieving them.

A variety of lecture course options are available to students. These include:

- Industrial robots – analysis of industrial robot variety; performance assessment; selection and financial justification.
- Programming robots – robot languages; problems and techniques.
- Robot applications – requirements for different applications; case studies; safety and maintenance aspects.
- Automation – handling and assembly; logic and sequencing; industrial, electrical and electronic systems; design, planning and implementation of flexible manufacturing systems.
- Numerical engineering – elements of numerical control systems; computer numerical control.
- CADCAM – commercially available CADCAM systems. State of the art in CADCAM. Computer-aided process planning.
- Management – organization and business policy; human and industrial relations; management accounting and economics.

The following list gives some typical student projects undertaken as part of the course:

- Performance assessment of industrial robots
- Product redesign for robot assembly
- Optimal robot selection for foundry applications
- Development of advanced robot hand for assembly tasks
- Solving robot programming problems in paint spraying applications
- Systems engineering requirements for a multi-robot handling system

A recent MSc group project has involved the development of a low-cost sophisticated robot arm, called CHARLIE. This is a five-axis device with a reach of half a metre, capable of lifting one kilogram. It was designed for industrial use as an assembly robot, but at a cost of less than £3,000 would clearly be suitable for use by universities or polytechnics as a teaching aid.

It should also be mentioned that CRAG has the largest university-based robotics research group in the UK. The group's interests extend to highly automated batch-producing or flexible manufacturing systems, and it undertakes a large volume of research and development work, much of it financed by industry. There is also a strong PhD programme.

Other CRAG activities include the preparation and circulation of the CRAG Bulletin, a free inter-university informal contact magazine. The group is also responsible for publication of *Robotics Technology Abstracts*.

MSc in Production Technology and Industrial Engineering – specialization in flexible manufacturing systems (FMS)

This one-year specialization was offered by the School of Production Studies from October 1982; it will be the world's first MSc course in FMS, and plans for it have been maturing over the last year, guided by an industrial steering committee. Whereas industry has in recent years introduced computer aids in a rather piecemeal manner, this course aims to provide a supply of engineers who are capable of co-ordinating all the computer-based activities in a company. The future of manufacturing will depend upon such co-ordination; it will allow greater flexibility, productivity and profitability than are at present attainable, and enable a company to be competitive in markets throughout the world.

The course has been designed to produce engineers who understand the basics of business as well as the application of computer-aided systems. It provides a blend of lectures, laboratory investigations, industrial visits, group and individual projects, and 'hands-on' experience with CADCAM and other systems.

The basic business-oriented lectures will cover human and industrial relations, management accounting, organization and business policy, corporate planning and economics, work organization, quality control, production and stock control, work study, factory layout and materials handling, systems analysis, data processing and computer programming.

The course will also deal with the applications of computers to design/drafting, manufacture (including NC, CNC, DNC, robotics and flexible manufacturing systems), process planning, factory layout, production and stock control, quality control, distribution and storage.

Computer control systems will also be discussed, as will the human and work organization aspects of CAM and flexible manufacturing systems. Special attention will be given to the management of a flexible manufacturing project, including the problems of linking systems and the calculation of cost effectiveness.

Students will work together on a group project, which will be a real-life project carried out in conjunction with a manufacturing company. The students will decide on the optimal configuration of flexible manufacture and computer-aided systems for the company concerned, and will submit their implementation proposals in a report prepared to professional standards.

Each student will also undertake an individual research project and will prepare a dissertation containing his findings. It is intended that the majority of these individual projects will involve practical problems posed by an industrial company.

To summarize the intentions of its organizers, this specialization is designed to produce a new breed of engineering managers who will take manufacturing industry into the 21st century.

The School of Production Studies already has a small FMS for teaching purposes, consisting of a model-maker's lathe and milling machine together with a robot arm and inspection station. The system is controlled by a PET computer. Larger projects are in prospect; a Herbert AL10 CNC lathe will be installed shortly, and the School will develop an associated work-handling system for it. Further CNC machines will be acquired in due course for FMS development.

Proposed MSc course in the design of CAD systems

This one-year course is currently in the detailed planning stage in the Centre of Engineering Design, which is aided in its deliberations by an academic/industrial steering committee. It is hoped that the course will take its first students in October 1983.

It is intended that graduates of this course will be systems designers who are capable of going out into industry and designing CAD systems tailored to the range of products made by their company. The course will adopt a top-down approach based on design methodology, systems methodology and information systems theory. There will be less bias towards mechanical engineering applications than in the case of the CAE Techniques course run by the Mathematics Department are described above.

The Centre of Engineering Design is a recently established unit whose primary activities have hitherto been in the centralized teaching of engineering design at

Cranfield, in running short courses and in consultancy work. A major research project on the human interface in graphics and modelling is now under way in the Centre, jointly with the Royal College of Art.

CADCAM short courses

Cranfield has for many years run a wide range of short courses, designed to broaden and update the knowledge and understanding, both of individuals and organizations, of their own and related technology. The bridging of the gap between theory and application is a critical aim of these courses. About half of all CADCAM short courses given in the Uk are organized by Cranfield.

Below is given a list of courses which are currently offered at Cranfield, with notes where appropriate. These are advertised courses which are run at regular intervals, but additionally many others are given in-house to a wide variety of organizations, each course tailored to the particular requirements of its sponsors.

Centre of Engineering Design:

- Computer-Aided Design for Design Engineers and Draftsmen (three days)
- Choosing, Commissioning and Exploiting a Turnkey CADD System (two days)
- Intensive Course in CAD (five days)

In addition to these courses the Centre of Engineering Design provides a CADCAM input to the eight-month EITB Fellowship course organized by the School of Production Studies, and regularly gives a six-day CADCAM course sponsored by the Irish government.

College of Aeronautics:

- Vehicle collisions (five days)

This course, organized by the Cranfield Impact Centre within the College of Aeronautics, is much concerned with structural deformation of road vehicles involved in collisions, and makes extensive use of CRASHD, a special-purpose finite element program developed at Cranfield for this application.

Department for the Design of Machine Systems:

- Robots in Manufacturing and Machine Systems (three days)
- NC – Wire – EDM in Toolmaking (three days)

This course combines an in-depth exposition of electro-discharge machining techology and numerical control technology and the essential programming. The orientation is very practical; each participant has the opportunity of programming and making a part on a Fanuc NC–wire–EDM machine.

Department of Mathematics:

- Computational Geometry for the Design and Manufacture of Complex Surfaces (five days)
- Geometric Modelling (five days)
- Fundamentals of Finite Element Methods (three days)

These three courses all combine theory and practice, making extensive use of the wide range of appropriate software available at Cranfield.

School of Electronic Systems Design:

- Modern Control Techniques in Practice (five days)

This course uses CAD techniques where appropriate. The School of Electronic Systems Design also runs courses in Automatic Test Systems (ATE) for various organizations.

School of Mechanical Engineering:

- Engineering Applications of the Finite Element Method

This new course, based on the School's wide experience in contract work using the Finite Element method, is running during 1982-83. It will be complementary to the more fundamental course run by the Department of Mathematics. The School of Mechanical Engineering also provides support to short courses in other departments by giving demonstrations of the full range of capabilities of a typical CADCAM system.

School of Production Studies:

- Robots for Top Management (two days)
- Effective Application of Industrial Robots (five days)
- Computer Applications in Industrial Engineering (five days)
- Computer-Aided Production Planning and Control (four weeks)

These two last courses make use of programs running on microcomputers to illustrate the solution of typical problems in production and industrial engineering.

- Engineering Manufacture in the Eighties (two weeks, one at Cranfield, one at Linköping, Sweden)

The School of Production Studies also runs numerous in-house courses on robotics. A course on Flexible Manufacturing Systems for Top Management is in preparation. A one-week vacation course on robotics for PhD students was given in December 1982, sponsored by the SERC.

As will be apparent from the preceding list, Cranfield has a fairly comprehensive range of short courses in the CADCAM area. This range is constantly being increased as advances take place in both hardware and software technology. Through its wide circle of international industrial and academic contacts Cranfield keeps abreast of the latest developments in CADCAM; short courses provide the medium through which relevant information is passed on to those sectors of British industry which will most benefit by its application.

Conclusions

It has been shown that Cranfield is making a major contribution to CADCAM education on three main fronts. Specialist MSc and PhD studies are available in various aspects of the subject. A broad range of short courses is offered for the benefit of industry. And finally, the teaching of CADCAM methods to students of more general engineering courses is becoming standard at Cranfield. All these activities will expand rapidly in the next few years. Our entry into this highly interdisciplinary field has brought the separate schools and departments of the Institute into closer contact with each other. It is to be hoped that CADCAM may correspondingly break down some of the barriers to communication in British industry.

Acknowledgments

The author would like to acknowledge the assistance of all those members of various departments at Cranfield who supplied information used in this paper. They are too numerous to mention by name.

Appendix 1 Computing and Software Facilities

The Cranfield Computer Centre operates two DEC VAX computers, an 11/780 and an 11/782, serving the whole Institute. Major items of CADCAM software currently available include:

CDM 300

Industry standard system for computer-aided design and manufacture, with two- and three-dimensional capabilities, marketed in Europe by Kongsberg, and in the USA (where it is known as ANVIL 4000) by MCS.

UNIGRAPHICS

Another powerful CADCAM system, developed by McDonnell Douglas Automation. The Cranfield implementation includes a version of the geometric modeller PADL 1, originally developed at the University of Rochester, USA.

MEDUSA

System for two- and three-dimensional mechanical engineering design, interfacing to a finite element program and a geometric modeller. This program, developed by Cambridge Interactive Systems, is implemented on a Prime 150 computer in the Cranfield Product Engineering Centre; one work station is available for teaching and demonstration purposes.

APTNMG

British Aerospace interactive graphical program for aircraft design, incorporating the Numerical Master Geometry surface defining system and the APT 140 system for generation of data for numerically controlled manufacture.

APT4 + SSR1

APT4 is a modern version of the APT part programming language for NC manufacture. This system incorporates the latest Sculptured Surfaces extensions, permitting the definition of very general patched surfaces. The VAX version of this system was prepared at Cranfield under a contract obtained by the Mathematics Department.

TIPS 1

A geometric modeller developed at Hokkaido University in Japan. Again, this program has NC facilities.

PAFEC

Finite Element analysis package for structural and heat transfer problems. This package includes facilities for dynamic and non-linear problems.

SUPERTAB/SUPERB/OUTPUT DISPLAY

SUPERB is another Finite Element analysis program for the solution of static, dynamic and heat transfer problems for complex structures. SUPERTAB is a pre-processor which aids the preparation of input to SUPERB by providing mesh generation facilities etc. OUTPUT DISPLAY is a graphical post-processor which permits the display of displacements, deformed geometry, mode shapes and contour plots of stress/strain or temperature.

GEOMOD

A pre-release version of SDRC's geometric modeller, currently undergoing user evaluation. Geometry defined using GEOMOD may be input directly into the SUPERTAB finite element pre-processor.

CADET

A stand-alone microprocessor-based system for printed circuit board design including an auto-routing facility. A Racal-Redac system.

SPICE

A general-purpose circuit simulation program for non-linear DC, non-linear transient and linear AC analyses. Circuits may contain any of the usual circuit elements, transmission lines and the most commonly used semi-conductor devices.

GINO-F/GINOGRAF/GINOSURF

General graphics packages. GINO-F contains basic FORTRAN sub-routines which are called by GINOGRAF for the generation of graphs, histograms, bar charts and pie charts. Similarly, GINOSURF calls sub-routines from GINO-F for the display of three-dimensional surfaces in various ways.

MOVIE/BYU

A suite of programs for the display and manipulation of data representing mathematical, topological or architectural models whose geometry may be defined in terms of polygonal elements or contour line definitiions. A major application is for the display of the results of finite element or finite difference analysis. Developed at Brigham Young University, Utah.

Other CAE programs can be accessed remotely, including:

BUILD

Advanced geometrical modeller under development at Cambridge University (on an IBM computer at Newcastle University).

POLYSURF/DUCT

Two surface-defining programs with numerical control facilities (on a Prime computer at Delta CAE, Birmingham).

Further programs which have been demonstrated at Cranfield by special arrangement include the geometric modellers ROMULUS and EUCLID, and the surface-defining system PROTEUS.

Cranfield undertakes a large volume of SERC research, and the Computer Centre accordingly has links to an IBM 360/195 at Rutherford Laboratory, a Prime 750 at UMIST and a DEC 10 at Edinburgh. The SERC has also sited one of its Multi-user Mini (MUM) computers in the Centre, a GEC 4085, which is available to local users as well as SERC grant-holders.

Appendix 2 Programme of CAE seminars for the academic year 1981–82

These seminars take place at 2.15 pm on Tuesdays during academic term time. Visitors are welcome; anyone wishing to have his name placed on the circulation list is invited to contact the author of this paper.

6.10.81   Mr M J Pratt (Cranfield)
Overview of computer-aided engineering
13.10.81   Dr J A Gregory (Brunel University)
Triangular and pentagonal surface patches
20.10.81   Dr A J Medland (Brunel University)
The impact of CAD on the design of mechanical elements for machines
27.10.81   Dr R G Bennetts (Cirrus Computers Ltd)
The present status of digital testing
3.11.81   Dr J Rooney (Open University)
Kinematic and geometric structure in robotics
10.11.81   Dr A Clarke (NC Graphics)
Polyapt: the numerical control part programming system
17.11.81   Mr E Kingsley and Dr M White (Compeda Ltd)
The SAMMIE system for computer-aided ergonomic analysis (talk and demonstration)
24.11.81   Mr R H Jackson (Baker Perkins Ltd)
CADCAM for general mechanical engineering
1.12.81   Mr D M P Davies (Rolls-Royce Ltd)
Overview of CADCAM developments at Rolls-Royce
8.12.81   Mr P Veenman (Shape Data Ltd)
Recursive sub-division techniques for the definition of surfaces
12. 1.82   Mr R W Mills (CALMA Interactive Graphic Systems)
Competence in software development
19. 1.82   Dr P Charrot (Pressed Steel Fisher)
Use of the NMG surface definition package at Pressed Steel Fisher
26. 1.82   Dr G F K Purchek (Cranfield)
A combinatorial approach to the planning of manufacturing cells
2. 2.82   Mr A J Scarr (Cranfield)
Computer aided quality control
9. 2.82   Mr A Griffiths (Kongsberg Data Systems Ltd)
Presentation and demonstration of Kongsberg CDM300 CADCAM system
16. 2.82   Dr P Wilson (Lucas Research Centre)
NC Machining of geometric models
23. 2.82   Mr M A Neads (SDRC Engineering Services Ltd)
A CAE design case study with strategic benefits
2. 3.82   Mr C A G Cary (Shape Data Ltd)
Parts classification
9. 3.82   Mr M Bonney (Nottingham University)
The 'Grasp' system for robotics simulation
16. 3.82   Mr D Ross-Turner Hume (Matra Datavision)
Presentation and demonstration of the 'Euclid' 3D CADCAM system
27. 4.82   Mr P E Galgut (Cranfield)
Computer-aided production planning and control
4. 5.82   Dr R G Newell (Cambridge Interactive Systems Ltd)
Presentation and demonstration of the Medusa CAD system
11. 5.82   Mr M Kellman (Lucas Research Centre, Shirley)
Proteus – a sculptured surface system for Design and Manufacture

18. 5.82   Mr R R Martin (Cambridge University)
             Lines of curvature and triangular patches
25. 5.82   Dr T Várady (Computer and Automation Institute, Budapest)
             A minicomputer-based CADCAM system for design and manufacture of free form surfaces
1. 6.82   Mr M J Bryan (Westland Helicopters Ltd, Yeovil)
             Design systems
8. 6.82   Mr G Mackie (General Electric Co, USA)
             The factory of the future
15. 6.82   Mr D B Trafford (Cranfield)
             An information systems approach to CAD system design
22. 6.82   Mr R S Scowen (National Physical Laboratory)
             IGES (Initial Graphics Exchange Specification) – a tool for computer-aided engineering?
29. 6.82   Mr J S Swift (Leeds University)
             The geometric modelling project at Leeds University

M Pratt, Department of Mathematics, Cranfield Institute of Technology, Cranfield, Bedford MK43 0AL.

Discussion

D M Brotton, UMIST:

Could you say whether your course provides your students with requisite experience for membership of the appropriate engineering institutions? It seems to be essentially practical in nature.

M J Pratt, Cranfield Institute of Technology:

We have not so far given any particular thought to that matter, but it will certainly be given consideration in the future. I am not quite sure at present which would be the appropriate engineering institution to approach, however. Perhaps the CADCAM Association may eventually develop into a professional institution, in which case it would certainly be the one most appropriate to our course. It is obviously a desirable feature of any course if it can lead to an extra qualification.

D G Smith, Loughborough University:

You said that an appreciation of how CADCAM systems fit into an engineering environment is given to the students. This is a very broad question, and perhaps one which may fall outside your response, but could you indicate in broad terms how this is done?

M J Pratt:

These lectures are given by David Farrar, the Director of the Centre of Engineering Design at Cranfield. His industrial experience is very extensive, and the lectures describe the organization of a typical engineering company, the roles of various people and departments in it, and the problems of introducing new CADCAM techniques into such an environment. The financial and economic benefits of the introduction of computer aids are also discussed.

R Metcalfe, Trent Polytechnic:

Is your course in computer-aided engineering broadly intended for the mathematician or the engineer?

M J Pratt:

The necessary mathematical background is not very extensive. We require a good grounding in basic calculus, matrix algebra and vector algebra (no initial knowledge of vector calculus is necessary). All these topics are covered to the required level in most first degree courses in a range of subjects including computer science, engineering, mathematics and physics; the phrase we usually employ is 'any numerate subject'. We can accept students from any such background into our course.

F Weeks, Newcastle Polytechnic:

You have shown an impressive list of software which is available at Cranfield, most of it commercial software. I wondered whether you had found it necessary to take any special steps to enable the relatively naïve student user to take advantage of the software as quickly as possible.

M J Pratt:

We find that our students are very self-motivating in this respect. If they know that the software is available and they have some reason to use it then they will want to use it. Our main problem here is that it is difficult to ensure that staff members are familiar with all the software; we often find that the students working with the handbook and a little bit of insight can be using a new system productively within a few days, before any staff member has had a chance to get to know the system. Our longest-established CADCAM system is the CDM300, and we currently have quite a number of students who are using this system productively despite having had very little formal tuition on it.

S A Abbas, Teesside Polytechnic:

MSc courses nationally are under-subscribed in most subjects. Could you indicate to us what numbers you have experienced in your two masters degree courses and what is the maximum number you could reasonably handle?

M J Pratt:

The Industrial Robotics course started in October 1981, and in the current academic year has had 23 students. This is very nearly that maximum which the School of Production Studies could cope with on that course. The Computer-Aided Engineering Techniques courses has been evolving over several years as we have incorporated more and more CAE, and current indications are that in the next academic year we will have about twice as many. The Mathematics Department could not comfortably handle more than 20 with our present level of resources.

38. Teaching computer graphics to mechanical engineers in Britain and the United States

P Cooley

Abstract: This paper reports some experiences of teaching computer graphics in the two countries. The courses described were intended for students of mechanical engineering with a small amount of practical experience with computers. Both courses had a significant level of practical work although the equipment used was very different. There were few significant differences between the students' initial abilities and aptitudes. The actual syllabus that was used is commented upon and details of practical work are given.

Introduction

The author recently spent one year as a visiting professor at the University of Connecticut. One of the courses offered was a one-semester three-credit class in Computer Graphics for students majoring in Mechanical Engineering. The course content was based upon material developed for an MSc level courses at the University of Aston in Birmingham. There were many differences between the two universities, particularly in terms of student intake and available equipment. The experience was valuable as a means to determine those topics and teaching techniques which are appropriate to courses at these levels.

It will be argued that, as so much of the material is totally new for most students, it is of minor importance when such a course is presented. The main prerequisites appear to be a good grounding in geometry and trigonometry. Students of mechanical engineering are interested in interactive computer graphics primarily as a means of geometrical definition for the purposes of design and analysis. For this reason the courses have been structured so as to concentrate upon geometrical modelling once certain fundamentals have been examined.

Syllabus

General

The Appendix details topics which were judged too important to be excluded from a course of this nature. There is, at the beginning, much ground to be covered concerning the hardware of interactive graphics. Familiarity with moderate-resolution raster-scan graphics displays (as typified by the Apple II personal computer) is evident in many American students. The same degree of familiarity cannot be taken for granted in this country although the situation is changing rapidly. Information Technology Year and the launching of several personal computers with colour graphics at under £200 are having a significant impact on awareness in this area. However, the direct view storage tube, which is ubiquitous in turnkey drafting systems and other graphics applications, does not have a domestic parallel. Therefore more syllabus time has to be devoted to the technology of this type of display.

Interactive graphic input devices are still rather rare on the domestic scene in both countries. Many students have heard of the 'light-pen' in much the same way that the general public has a vague idea of the nature of a laser. The other methods of input also require detailed explanation and, when possible, demonstration. On this score it was comparatively easy to arrange demonstrations at industrial installations within a state such as Connecticut with considerable investment in advanced technology.

Two dimensions

It is convenient to introduce students to graphics software via graph-plotting. This is a function that is both familiar and frequently required in engineering. It introduces the concepts of plotting and scaling and the need to display character strings in a graphics environment. From these basic concepts one can move on to that of windows, clipping and zooming. It is when one considers a topic such as clipping that many students realize one basic truth about this subject. What appears to be a simple operation is the result of considerable attention to detail and some ingenuity by those who have laboured in this field (Jarvis, 1975). Another topic shown in Figure 1 also falls into this category. Most students are surprised to discover that the simple action of drawing a line on a raster-scan display requires a method of the complexity and elegance of Bresenham's algorithm (Bresenham, 1965). It was included in the syllabus not only for that reason but also because it may have to be implemented by software in several of the more sophisticated desktop machines available to engineers. (SuperBrain and Sirius are examples.)

The other topics considered in this section are obvious choices. The 2D transformations of translation, scaling, rotation and reflection can be treated descriptively at this time, postponing matrix methods until the 3D transformations are presented. It has been found to be particularly useful to introduce Homogeneous co-ordinates at this stage (Maxwell, 1946). Very few students on either side of the Atlantic have met this concept before but most can readily appreciate the powerful nature of a new tool introduced at this time for representing points and lines. Later extension into three dimensions is a natural progression.

The Dirichlet (Voronoi or Thiessen) tessellation (Green and Sibson, 1978) has been included in this section because, in the author's view, efficient methods of geometric sorting and searching are of vital importance in computer graphics. Knuth (Knuth, 1973) refers to this search as the 'Post Office' problem: given the location of post offices in a city and the location of someone wishing to use a post office, determine the nearest post office. Its application to computer graphics is obvious and the methods of sorting and searching used form an excellent introduction to more advanced procedures.

Three dimensions

The mathematics of planar geometric projections is a topic more sinned against than sinning. Perspective projections have been excluded from a course designed for engineering students. Orthographic projections, in the author's experience, are readily understood provided the conventions are clearly defined and the rotations are treated one axis at a time. Teaching aids in the form of wire and string models are essential. Matrix methods provide a convenient means for recording the transformations and may be used in some applications software but the basic concept (the precise co-ordinates of lines projected onto planes) should be treated as elementary trigonometry.

One is then ready to consider the large body of knowledge dubbed 'geometric modelling'. In courses at the levels described only 'wire-frame' modelling and parameterized shapes were considered suitable for inclusion. The former concept is readily grasped but consideration of details concerning well-formed shapes (Baer, Eastman and Henrion, 1979) is a minefield which instructors might be well advised not to cross. Apparently only active researchers in this area can really appreciate the importance of such concepts.

It has been found that many students are familiar with the operations of Union, Intersection and Difference that are of fundamental importance in modelling by means of parameterized shapes. This familiarity came of course, from new school mathematics curricula which include set-theory illustrated by means of Venn diagrams. Hence the above operations can be presented in two dimensions and extended to three dimensions. Requicha (Requicha, 1977) stresses the vital importance of regularized shape operations for applications: hence the inclusion of this topic.

The last topic in this section, hidden edge and surface removal, are included for two main reasons. First, they are important at the applications level. Technical illustration and visual realism demand such algorithms. Second, the particular methods described illustrate several important features of work in this area. Warnock's algorithm (Warnock, 1969) is a delightful example of the 'divide and conquer' approach which all students can learn to emulate. Scan line algorithms (Sutherland, Sproull and Schumacker, 1974) illustrate the benefits to be gained by marrying features of the hardware to the software techniques that are devised.

Curved surfaces

No course on this subject could omit three important names. There are, of course, many ways to define a cubic parametric curve (one for which x, y and z are each represented as a third-order polynomial of some parameter t). Time constraints limit discussion to consideration of the Coons' patch (Coons, 1967), Bezier curves (Bezier, 1974) and the B-spline form (Reisenfeld, 1973). It is interesting to observe that the late Steven Coons is a familiar name to many American students who have but scant knowledge of his actual work.

Practical work

Unlike many other subjects, practical work in computer graphics is subject to (at least) four major constraints:

1. Availability of suitable hardware
2. Availability of suitable software
3. Student familiarity with the available resources
4. Student time allocations.

Students taking a course of this nature are apt to be unfamiliar with both hardware and software. Therefore it has proved useful to set assignments which require a simple program to be written from scratch. Interaction is limited to keyboard inputs because many students will be mastering the operating system, editor, compiler, graphics sub-routines, peripheral interchange software, loader etc, at this time. In addition there may be students with but a rudimentary knowledge of the high-level language they are required to use and such an assignment provides the opportunity to revise and progress.

The second assignment is much more challenging and is posed to students in the form of a flow chart for an algorithm which is known to work. Students are also provided with sub-routines for some of the necessary computations, eg. test that point lies within bounds of line, ordering of points that lie on one line, etc. To reinforce the lecture on homogeneous co-ordinates students are required to write their own sub-routine to compute the intersection of two lines. This assignment also introduces whatever techniques are available for interaction with the graphics display. As the aim of all the exercises is to reinforce lecture material, the details of the assignments specify some foreseeable pitfalls in the hope that students will manage to avoid them. In the cross-hatching exercise students are warned that there may sometimes appear to be an odd number of intersections (when a line passes through a vertex) and that this result should be trapped out before the hatching lines are plotted.

The third assignment takes the form of an existing program which is to be modified. It introduces three dimensions in the form of a wire frame model and the program supplied allows the user to rotate a simple object around the axes and to become familiar with the conventions that have been adopted. Students are required to enhance the program so that the user may delete those edges which he judges to be hidden. The assignment may be tackled in several ways according to the student's mastery of the equipment and the geometrical problems posed. The soft option is to have the program write numbers alongside the lines for identification and to let the user type in the line numbers to be deleted. A harder option, which appeals to some students, is to move a graphics cursor unambiguously close to the line to be deleted and have the program process the geometry in order to identify the line that is to be removed. This exercise also introduces the important concept of updating a data structure. It is quite deliberately a challenging assignment which very few students complete without significant tutorial assistance.

The final assignment is designed to reinforce class-work on curved surfaces. Students are presented with a program which draws a pre-defined Bezier surface using sixteen control points. A Bezier surface is chosen because the control points can be easily manipulated to change the shape of the surface patch. The students are asked to devise a means whereby the program user can interactively manipulate the control points in three dimensions and have the modified patch computed and displayed. Many of the ideas developed for assignment no. 3 can be used here and, without a good attempt at the previous job, few students can complete this task unaided. This type of exercise also serves to reinforce work done on transformation matrices.

Assignment	Description	Syllabus topics
1.	A simple function of two independent variables is given. One variable can be made the subject of the equation. Plot contours for the dependent variable with suitable labels for the axes, contours etc.	Graphics sub-routines. Scaling. Character strings. Use of plotter. Use of display. Clipping.
2.	Write a program to allow the user to draw a polygon interactively on the display device. Close the polygon. Given a suitable algorithm write a program to let the user have the polygon cross-hatched at an angle and pitch which he specifies.	Input devices. Graphics sub-routines. Homogeneous co-ordinates. Sorting. Line generation.
3.	Given a program which allows the user to input rotations of a wire frame model and have the results displayed, enhance the program to allow the user to delete edges judged to be hidden.	Geometrical projections. Transformation matrices. Wire frame models. Data structures. Conventions. Hidden lines.
4.	Test a program which displays a Bezier surface. Enhance the program to enable the user to change the control points and have the modified patch computed and displayed.	Parametric cubic curves. Transformations. Interactive techniques.

Figure 1 *Outline syllabus in computer graphics for mechanical engineers* *(see page 9)*

Software

The software used for the course described above consists mainly of source programs in FORTRAN, supplied by the author and library graphics routines that are commercially available at a reasonable cost. One major problem in mounting a course of this nature is the prohibitive cost of more sophisticated software. A 2D drafting package developed by the author (Cooley, 1979) is available to Aston students and this was adapted to run on very different hardware in Connecticut. No geometric modelling software nor, for example, finite element mesh generation software could be made accessible to students due to cost constraints. However, many suppliers are willing to demonstrate CADCAM systems to small numbers of students on their own premises.

The subject of computer graphics is still in its infancy despite enormous progress since birth. A course of the type described here can only be a window-opening exercise and is certainly not intended as a training course for professional workers in this field. The major deficiency of the courses run to date has been the lack of exposure to a variety of advanced software.

Conclusions

The teaching of a subject which barely existed 30 years ago and which has developed in an astonishing number of directions must be based primarily on the requirements of the participants. Students of mechanical engineering will probably utilize interactive graphics as a method of presenting data, defining and processing solid objects and a means of communicating design proposals. It is therefore important to lay emphasis upon the state of the art concerning both hardware and software. However, it is also important to treat those topics which, at this time, can be identified as fundamental principles of the subject. Practical work is an important constituent of such a course. By this means students attain a deeper understanding of the subject and an insight into practical problems which would otherwise remain mere irritations in their professional careers.

Appendix: Syllabus

1. Graphics Hardware.
    Displays: Rastor-scan. Vector systems. Direct view storage tube. Shadow-mask colour tube. Plasma panel display.
    Input Devices: Keyboards. Inching buttons. Tablets. Light pen. Thumb wheels. Joysticks.
    Hardcopy devices: Pen plotters. Electrostatic plotters. Dot matrix printer-plotters. Light/heat sensitive devices.

2. Two-dimensional graphics.
    Graph plotting. Windows. Viewpoints. Clipping. Scaling. Character strings. Zooming.
    2D transformations: Rotate, Scale, Translate, Reflect.
    Raster algorithms for line and circle generation.
    Homogeneous co-ordinates. The Dirichlet tessellation.

3. Three-dimensional graphics.
    Planar geometric projections. Transformation matrices. Conventions.
    3D homogeneous co-ordinates.
    Geometrical modelling: Wire frame models. Well-formedness. Parameterized shapes. Regularized Shape operations – union, intersection, difference.
    Hidden line and surface removal. Warnock's algorithm. Scan-line algorithms.

4. Curved Surfaces.
    Parametric cubic curves. Bezier curves. B-spline. Hermite.
    Parametric cubic surfaces. Coons' patches. Transformations.
    Speherical products. Superquadrics.

References

Baer, A, Eastman, C and Henrion, M (September 1979) Geometric modelling: A survey. *Computer-aided Design* **11** 5: 253–72

Bezier, P (1974) *Mathematical and practical possibilities of UNISURF*. Barnhill, R E and Riesenfeld, R F (eds) Academic; New York Computer Aided Geometric Design

Bresenham, J E (1965) Algorithm for computer control of digital plotter. *IBM Systems Journal* **4** 1: 25–30

Cooley, P (March 1979) Mechanical drafting on a desktop computer. *Computer-aided Design* **11** 2: 79–84

Coons, S A (June 1967) Surfaces for computer aided design of space forms. Institute of Technology; Massachusetts Technical Report TR-41

Green, P J and Sibson, R (1978) Computing Dirichlet Tessellations in the plane. *Computer journal* **21** 2: 168–73

Jarvis, J F (1975) Two simple windowing algorithms. *Software Practice and Experience* **5**: 115–22

Knuth, D E (1973) *The art of computer programming*. Addison-Wesley: Reading, Mass. Fundamental Algorithms 1

Maxwell, E A (1946) *Methods of plane projective geometry based on the use of general homogenous coordinates*. Cambridge University Press: Cambridge

Requicha, A (November 1977) *Mathematical models of rigid solid objects*. College of Engineering and Applied Science; University of Rochester Technical Memorandum **28**: 6–7

Riesenfeld, R F (March 1973) Applications of B-spline approximations to geometric problems of computer-aided design. Computer science department; University of Utah; Salt Lake City Report UTEC-CSc-73-126

Sutherland, I E, Sproull, R F and Schumacker, R A (March 1974) A characterization of ten hidden-surface algorithms. *Computing Surveys* **6**: 32–4

Warnock, J E (1969) *A hidden line algorithm for halftone picture representation*. University of Utah Computer Science Technical Report TR4-15

P Cooley, University of Aston, Department of Mechanical Engineering, Gosta Green, Birmingham B4 7ET

The teaching of CAE in a polytechnic engineering department. Huddersfield experience
T Hargreaves (see pp. 360–67)
Teaching of CAD and CAM, G Cockerham (see pp. 368–77)

39. The teaching of CAE in a polytechnic engineering department – Huddersfield experience

T Hargreaves

Abstract: Experience in the teaching of CAD in a highly design-orientated mechanical engineering department is described and proposals for the future are offered.

Integration is seen as the greatest need – CAE instead of CAD, with emphasis placed on the everyday use of the computer as a tool in all 'traditional' engineering areas.

The object is not just to teach students how to operate a computer, but to equip them to face the challenges of computer-aided engineering which they will meet in their future careers.

Introduction

The Department of Mechanical and Production Engineering of Huddersfield Polytechnic has had a long involvement in engineering design. It has been said that design is the heart of engineering, and the Department's endorsement of this statement is reflected in the importance which it attaches to the subject in all its courses. By its nature, design must take advantage of the latest technological innovations and it was thus inevitable that computers should find early applications in the design field.

This prospect was recognized when the Polytechnic's first computer was installed, and prompted the introduction of computing elements into the Department's courses. When the degree course (BSc Mechanical Engineering) was approved in 1974, it included, within its extensive design content, syllabuses in CAD and concurrently CAD elements were introduced, under the design heading, in the final year of the Higher National Diploma course.

The recent 'computer explosion' led to a reappraisal, some 15 months ago, of the teaching of CAD, with particular regard to future industrial needs, resulting in a decision to expand considerably computer-related work in the Department.

It seems to be a matter of historical fact that the early development of computer applications in engineering was highly compartmented: the acronym CAD was coined to describe the application of computers in the design area – at first analysis, later graphics. CAM was another area entirely. Although the benefits (or horrors, depending on viewpoint) of the 'automated' factory were being discussed 25 years ago, the possibility of a marriage between CAD and CAM as a means of approaching this objective does not seem to have been recognized until comparatively recently. The dramatic advances which have been seen in the last few years have made the concept at least technically feasible, and almost at a stroke, have significantly modified the designer's potential role. No longer will he necessarily just use the computer as a tool to speed up the design process or to produce drawings – indeed, with the availability of direct computer control, drawings may become superfluous. In designing for such systems the designer will need knowledge of how the machines work, their capabilities and limitations, their special requirements: he may indirectly become, to some extent, their controller.

With this prospect in view it became apparent that teaching under the title of CAD was likely to be far too restrictive and a wider, more integrated approach was sought.

The increasing availability of what the computer sales people lightly refer to as 'all singing, all dancing' packages has led to the introduction of a new acronym – CAE (computer-aided engineering) – typically embracing computer-aided design (CAD), computer-aided manufacture (CAM) and computer-aided drafting (CADr).

It was felt that the general term was more in keeping with the teaching objectives in the Department and under the heading of CAE a team was set up with a mandate to develop teaching and project material on as broad a base as possible, with a view to providing comprehensive courses in computer applications in engineering.

Background

Computing elements were introduced into courses soon after the Polytechnic acquired its first computer in 1969. The objective was the teaching of a programming language – initially ALGOL and later, with a change of computer, FORTRAN. In what seemed the natural order of things at the time, computing was seen as a branch of mathematics and was taught by the Mathematics Department.

Justification for the subject, if justification were needed, was that, with the availability of computers, engineers would need to be able to program. Further it was held in some quarters that high-level computer languages might replace normal algebra in technical communication, so that the ability to write and read a programming language would become essential.

At that time the Department's 'main' courses were a Higher National Diploma in Mechanical Engineering (three-year sandwich) and a one-year full-time post-Diploma Polytechnic Associateship course. It was on this latter course that the Department first introduced work in computer applications. As part of the course each student was required to undertake a major project, and, as staff experience and expertise developed, a number of these were in the computing area. Emphasis in the computer projects was on the use of the computer to solve an engineering problem: the student was required to develop the necessary theory, write the program and (hopefully) show that it provided satisfactory results. In the early 1970s there was significant staff research in the then relatively new area of finite element techniques, and, as a result, a number of students took, as their projects, exercises in developing finite element programs. By today's standards such projects might be considered almost trivial; at the time however, they presented a considerable challenge to the student. Not least of the difficulties was the use of a 'small' computer (an IBM 1130). The core size of 32K made finite element projects major exercises in storing and retrieving data, whilst run times were often very long – two hours was not unusual for a very modest program!

Whilst the projects were, admittedly, of limited scope, the experience gained in their supervision was extremely valuable in the light of future developments. Figure 1 lists a few of the projects undertaken.

Finite element analysis of stepped shafts.

Finite element program for the solution of plane stress problems.

Hand programming of an NC machine tool.

Computer aided design of axial flow compressor blades.

Figure 1 *Early Project Work*

The belief that computing could be equally important at other levels led, concurrently, to the offering of Computer Programming as an endorsement subject on the Higher National Certificate course.

With the approval of the degree course, the subject of Computer-Aided Design appeared for the first time as a course element in its own right. Computing (ie programming) and Computer-Aided Design were offered as separate subject, the former, as by now traditionally, taught by the Mathematics Department. Students would acquire and develop programming skills in the first two years of the course, only starting CAD, taught in the Mechanical Engineering Department, at the third year.

It was noticeable that a significant number of students found computing a 'necessary evil' and if they could avoid it they would. At that time all computing was in batch-mode and it seemed that the attitude of those students possibly resulted from a fear of using the computer – perhaps based on unfamiliarity and engengered by problems met at an early stage – typically in using the job control system.

The commissioning of a new machine (an ICL 2960), with terminal access available, promoted considerable interest throughout the Polytechnic and there was some evidence of an improvement in student use of the machine. Subsequent discussion with students has suggested that a major problem was the 'remoteness' of batch operation: with a terminal the student feels that he is more 'in control' and 'can see what he is doing'. With the new computer came the availability of BASIC: the simplicity of the language, coupled with its free format makes it ideal for terminal use, particularly for interactive work and it was decided that students should be taught both BASIC and FORTRAN.

In view of the earlier problems there were fears that, if as seemed logical, BASIC was taught first, a number of students would choose to ignore FORTRAN on the basis of 'why learn more than one language?', so it was proposed that FORTRAN should be taught first, when a conversion to BASIC could be achieved fairly quickly and easily. In view of the recent rapid development and availability of microcomputers it seems that this decision is perhaps no longer relevant. If, as seems to have been the case, the biggest initial problem is introducing the computer – 'breaking down the barriers', so to speak – it may be a positive advantage to start students on microcomputers using BASIC. The simplicity of operation of the micro will then permit them to obtain initial computer experience relatively painlessly and enable them to realize the benefits of the computer without experiencing many of the problems. Hopefully, by the time they graduate to use of the main frame (in either batch or interactive mode) they will have acquired sufficient confidence to be able to cope with problems as they arise.

It can be expected that, with the increasing teaching of computing in schools and the promotion of computers in leisure areas, these types of problems, encountered in the relatively early days of computers in education, will largely disappear. We are currently in a transition stage and evidence with Huddersfield students is that the transition is occurring fairly rapidly. Some students are already 'hooked' on computing when they join the course; a result of this is seen in their willingness to use the machine and to experiment with programming techniques which, five years ago, would have been considered 'too difficult'.

The writing of a CAD syllabus for the 1974 submission presented some problems. The difficulties of writing such a syllabus are readily apparent even now. Most subjects taught on engineering courses are well developed and fairly stable and syllabuses can be readily devised out of staff experience. CAD however has no tradition: developments in industry have taken place in diverse directions in which, often, the only common factor is the computer. There are no general textbooks to provide 'pegs' on which a syllabus might be hung: there are no guidelines on what should be taught, or even, given that decision, how to teach it. This is true today: it was all the more true in 1974.

It was recognized that such a syllabus must go further than simply programming and a range of topics was offered which reflected the then current state of art.

The first year of CAD was seen as essentially a period of consolidation. The student would undertake a number of projects designed not only to extend his programming ability, but to establish concepts of program and package design. In practical engineering applications programs are rarely written for the sole use of the writer: emphasis was therefore placed on the necessity for back-up documentation to enable another engineer to use the program as a working tool; if necessary to be able to adapt it to his particular needs. A marking scheme was provided, indicating the relevant areas, typically as shown in Figure 2. This basic format, with some variations, is used for marking all CAD project work.

Brief description of program	5
Flow chart	5
User guide	5
Program	20
Specimen output and longhand checks	10
Presentation	5
Total	50

Figure 2 *CAD Marking Scheme*

When terminals became available the way was opened up for interactive work and now virtually all projects are designed for interactive use. Figure 3 lists typical class projects which have been used in the first year of CAD.

Program to determine the diameter of a shaft subjected to combinations of bending, torsion and axial load.
Program to analyse readings from a strain gauge rosette.
Program to calculate and plot deflections of a beam subjected to combinations of concentrated and distributed loads.
Program to calculate dimensions or stresses for a thick cylinder subjected to internal and/or external pressure.
A simple pseudo-interactive drafting package using GHOST out of BASIC.

Figure 3 *Class Projects*

The final year syllabus contained topics (particularly in the graphics area) which, whilst they could be lectured on, would not be demonstrable because of lack of resources. It was decided, therefore, to cover such areas by 'state of art' lectures and to concentrate practical work in the finite element area. Commissioning of the ICL 2960 coincident with the start of the first final year, provided the necessary scope. The larger machine made finite element work a much more practicable proposition and it was decided to run a lecture course, for all final year students, on the finite element method, with every student preparing his own program. So as not to overface the less able student it was decided to confine the work to development of a beam program, working on a lecture/tutorial basis. Typically programs are prepared piecemeal, following lectures on the appropriate sections, and gradually built up to provide a complete package. The beam program has the advantage that the concept of the stiffness matrix can be taught by direct relation to Mohr's theorems and the student is thus able to generate it longhand, term by term. An inversion routine from the NAG library is used, but the student is invited to develop his own algorithms for structural assembly and basic matrix operations. Figure 4 shows a typical breakdown of lecture material.

The system has been successful in that virtually every student has developed a program and some of these have been of very high quality.

The degree course has a large proportion of design project work, culminating in individual major projects in the final year and the availability of terminals and

Introduction and general description
Element stiffness matrix
Structural stiffness matrix
Load vector and boundary conditions
Inversion and displacement calculations
Calculation of stresses
Output layout and support documentation

Figure 4 *Finite Element Method Lectures*

microcomputers has led to an increasing application of computers in this area. Each year a small number of the final year projects have been computer-based, whilst, from time to time, design projects with a strong computer 'flavour' have been introduced in earlier years. Figure 5 shows typical final year project titles.

Program for pressure vessel design to BS 5500.
Use of PAFEC for the design of a high speed chuck.
Design of plane mechanisms on a 4051 Graphics Terminal.
Feasibility study of an integrated CADCAM system for turbine blade manufacture.
An elementary drafting package based on the 4051 Graphics Terminal.
Design and programming of a 'pick and place' robot arm.

Figure 5 *BSc IV Projects*

Discussion so far has largely centred on the degree course, but there has, of course, been other work. Computer-aided design has appeared for a number of years in the final year of the Higher National Diploma course (following 'Computing' in the first two years) and its pattern has been essentially similar to the CAD in the third year of the degree. It would be true to say that, as a 'required' subject it has not acquired the popularity which seems to have developed on the degree course and in a recent re-submission it was decided that, in keeping with the 'technician flavour' of the course, it should be replaced by project work across the engineering field. This has permitted students with a particular interest in CAD to undertake project work in the area, whilst allowing others, either less able or less committed to computing, to take up projects in other areas. In the past year about a third of the projects were computer-based.

The relatively new Highter TEC certificate course has computing elements and in the Project unit a number of projects have been in the computing/CAD area, a selection of which are shown in Figure 6.

Microcomputer program for stores control.
Program for costing turned components.
Hardware and software for an X-Y drive system based on the AIM-65.
A 'brief-case' computer system based on the ZX81.

Figure 6 *HND/TEC Projects*

A new college-devised HTEC in Engineering Design had its first intake this year and this has both computing and CAD elements incorporated. Since these units occur in the second year of the course, no experience has yet been gained.

Figure 7 summarizes computing and CAD activities on the Department's courses to date.

Figure 7 *(see page 9)*

Problems in developing CAE

Learning computer-aided engineering must ultimately be a 'hands-on' exercise and an immediate problem in any educational establishment is one of resources. The difficulty lies in giving a large number of students access to a limited amount of hardware. No great problem arises in analytical work, where access to a large time-sharing machine will meet most requirements, but graphics is much more of a problem. Graphics terminals are still relatively expensive and it would be difficult to justify purchase of a large number of terminals if they were to be used for only a relatively small proportion of time on any course. A similar, though slightly less expensive problem exists, of course, if students are to be required to make extensive use of micros. The problem can be met to some extent by use of booking systems – the Department has recently run a project with a class of 30 students, using five microcomputers. It has to be accepted, however, that this imposes a time penalty on such work and had 30 micros been available, the project would have been completed much more quickly (but probably not six times as quickly!). Similarly it has been possible to operate group projects (four students to a group) on a single graphics terminal, but, clearly, the availability of more terminals would have permitted a faster turnround.

Hardware resources are not however the only source of difficulty. Ideally it would be desirable to be able to provide extensive experience reflecting the industrial use of computers and there are those who, in support of this approach, advocate teaching the use of commercial packages. Whilst there is undoubtedly a place for this type of work it has significant limitations. When industry chooses to enter the CAE field it usually does so with clearly defined and limited objectives: for an educational institution to acquire a sufficiently wide range of packages would be prohibitively expensive. Given limited resources of this type, there is little educational benefit in teaching students to use a specific package any more than there is in teaching them to 'press the buttons' on a specific computer system. The end objective must be to equip the student so that he can readily adapt to any system: to enable him to cope with whatever developments may occur.

The most important resource, of course is, ultimately, the teacher. It is a fortunate institution which is able to 'buy in' the expertise necessary to teach CAE. In the main staff have to acquire expertise; even in some cases to be persuaded of the value of getting involved in the subject at all. Those involved need time, perhaps more so than in other subjects, for the preparation of teaching material, writing of software and so on.

It was the brief of the CAE team to address itself to problems such as these.

Future proposals

When CAD was first introduced into the Department's courses, it was seen as something of which the engineer should be aware: he might meet CAD in industry, but the chances were that he probably would not and, in general, he would require little more than an awareness of its existence. It was felt at the time that the work undertaken more than satisfied this requirement. In order to meet current needs, a much broader approach was considered necessary and five areas were identified for development, namely:

Analytical programs (Main frame and micro)
Finite element
Graphics/drafting
Microprocessor applications
Computer-aided manufacture

The titles are, of course, arbitrary: in practice, significant overlap occurs and whilst theoretically one member of the team is responsible for each area there is, of necessity, considerable inter-relations. Reasonable progress has been made and teaching programmes has been developed in all areas, which will come into operation in 1980. Classes will be split into groups and taught on a rota system to optimize the opportunity for 'hands-on' experience. This is only an interim programme: until other changes are implemented the full scheme will not come into operation.

One such change is the decision to teach all computing in the Department. This is no reflection on the Mathematics Department; it is considered preferable however to introduce students to computing at year one, via engineering applications and such a change is seen as a means of accelerating the students' development in CAE. Until this change filters through the system the CAE programme in the third and fourth year will not be fully developed.

Integration of computer use into other subject areas is seen as extremely important and staff have been asked to incorporate problems utilizing computers in their subject areas. This will open up the possibility of dealing with significantly more complex problems than heretofore and hence extending the students' experience into more realistic applications.

As has been stated, the Department's courses are extremely design-orientated. Apart from the specific computer-based projects which are set, day-to-day computer use in the design area is largely calculation work in support of design proposals and this must be extended. An early priority is a computerized data retrieval system – initially a design studio 'standards book' embodying, for example, information on available materials and section sizes, machining processes and stock cutting tools, operating costs, and so on. Ideally such a system should be based on a dedicated minicomputer and it is very tempting to await the availability of such a machine before starting. There is, however, the attendant risk that the project will never get off the ground. It is therefore envisaged that the system will be initially developed in a limited form for microcomputers with material stored on floppy disc.

A more ambitious proposal relates to the use of computer graphics in design. On the degree course students spend a great deal of time actually designing and every student has his own 'permanent' work station in a design studio. If, as might be anticipated, the design process is to become increasingly computerized, students should have experience of the use of graphics terminals. A scheme is currently being examined for the installation of networked terminals in the design studios – ideally one per student, but, perhaps more realistically, one terminal per four students. Turnkey-type systems are of course out of the question on financial grounds, but a system is envisaged employing microcomputers, with good graphics facilities, utilizing a rudimentary drafting package, and networked to a minicomputer data-base. Such a system would allow the student to produce and store design schemes which could then be developed to produce 'selected' final drawings on a high level graphics terminal, linked to the minicomputer, using a commercial drafting package. This is, of course, a costly exercise and its implementation will depend on the availability of financial resources. If, as is expected, relatively low cost 'graphics computers' become available, the project could become feasible in the near future.

Thus far little has been said of computer-aided manufacture. Traditionally this area has been dealt with under the umbrella of Production Technology – students have undertaken an exercise on an NC machine, producing a tape and machining a simple component. We must now go further than this. Students must have experience of NC and CNC machines; they must however learn more than just 'which buttons to press'. They must have experience of computer preparation and verification of control tapes and facilities are now available for this, but, more important, they must be made sufficiently aware of the possibilities afforded by computer control to be able to comprehend, develop and use systems employing the most advanced techniques available. To install such a system, incorporating CNC, DNC and robotics would be prohibitively expensive and it was to go some way to meeting these educational needs that a microprocessor applications laboratory was set up. Here the student can acquire the principles of computer control with 'hands-on' experience – driving an X-Y plotter by computer is no different in principle from controlling the movements of a machine table; the pick-and-place robot arm under microprocessor control operates in exactly the same way as its industrial counterpart. Linking a number of such applications simulates the machining centre.

The direct linking of CAD and CAM is more difficult in the educational environment. Given the developments in the CAD area however, it is not unreasonable that, using the minicomputer data-base, direct connection to at least one machine tool could be achieved and this must be another objective for the future.

Computers are increasingly being used in logging and analysing test data and this is another area under development. Current work has ranged from computer curve fitting of experimental results to the recording and analysis of strain gauge readings. There is scope for much more – 'computerized' instrumentation on engine tests and prototype investigations. When these results feed back into the design process we 'complete the circle'.

Conclusions

In developing a CAE policy the predominating theme has been integration: the use of computers in the traditional engineering areas and the interrelation of the various branches of CAE as tools in the overall engineering process.

Computer-aided engineering is not a subject to be studied as an end in itself: its only value is in its application. It may be necessary to teach current state-of-the-art material to bring students up to date, but, given the present rate of development of computer applications, we must remember that such material may be outdated tomorrow.

Our students must be fitted to face up to this situation: we must ensure that they will be capable of meeting any challenge which the computer may present in their future careers.

40. Teaching CAD and CAM

G Cockerham

Abstract: There are many problems to be faced by those wishing to introduce CADCAM into the syllabus. Not only has the problem failed to incorporate into an already overall syllabus to be considered but also the problems of presenting and illustrating the integrated nature of CADCAM with the very high cost of such systems. This paper details the approach adopted by one polytechnic.

Introduction

Recent developments in micro-electronic technology have provided a range of automatic machine tools for a large number of manufacturing processes. Similarly, computers capable of mathematical manipulation and display of data are available in a wide variety of shapes and sizes. The combination of these two has resulted in a capability for integrated geometric definition, functional analysis and automatic manufacture which, when successfully applied in industry, will improve the quality and lead time of design and improve the quality cost and delivery of manufactured goods.

In teaching such a subject to mechanical and/or production engineering undergraduates, the difficulty is deciding what to include in curricula that are already considered overfull with increasing pressure being applied to reduce syllabuses. To become a real expert in the area requires high levels of knowledge in a wide variety of topics such as mathematical geometry, computer science, machine tool technology and production management.

In order to reflect accurately the total system approach requires enormous investment in hardware and software which may not be practical in an educational environment. In addition, limited use of computers within small companies can be the best solution in many cases.

This suggests that teaching the individual elements of CADCAM may be an acceptable activity provided information is supplied to students on integrated systems, preferably in an industrial context.

The approach at Sheffield City Polytechnic is to incorporate the subject within two four-hour modules at final year level of the BSc Engineering; one module as part of the Mechanical Engineering option emphasizes engineering analysis (Cockerham, 1977), whilst object definition and automatic manufacture is the dominant part of the module for the Production Engineering option.

Philosophy

The objective of the courses is to provide a broad appreciation of the technical and economic benefits, the range of available hardware and software and an understanding of the underlying principles sufficient to provide a starting-point for any original work that might be contemplated. Additionally, case study work is an important feature, being based upon a number of industrial consultancies carried out within the

Department with local industry. These case studies provide students with the opportunity to appreciate the problems faced by industry in deciding if and what computer-aided design and manufacture facilities can be used to good effect within the company. Discussions with company personnel are encouraged wherever possible. Figure 1 shows details of a block diagram representation of areas of application of CAD and CAM. Emphasis is placed upon the possibilities for using microcomputers within the process having earlier clarified the demarcation between different sales of computer systems.

Figure 1 *CADCAM (see page 9)*

Mechanical option

The 40-hour CAD module on this course is taught within the subject of Engineering Design with emphasis being placed on the benefits brought to the design process by computer aids. Figure 2 shows a block diagram representation of the design process and the extent to which computer methods can be used to evaluate and optimize any alternative concepts that the designer proposes. Some reference to computers in manufacturing is included but is not covered in great detail. As mentioned earlier, emphasis is placed on the wide variety of computer hardware that can provide aid to design and project activities are used extensively.

Figure 2: *The design process (see page 9)*

Large systems

Figure 3 shows output from a large suite of linkage software requiring a large main frame computer (IBM 4350) and graphics terminal (Tektronix 4010). This provides possibilities for several design projects and demonstrates quite clearly the benefits to be gained in the design process using high-speed functional analysis. Stress analysis by finite element or finite difference methods are two other programs that are used in this area.

Figure 3: *Typical linkage analysis (see page 9)*

Small systems

Recent developments in desktop personal computers have presented an opportunity for the spread of computer-aided design to large numbers of small companies. A gear analysis program originally designed for a main frame facility (Cockerham and Waite, 1976), has been squeezed on to a 48K Apple II and shows the possibilities of such small computers. Again many projects can be built around this item of software. A typical student effort to an automatic gearbox design is shown in Figure 4.

Figure 4: *Automatic gearbox design carried out using computer analysis (see page 9)*

Production option

Again the 40-hour module in CADCAM is contained within another subject called Manufacturing Plant Automation which contains other subject matter of a general nature. The contrast between large and small computer systems is continued and the project theme maintained. In addition, the types of machine tools available and methods of programming them are also included to provide an awareness of the ultimate objective of any computer-aided design and manufacture system; ie the production, storage and transmission either directly or indirectly, of instructions to cause the automatic manufacture of a required object.

Large systems

Many large firms have been involved in CADCAM for more than a decade, mainly in high technology areas solving complex design problems, quite often using large computers together with sophisticated computer programs. The areas of application are two-fold; three-dimensional regular geometry (ie combinations of straight lines and circles) and free form three-dimensional curved surfaces. Typical of these is the APT/FMILL (IBM Technical Publications) suite of programs developed by Illinois Institute of Technology for IBM which are available on the IBM 4350 main frame computer at Sheffield City Polytechnic. They have been superseded, expanded, contracted in one form or another in recent years, but are still extremely useful in establishing some of the basic principles of regular or free form surface definition which have been used in subsequent application programs such as NELAPT, GNC, POLYSURF, COMPACT etc. An understanding of the unambiguous way in which APT enables points, lines, circles, planes, etc to be defined and a tool to be driven around them in combination to produce a component, can provide a fundamental understanding of CADCAM for regular objects which is transferable to any other software package. After several hours using such a program, it is not difficult to appreciate the calculations and logic contained within such a program and the possibility of producing a purpose-built facility with limited facility on a small computer is not lost on the students.

This is not the case when using the free form surface definition section (FMILL) of this, and no doubt other programs. In this case, some basic mathematical background is essential in comparing and contrasting the various curve and surface fitting techniques commonly used. In particular, the distinction between parametric and non-parametric curves (Faux and Pratt) is highlighted. Figure 5 shows output on a Tektronix 4010 graphics terminal of a parametric cubic spline computer program which can be used to gain an appreciation of the ways in which surfaces defined in such a way can be modified and adjusted.

Figure 5 *Typical parametric cubic spline curve fitting (see page 9)*

Small systems

The particular piece of software used to demonstrate the possibilities of small computers has been developed within the Department of Mechanical Engineering at Sheffield City Polytechnic as part of a consultancy project carried out for Precision Manufacturing Services of Sheffield and reported elsewhere (Cockerham and Panter, July 1981; Cockerham and Panter, October 1981). The company is using spark erosion techniques with the electrodes made on NC machine tools, for the manufacture of precision tooling, typically as shown in Figures 6 and 7 for a die-cast steam iron base. The company concerned is small and has only two NC machining centres which limits the investment possible in CAD and CAM facilities. The computer used to define the shape and produce the machining instructions is a 48K Apple II with single disc drive and monitor and indicates the possibilities for introducing low-cost CADCAM into industry. The software produced has four particular features:

    (i)   two-dimensional geometry definition
   (ii)   graphical representation feature by feature
  (iii)   storage and editing of machine tool instructions
  (iv)   automatic production of punched paper tape to drive the machine tools.

Figure 8 shows one of several geometric configurations selectable from a menu to allow the definition of the necessary geometric intersection data as shown in Figure 9. It can be seen that these geometric features are essentially based upon the ideas of APT.

The graphical layout of the shape is shown in Figure 10 and although the resolution of the picture is limited, it is considered by the company using it to be sufficient to

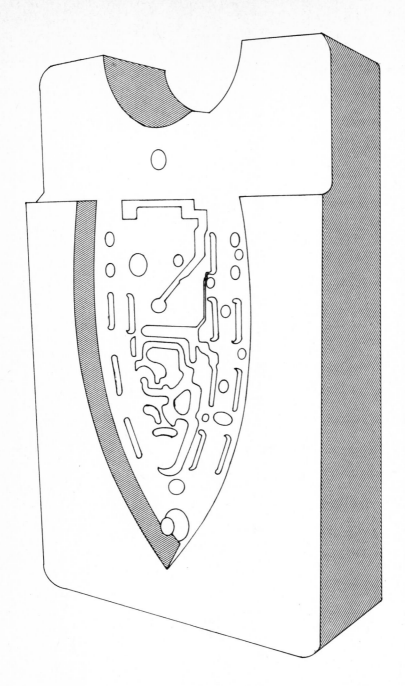

Figure 6 *Spark-eroded die block for die-cast steam iron base*

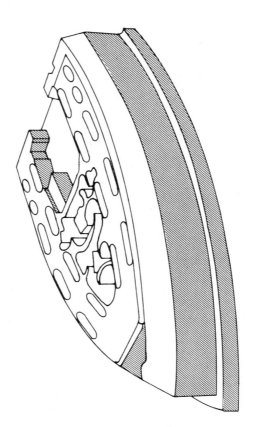

Figure 7 *Graphite electrode used to produce die block*

identify any obvious errors or mismatching between features. Rubout and redraw features are also included.

Figures 8 and 9 show the machine tool instructions automatically produced and the capability of the system for editing in or out additional instructions or commands.

Figure 8 *Geometric configuration for intersecting circles with blend radii (see page 9)*

Punched paper tape is produced automatically by interfacing the Apple to a teletype terminal with punch or a punch itself. Thus, a real illustration is given of the productivity benefits to be gained by applying small-scale CAM. As with the large-scale system, all exercises and projects lead to the manufacture of components with some attempt at a cost breakdown and comparison with alternative conventional methods of production.

Reactions

The modules as described have been in operation for several years and student reaction suggests that the 'global coverage' provides them with an appreciation of the possibilities of CADCAM and an awareness of the currently available hardware and software systems. With a large proportion of these systems having been developed over hundreds of man years their successful implementation is clearly the sort of task that today's undergraduate engineers are likely to face rather than an *ab initio* development. The information they require in such circumstances certainly is *not* just a commercial catalogue, but sufficient information on the basis principles of the software to be able to ask the right questions and evaluate the answers. To this end, discussions with personnel within local companies having made the move into CADCAM are seen as vital in providing the students with an appreciation of the complex nature of decision-making in this area. The real activities of manufacturing piece parts, visiting large installations such as Delta Metals, Birmingham are stated by the students as invaluable in completing the total scene.

Similarly, the idea of using small systems seems to be particularly well received and several recent graduates have initiated activities within their own companies. Perhaps the nature of Sheffield industry, involving a large number of small to medium firms is significant in that respect.

Having operated in this manner on the two distinct options of the BSc Engineering it is now felt that a third cross-option concerned with design and manufacture would represent a course of value to industry. This is currently being proposed and will consist of the two subjects of engineering design and manufacturing plant automation, together with three others from a menu of both mechanical and production subjects. Thus, even with CADCAM finding its place within a single course it is felt that it is better located within more fundamental subjects as an appropriate technique rather than being a single identifiable subject.

References

Cockerham, G (July 1977) Fitting computers into design. IPC Science and Technology Press CAD ED.

Cockerham, G and Panter, N (July 1981) Microcomputers in numerical control. Sheffield symposium on micro-electronics.

Cockerham, G and Panter, N (October 1981) CADCAM based on a microcomputer aids production of precision tooling. Numerical Engineering.

Cockerham, G and Waite, D (1976) Computer-aided design of sput and helical gear train. *Computer Aided Design* 8 2.

Faux, I D and Pratt, M J *Computational geometry for design and manufacture.* Ellis Horwood.

System/360 APT numerical control processor IBM Technical Publications.

G Cockerham, Sheffield City Polytechnic, Pond Street Site, Sheffield.

```
]RUN CONTROL

THIS PROGRAM PERFORMS N-C CALCULATIONS,
PLOTS THE RESULT & CREATES A M/C TOOL
LANGUAGE FILE. SELECT YOUR CHOICE BY
TYPING THE APPROPIATE NUMBER.

     1-TWO CIRCLES AND AN INSIDE
       BLEND RADIUS
     2-TWO CIRCLES AND AN OUTSIDE OR
       OUTSIDE/INSIDE BLEND RADIUS
     3-TWO LINES AND A BLEND RADIUS
     4-A LINE AND A CIRCLE AND A
       BLEND RADIUS
     5-TWO CIRCLES AND A COMMON TANGENT
     6-A CIRCLE AND A TANGENT OF A
       KNOWN INCLINATION
     7-A CIRCLE AND A TANGENT FROM A
       KNOWN POINT
     8-CREATE A N-C FILE
     9-DRAW AND CREATE A M/C TOOL FILE


     88-CATALOG
     99-FINISH
?1

THIS SECTION DETERMINES THE
INTERSECTION POINTS FOR TWO CIRCLES
AND AN INSIDE BLEND RADIUS

TYPE X1,Y1,R1,X2,Y2,R2
?2,2,2,5,2,2

INTERSECTIONS ARE:
X1=3.5              X2=3.5
Y1=3.32287565    Y2=.677124345

TYPE INTERSECTION NO.(1 OR 2)
?1
TYPE BLEND SECTOR
?1
TYPE BLEND RADIUS
?1
CENTRE OF BLEND RADIUS
X=3.5
Y=4.59807622

INTERSECTION OF BLEND & CIRCLES
X1=3               X2=4
Y1=3.73205081    Y2=3.73205081

DO YOU WISH TO FILE THESE FIGURES
?N
```

Figure 9 *Typical Intersection program*

Figure 10 *Steam iron profile on an Apple monitor*

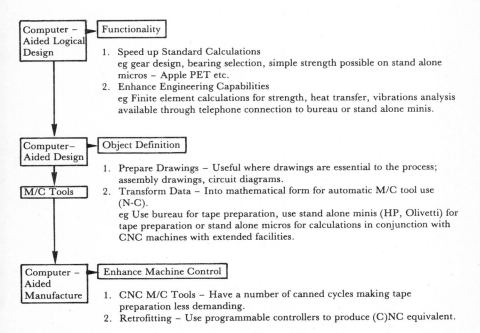

Computer – Aided Logical Design → Functionality

1. Speed up Standard Calculations
   eg gear design, bearing selection, simple strength possible on stand alone micros – Apple PET etc.
2. Enhance Engineering Capabilities
   eg Finite element calculations for strength, heat transfer, vibrations analysis available through telephone connection to bureau or stand alone minis.

Computer–Aided Design → Object Definition

M/C Tools

1. Prepare Drawings – Useful where drawings are essential to the process; assembly drawings, circuit diagrams.
2. Transform Data – Into mathematical form for automatic M/C tool use (N-C).
   eg Use bureau for tape preparation, use stand alone minis (HP, Olivetti) for tape preparation or stand alone micros for calculations in conjunction with CNC machines with extended facilities.

Computer – Aided Manufacture → Enhance Machine Control

1. CNC M/C Tools – Have a number of canned cycles making tape preparation less demanding.
2. Retrofitting – Use programmable controllers to produce (C)NC equivalent.

Discussion

Chairman, B T David, Laboratoire Imag, France:

The last speaker introduced a separation between CAD and CAM but it seems to me that it is not possible to identify precisely the frontier between CAD and CAM. Usually the problem is to put together all aspects of CAD and CAM; graphics, simulation, finite elements calculation and so on, *to integrate* rather than separate. Then the problem of modelling becomes very important. In this case we need a model of an object in CAD and in CAM for different operations and we must use a model which is able to integrate all CAD and CAM aspects of the elaboration process.

G Cockerham, Sheffield City Polytechnic:

Going back to the point I made earlier, an awful lot of engineers wanted to use computers for systems analysis; the analysis of components in systems and how they react to external stimuli. CADCAM as an integrated subject relies upon single component manufacture. One can accept that object definition and manufacturing data generation are tied together but the example I gave of the analysis of a linkage system seems to be a fairly long way away from the manufacture and assembly of that particular system. I am not sure whether CADCAM is really coming to mean object definition and subsequent manufacture rather than systems analysis, which is still an important part of the CAD activity.

T Hargreaves, Huddersfield Polytechnic:

I think the disagreement is possibly more apparent than real. I would agree that, in an educational situation, it's extremely difficult to get the connection between CAD and CAM. It's a real problem and it's rather like the other problems I mentioned. When industry sets out to solve this problem, it knows what problem it's solving, so it knows exactly what it wants to do. If we tried to do the same, I agree entirely that we are down to a single component generation at least. If we get that far. I don't think we have fundamental disagreement on what can be done, and in terms of integration, we both have it in our own ways. The integration that I am talking about educationally is in getting into the different subject areas. I would love to be able to argue that you can integrate CAD and CAM inside our institution. I would tell you quite categorically it's impossible. We couldn't afford the machinery. And this is one of the reasons why the microprocessor lab is there: hopefully we can do some simulation and demonstration in that sort of way in CADCAM.

H C Ward, Teesside Polytechnic:

Having just heard those two statements, I would like to contradict them. You can integrate CAD and CAM quite successfully. We usually teach CAM in second year and we start CAD in the third year, at least the deeper aspects of CAD, but it is quite possible then, that having developed the system, to define the object geometry and then go on to integrate the design into the manufacturing field. We are fortunate in having a Prime 250, where we can run GDS initially to go as system development or the definition form. We can then interface this with GNC and produce the tapes for manufacture. The biggest danger in integrating these is that the designer is not a production engineer and will not in fact produce the best manufacturing method.

Unidentified contributor:

I am not familiar with GDS. Is it a drawing routine rather than an analysis routine, are you doing engineering systems analysis and then manufacture, or are you really doing graphics, object definition and then manufacture?

H C Ward:

We do a little bit of both. We are again rather fortunate, in that as well as the Prime 250, we have a UNIVAC 1108 and Prime 750. So between the three (the three are interlinked) we can do the whole act.

Contributor:

How much did they cost?

H C Ward:

We spent I think £750,000 on the UNIVAC, which was a cut price rate: I believe its actually worth twice that. The Prime 250 system which consists of five work stations with ten terminals on it is around £80,000. The Prime 750 I must admit I don't know. The UNIVAC I must say has proved a very disappointing buy and will probably be scrapped within two years.

G Cockerham:

Do you not feel that by pushing integration on to students that you may give them a false impression? Perhaps they ought to be able to decide for themselves. Now, we don't worry about integration in the sense that we say it's not necessary, or shouldn't be done. We have a self-written program for designing polydine cams and then producing the NC tape and the CAM of intergrated design and manufacture. That's a simple example but I am trying to get further than just the simple examples and say that there are circumstances in which a student can leave an institution saying, 'If you don't do CADCAM it's not worth doing anything,' which may give a false impression. We take our students on an industrial visit to the Department of Industry's expert centres. This year it was Delta Metals in Birmingham and the students there have seen the best software in this country of an integrated nature. But most of them said, 'Well, I am working with a company with 25 employees and the managing director is the chief engineer, and perhaps he is not going to be too happy at me saying that half a million pounds is what he needs to make his firm profitable.' So I am not suggesting that we are telling them that they shouldn't integrate design and manufacture, but that integration is not essential to the successful use of computers within both the design and manufacturing sphere.

L A Grantham, Manpower Services Commission:

Mr Hargreaves made the point that the Polytechnic teaches students what the Polytechnic has decided they need to know. I would like to know on what research this decision was taken and how closely those needs are related to what industry thinks students need to know?

T Hargreaves:

I suppose the answer to that question is 'In our considered opinion'. On the basis that most of the people involved in it are in contact with industry or are recently from industry. We don't make our decisions in a vacuum.

G Cockerham:

That's one of the reasons why we concentrated on project activities. It may not be exactly what everybody in industry wants but certainly from our point of view it's what half a dozen people in local industry have asked us to do, and we are trying to pass on that experience to the students through the project work.

41. A practical approach to CADCAM training

H Holland

Abstract: This paper describes the CADCAM courses that are being run at and by Counting House Computer Systems Limited. It looks at the course programmes and defines the training objectives and methodology as well as making a comment about the divergent characteristics of the average trainee attending the CAM as distinct from the CAD course.

Special emphasis is given to the importance of proper implementation of the total project (as well as the training element) which should take full account of the impact of change on the organization and the individual people who work within that organization.

Background

Counting House Computer Systems Limited was started in 1976. It was situated first in Haverhill in Suffolk and then three years ago moved to its present headquarters at Bury St Edmunds. The main activities are centred round IBS, ITS, and the Training Centre. An associate company manufactures the high resolution graphics terminals which are used by the company.

IBS – integrated business systems

This is an integrated suite of stand-alone modules, designed to satisfy the commercial, accounting and control needs of engineering and other companies. The system has been developed by Counting House.

ITS – integrated technical systems (on which I will be concentrating in this session)

This is a turnkey CADCAM system based on Prime minicomputers and the new graphics work stations. It includes Graphical Numerical Control (GNC) and General Drafting System (GDS). In its most comprehensive form ITS integrates design, analysis, drafting, planning and manufacture using a common data-base accessed via a number of work stations.

Training Centre

This is where the training courses for IBS, GNC, and GDS take place. It has 21 single bedrooms, a main conference room, a fully equipped demonstration room, three syndicate rooms, dining-room, lounge and bar, as well as a squash court and a 4½-acre garden.

The CAM training

The Graphical Numerical Control system which has been marketed and supported by us for some years was developed by the Computer-Aided Design Centre, Cambridge.

It is an NC part-programming system employing interactive graphics which allows the following operations to be submitted visually: turning, milling and drilling, shape nesting, flame cutting, nibbling, and punching.

Any machining errors can be seen on the screen in simulated form, and rectified before control tapes are produced.

Main advantages

Interactive graphics simplifies part-programming leaving the programmer free to concentrate on machining language.

Visualization gives less opportunity for part-programming errors.

– drastically reduces lead time in producing control tapes.

– increases machine productivity owing to less prove-out time on the shop floor.

– improves estimating accuracy and timing.

The CAD training

The General Drafting System was developed by Applied Research of Cambridge Limited and is marketed and supported by us.

The drafting system is a computerized set of drawing facilities and is used to produce and revise two-dimensional drawings in much the same way as they are produced normally.

The designer or draftsman tells GDS what is to be drawn using simple instructions that relate closely to the concepts, techniques and accuracy of manual drawings.

The system produces drawings that look exactly as the draftsman would wish to see them, but all the benefits of using a computer, such as speed, accuracy and automation are exploited to the full, all of which results in overall time-saving and quality increases.

Main advantages

– produces top-quality drawings, accurately and consistently.

– substantially increases drawing office productivity primarily through fast repetition of standard details.

Training objectives

The objectives on each of the courses are to enable the trainee to operate the system, and give him/her sufficient understanding of the basic commands. In GDS it is to create drawings, and in GNC it is to be able to use the system as a tool to generate NC tapes.

The training continues on site as the trainee builds up expertise and confidence with the active help of the support engineer.

Learner characteristics

It is not usually too difficult to identify quite considerable divergencies between those who come on a CAM course and those who attend a CAD course. On the CAM course the trainees typically come from organizations who have got sophisticated machines, and they attend the course to enhance and to build on skills they already have. Their motivation is high and the training they are getting is further evidence of the company's investment in their future. The CAD trainees will frequently be draftsmen or designers who are being brought into contact for the first time with the new technology and for whom all the normal problems of introducing technological change and the difficulties of retraining are exacerbated by well-founded fears that the system is going to result in the redundancy of some traditional jobs.

Training methodology

The prospective user and the sales executive determine the installation timetable and the specific training requirements including the pre-course briefing for the selected candidates for the CAD or CAM course.

The five-day GNC course (CAM)

Systems

- filing structure in the Prime system
- log-in and log-out procedures
- use of the editor.

K curves

- defining points, lines, circles
- binding up geometric elements to form the part shape
- practical exercises.

GNC viewing mode

- handling the various shapes – rotating, copying, mirroring, etc.

Turning or milling or nesting, etc

- thorough review of the turning, milling, nesting or other procedure together with practical examples.

Abstract:

- procedures of running a job from beginning to end
- how to run a post-processor and how to output paper tape for the machine tool.

The five-day GDS course (CAD):

Introduction

- use and management of the system
- production or revision of individual drawings
- organization, storage and issue of whole sets of drawings
- appreciation of filing structures
- log-in and log-out procedures
- appreciation of the use of the editor.

Basic drawing techniques

Computer facilities

GDS control

High emphasis is placed on 'hands-on' experience – at least 75 per cent of the course – with the remainder spent on demonstration and lectures.

There is also a high tutor-to-candidate ratio which is especially helpful when coping with the different ability levels of people attending and the wide variety of age groups.

After the five-day course the responsibility for the successful implementation is in the hands of the support systems. For example, when the post-processor is installed, the support engineer will go to the customer's site to show them exactly how to format GNC to produce a working program.

Important considerations for any CADCAM installation

Co-ordinated planning

Specifically of the training requirements of those who are principally and in any other closely related way involved with the planning of the total project.

Selection and training of a system manager

With a properly defined role which will include responsibilities (among others) of devising and controlling the management of the system with respect to back-up, security, etc; organizing the recruitment and training of operators; ensuring that quality is maintained and there is proper planning and control of the work through the system.

Early involvement

Participation of everyone involved in the proposed system.

Detailed preparation

For example, decision-making on the environment, on consequential changes, on the timetable, on the work needed to be done before implementation, on the definition of the roles of everyone who will be involved.

Integrating the planning of the training

First, identify who is to be trained, in what, by whom, where it will be done and when. This may well include awareness briefings in plant, pre-course briefing of all those who are going to receive specific training in CADCAM techniques, the course itself, implementation and on-going support back in the company and the retraining of others who are going to be affected by the system going live.

Timing

This is important, and can be particularly crucial when the delay in one element has a detrimental bearing on another. For example a successful five-day course can lose a great deal of its value if the system is not ready to run on the following Monday or if the course has been arranged to be held the week before Christmas or immediately before an annual shutdown.

Understanding how we bring about change

Translate training needs into action. There are four factors essential for change to happen and they are as follows:

    (i) dissatisfaction with the present state
   (ii) a vision of something better
  (iii) ease in taking the first step
  (iv) the cost of change (direct costs, indirect costs of upheaval and disturbance and/or the cost of not changing) must not exceed the sum total of (i) + (ii) + (iii).

If one consciously works through this formula the appropriate strategy for achieving change for one's own business situation becomes apparent. The greater the dissatisfaction with the present, the more clearly defined, quantified and pinpointed, the greater the willingness to change. If the business is too 'fat and happy' there will be reluctance to acknowledge any training need and the 'change agent' will need to create dissidence and shake complacency, probably by stressing the external pressures for good or evil which will apply even if there is little cause for complaint within the organization. However, dissatisfaction alone is not enough. Indeed too much can create desperation and a 'better the devil you know' attitude. For the 'lean and miserable' organization emphasis must be on the second factor – building the vision of something better and the desire for the new situation. In any case the all-important requirement is what to do first and unless it is easily understood and accepted, change will not take place.

Conclusion

What I have concentrated on is my experience at Counting House and some of the thoughts that spring from the experience we have gained. In doing so one should not lose sight of the broad picture that is about us.

The rapid development of computing from simple number crunching, to include data handling, on-line techniques and then more recently interactive techniques, are all coming together to form a linked hierachy of distributed computing power which will evolve into the flexible design and manufacturing systems of the future.

The central feature is the interactive work station and the CADCAM systems. These provide the means of linking application tasks and separate processes in industry, and of gaining access to further computer power and data banks of engineering and business information in the background.

And because CADCAM systems provide the means of communication and control of ideas and judgements man to man as well as man to machine, they are destined to play a key role in the shape and form of future industry (Report on CADCAM Training and Education by the CADCAM Association).

For this reason and also because they form the human face of computing, it is important that the whole concept of integration of team work and the man/machine relationship be taken fully into account in education and training, quite apart from the need to produce the necessary skills to sustain the technology.

The application of CADCAM is being recognized more and more in the world as the key to productivity and the stepping stone to the future regeneration of industry. Effective training, implementation and use of this technology is essential if we are to compete and survive.

H Holland, Training Department, Counting House, Computer Systems Ltd, Fornham House, Fornham St Martins, Bury St Edmunds, Suffolk.

42. Promoting industrial awareness of CADCAM*

M W Looney

Abstract: This paper will describe the work being carried out at UMIST and MTIRA to encourage the industrial application of CADCAM. Early in 1981 UMIST installed two turnkey systems to help educate industry and answer the questions companies ask in assessing possible application of CADCAM in their organizations. This is achieved by providing practical demonstrations and giving company staff training and the opportunity to gain 'hands-on' experience of CADCAM. The method of operation of the facility and the nature of companies using it will be described, together with their objectives.

MTIRA initially held seminars for its members to assess the level of interest in CADCAM, to try to identify application areas and to determine how the Association could best assist industry in this area. These culminated in a formal agreement between MTIRA and UMIST for co-operation and selection by the Department of Industry as a joint Practical Experience Centre. The conclusions of the seminars and the work now in progress will be discussed.

*This paper is published by courtesy of UMIST and of the Director of the Machine Tool Industry Research Association.

Introduction

The objectives of this paper are to describe activities at the Machine Tool Industry Research Association (MTIRA), Macclesfield, and the University of Manchester Institute of Science and Technology (UMIST) aimed at encouraging industry to use CADCAM and in particular those leading up to the establishment of a Practical Experience Centre under the Department of Industry's CADCAM Awareness Scheme. Both the Scheme and the Experience Centre are also discussed.

CADCAM at UMIST

Background

UMIST has been involved in the industrial application of CADCAM for approximately six years. In 1979 a major project was started with a company manufacturing textile machinery under the Teaching Company Scheme, to carry out a feasibility study into the potential application of CADCAM, to select and justify a system, to plan for its installation and carry out its implementation. Government funding was provided for the purchase of the initial system which was to be installed in the Company's premises. Unfortunately between the order being placed and installation the Company was severely hit by the recession and late in 1980 had to withdraw from the project.

Rather than cancel the order and incur cancellation charges discussions were held with a number of companies in the North-West to determine whether and if so, how, the system could be used to assist and encourage industry in using this technology. Discussions with about 30 major companies who were not users of CADCAM revealed a high level of general interest, but a lack of knowledge of what systems could do and in particular how this related to their own activities. There was agreement that the best way of overcoming this situation was for company staff to obtain some 'hands-on' experience of technology. In general the companies were open about this lack of knowledge and realized that this was something that they would have to find out about and assess, if not immediately, then in the next two or three years.

Industrial use of UMIST's CADCAM facility

It was concluded from these discussions that there was sufficient support to operate a scheme to assist these companies. The above-mentioned CADCAM system, shortly followed by a second were therefore installed at UMIST. It was intended to work with companies who wanted to carry out an in-depth investigation of CADCAM into, for example, how it would help in its particular type of work, how much effort would be needed to build up libraries etc, how standards and macros could be used, or how to give their staff some experience of CADCAM equipment.

A scheme was established which enabled interested companies to come for initial discussions and demonstrations of CADCAM. However, as stated above, it had been concluded that the best way for companies to gain a full appreciation of CADCAM was to carry out work on a system.

Following these initial activities companies were therefore offered a package of training, 'hands-on' time and support. A one-week operator training course for two people was offered, followed by 100 hours work station time, during which trained operators were available to help overcome problems which would inevitably occur. A fee of £3,000 was charged for this package.

In order for this limited time to be used effectively a number of guides were established. First, before work started a set of objectives must be established and an exercise to meet them defined. This must be realistic in terms of both the amount of work involved and its degree of difficulty. Obviously companies need to appreciate more than a system's basic facilities; which is all the training can cover. Demonstrations are therefore given to support the 'hands-on' exercise.

The intensity of system use has been found to be important. The operators need to start using the system as soon as possible after training and must use the 100 hours in a maximum of two or three months, ie use the system for between one and two days a week. If this level is not reached, it has been found necessary to provide refresher training courses.

This Scheme has been in operation for 12 months now and has been used by a range of medium and large engineering companies. There is a steady level of interest in it with companies initially welcoming contact for discussions and demonstrations with an impartial, experienced user and leading on to use of one of UMIST's systems.

Student involvement with CADCAM

Although the systems at UMIST were installed primarily to assist industry, the Institute obviously also has a responsibility to send out graduate engineers who are aware of CADCAM. System time is a major constraint on giving the large number of mechanical engineering undergraduate students at UMIST real experience of the systems. However, demonstrations are given to students in small seminar groups and a limited number have the opportunity to carry out projects on the systems.

Similarly, there is a limited amount of normal post-graduate work carried out using the systems. However UMIST has Teaching Company Schemes with a number of companies in the CADCAM field and research associates on these schemes are major system users.

Teaching Company Schemes

Teaching Company Schemes are joint ventures between a university and a company to encourage exceptional young graduates into manufacturing industry and to stimulate interchange between staff of the company and the university involved. The graduates are appointed as research associates and carry out a substantial project in the company which has significant industrial importance. They are jointly supervised by company management and university staff.

UMIST has a number of Teaching Company Schemes in the CADCAM field, and a new multi-company scheme has recently been approved which will significantly expand the number of companies involved. The Schemes are either in companies who already have CADCAM equipment installed, in which case the projects are concerned with expanding its use into new areas, or in companies who are planning for its implementation. In general the research associates work on a project for two years and submit their work as an MSc thesis.

CADCAM at MTIRA

MTIRA has been a user of computers for engineering design and analysis since 1967 and a large number of programs have been written for use in the machine tool industry and elsewhere. One of the Association's objectives throughout has been to help and encourage its members and industry at large to make effective use of such methods. It was a natural development to initiate activity in the CADCAM field. A seminar was therefore organized in March 1980 for invited delegates to discuss CADCAM and the role MTIRA might play.

First seminar – March 1980

This first seminar was attended by 30 delegates from the machine tool industry, nearly all senior members of design departments. After initial presentations on CADCAM, the seminar took the form of a structured discussion.

It was apparent that the majority of companies had given the design aspects of CADCAM some consideration and had identified three major problems it could help solve:

(i) shortage of skilled draftsmen
(ii) design and quotation lead times too long
(iii) drawing quality and consistency.

A number of application areas had also been identified and again there was consistency in the delegates' ideas. The main applications were seen as:

(i) tendering and tooling design
(ii) electrical and hydraulic schematics
(iii) parts listing.

Most people had difficulty seeing how CAD would help in new machine design in the short term because of the problem of data-base creation and the only use of 3D identified was in detecting collisions on tooling layouts. It was anticipated that manufacturing could be assisted by more easily dimensioning drawings as they were required for NC programming etc. Process planning and NC programming were thought to be longer-term applications. The consensus was that costs were at that time too great for the benefits that had been identified; however if costs could be reduced there would be considerable real interest.

Thus, although the delegates had obviously some knowledge of the potentials offered by CADCAM, they were not aware of the full range of applications, for example the use of macros, standards, analysis programs or the range of manufacturing applications.

This lack of awareness was in part inhibiting further progress.

The conclusions of this seminar were:

(i) that the possibility of setting up a 'satellite' system based on a processor at MTIRA should be considered. This would have a number of remote work stations that could be installed in members' premises for short periods so that they could carry out some 'hands-on' exercises to assess the technology more thoroughly. Members were asked to consider contributing to the cost of setting up such a system;

(ii) companies should consider further the application of CADCAM to their operations;

(iii) a second seminar should be held to discuss the subject further.

Second seminar – May 1980

The satellite scheme was initially discussed, but could not be advanced because the industry could not at that time commit itself to making the contributions estimated at £10,000 per company that would be necessary as there would be few tangible benefits. This did not imply a lack of interest in CADCAM, but was brought about by the recessed state of the majority of machine tool companies. There was now general belief that CADCAM would be essential in the future and that progress towards its implementation should be made in whatever ways were possible. In particular, companies wanted to prepare themselves so that when money was available they would be fully aware of how CADCAM could assist them and the range of equipment available; such that they could take advantage of it as rapidly as possible. It was agreed that a number of group visits should be organized to suppliers of CADCAM systems to start this programme. These took place at regular intervals throughout 1980 and 1981.

It was recognized, however, that the key to a real understanding was not just seeing and talking about CADCAM, but 'hands-on' experience. It was further recognized that there were many other companies outside the machine tool industry in exactly the same position and with the same concerns as the delegates. It was therefore agreed that the scope of this activity should be widened and that on this basis the Department of Industry should be approached to discuss the possibility of support for setting up a facility to enable companies to obtain this experience.

Submission to the Department of Industry

After initial discussions a submission as outlined above was made to the Department of Industry. However no quick answer was expected because it was still considering its response to the ACARD report on CADCAM (ACARD, 1980) and was itself at that time being reorganized.

Shortly after this, the possibility of UMIST establishing a CADCAM facility arose. The two organizations have work together on many projects in the past and agreement was reached for MTIRA to use a substantial amount of time on the UMIST system to build up operating experience. It was also agreed that they should work together to run CADCAM appreciation courses.

When the Department of Industry picked up MTIRA's earlier submission a working relationship in this field had been well established. It was agreed that the submission should be modified to take account of this relationship, the equipment already available at UMIST, and also the complementary areas of expertise of the two organizations.

The Department of Industry CADCAM Awareness Scheme

The submission now requested by the Department of Industry was for MITRA in conjunction with UMIST to operate a Practical Experience Centre which was to be one part of a national CADCAM Awareness Scheme. This submission was accepted by the Department in late 1981.

The basic objective of the Awareness Scheme is to encourage industry to take up CADCAM more rapidly than it would otherwise. Before discussing the activities of the Practical Experience Centre, the various parts of the Scheme will briefly be described.

Management seminars

A series of about 120 seminars aimed at the chief executives of small and medium companies not using CADCAM is being organized over the next three years by the Institutions of Mechanical Engineers and Electrical Engineers. They are designed to illustrate the advantages of CADCAM so that delegates can appreciate what can and cannot be achieved. They are non-technical. It is intended that they will stimulate interest in the subject and lead to company participation in the more detailed aspects of the Scheme.

Practical Experience Centre

The Centres will be described in the next section.

CADCAM consultancy

The Department of Industry will provide a grant covering the first £2,000 of the cost of a feasibility study to assess the economic or technical feasibility of using CADCAM or to advise on its application. The project must be carried out by an approved consultant. This scheme is administered by Warren Spring Laboratory.

Visits to user firms

The Department of Industry is arranging for some of the most experienced CADCAM users in the country in as many different fields as possible to show their installation to, and discuss its operation with, other companies.

CADCAM information point

An enquiry point is being established to provide interested companies with information on courses, consultants, equipment, software etc.

MTIRA/UMIST Practical Experience Centre

The MTIRA/UMIST Centre is one of five, the others being at British Ship Research Association, CAD Centre, Delta CAE Ltd and National Engineering Laboratory. They are geographically spread, but the distinction between them is not only geographical, they also have different particular areas of expertise. MTIRA and UMIST are both primarily involved with light and medium engineering. Their experience covers all aspects of the application of CADCAM from design and drafting through to manufacture.

The Centres have a range of CADCAM equipment and staff who are both experienced in the use of that equipment and also generally knowledgeable in the industrial application of CADCAM and the range of equipment available. The MTIRA/UMIST Centre is based on the two systems already described at UMIST and two further systems more recently installed at MTIRA.

The role of the Centres is to allow key staff from potential user firms to study the techniques, try out the equipment and obtain expert and objective advice. People coming to a Centre will either already have received a basic introduction to CADCAM, either in the earlier stages of the Awareness Scheme or elsewhere, or they will be coming for their first contact with the technology. They therefore provide a range of services.

The Centre provide assistance to all mechanical engineering companies with a UK manufacturing facility. However, particular efforts are being made to assist smaller companies, who perhaps have automatically assumed that CADCAM could only be justified by larger organizations. A Centre might for example explain to them first, the wide range of system types available today and second, the need to consider CADCAM against the background of the company's whole operation rather than from within a particular department. The MTIRA/UMIST Centre expects to assist approximately 120 companies per annum during the three-year life of the Awareness Scheme.

Initially a company will come for discussions and perhaps a demonstration of CADCAM. As described above this may be a company's introduction to the technology. In either case, in part the objective is to explore how the Centre can further assist the company in its investigation of CADCAM. There are many opportunities today for companies to learn about CADCAM in general and hear the benefits it offers. The Centres provide the opportunity for more specific work to relate the technology to a company's own operation. After this initial stage exercises are planned with well-defined objectives. The exercises may consist of a series of detailed demonstrations identified as of particular relevance to the company. These will be carried out in order to show both what the systems are capable of and also what is involved in carrying out the work. This will enable the company to gain a realistic assessment of the work that would be needed to use CADCAM effectively in that area.

Alternatively or in addition to this, company staff can carry out a 'hands-on' exercise. This will inevitably be limited to the more basic applications of CADCAM, but will lead to a greater understanding of the techniques. Basic training is provided to company staff using this facility and Centre staff are on hand to help overcome the difficulties that may be encountered. The training is structured to enable the specified exercise to be carried out.

Initial activity of up to one day's duration is funded by the Department of Industry. This includes both staff time and system time. Beyond this charges are made, but a subsidy of £1,000 is available on the first £2,000 of expenditure at a Centre.

Reference

ACARD (Advisory Council for Applied Research and Development) (1980) *Computer-aided design and manufacture.* HMSO: London.

M W Looney, Department of Mechanical Engineering, Manufacturing and Tools Division, UMIST, PO Box 88, Manchester M60 1QD

Discussion

P N Smith, Dudley College of Technology:

Can Mr Holland tell me the cost of his five-day courses?

H Holland, Counting House Computer Systems Ltd:

The five-day courses are fully residential, including all accommodation, course notes, etc. For five days from Sunday night to Friday afternoon they are £100 per day. People who normally attend the courses have actually bought the system, and this is part of the deal. We do have some people who come to evaluate a system but it's generally for people who have bought a system and are now ready to start using it. It's probably something like £70 for the actual course, tuition fees, and £30 for the 24-hour rate to stay on the premises and use of all the facilities.

J Kinsler, Paisley College:

I wonder if both speakers can tell us why their courses are of five-day duration?

M W Looney, MTIRA UMIST:

In our case we wanted to provide people with enough training to be able to carry out a meaningful exercise in the shortest possible time, and we initially thought that five days would enable us to provide just enough training, and our experience over the last year has shown us to be right in that assumption.

J Kinsler:

So what's your answer to Dr Abbas, who wants to run a year-long course?

M W Looney:

Well, we are not training people to be experts in the system, we are providing them with absolutely the minimum tool to do something meaningful with somebody skilled and experienced looking over their shoulders.

J Kinsler:

Do you actually think one week is sufficient?

M W Looney:

Yes, we have trained a lot of people in this way.

H Holland:

In our case the five days is essentially the basic course, a launching pad on which to start, but the training then continues on site as they start to use the system. That's when, together with the relevant person from our company, they build up the experience, but the launch is put over in the five days. Essentially in the five days we can have a very thorough grounding in the basic commands and how to operate the system.

A L Johnson, Cambridge University:

Can Mr Looney give us some indication of the range of packages which these practical experience centres offer to industry? Obviously some of the best CAD packages are extremely expensive to buy. Do these practical experience centres have special access to these packages so that industry can appreciate them?

M W Looney:

The various centres do have an example of every type of system, from the lowest cost, to the drafting system up to the most sophisticated turnkey systems, so yes, we do have that range between us. In our particular case, our systems are all at the top end, but we have structured the demonstrations that we do in such a way that we can identify at what point the capabilities on the lower-cost systems end. In most cases the methods of operation are similar enough for us to be able to identify what the differences are and point them out to people.

P de Santos, Preston Polytechnic:

We have a lot of experience in the running of some of the systems described. When you are talking about training, you are talking training on specific items, a specific piece of equipment or a specific method. I am suggesting that the subject must have wider parallels which we can identify, and therefore we can talk in terms of education rather than training.

M W Looney:

I would agree with that. But having said that, the real essence of what we are trying to do here is to give people 'hands-on' experience. In talking to a large number of people and to engineers in general, there is nothing like getting their hands dirty to come to grips with and understand the principles involved, and it does necessarily involve a particular technique.

Unidentified contributor:

What you are saying is when we are looking at it as an educational institution or a training centre, we are looking at things too narrowly and that in the interests of technology, we ought to think a bit more broadly.

P de Santos:

I think there is a need for both. You are talking mainly to educationalists who haven't got access to the sort of facilities you have. How do we build some of this into some of our undergraduate courses in particular as an educational medium rather than as a training necessity?

Unidentified contributor:

Mr Santos has highlighted the aims as being different. Some papers presented at this conference were looking essentially at education, at CADCAM education. We were not concerned with product-specific training or system-specific training. Now the two papers you heard today were product-specific/system-specific, and I do think they make no bones about it, that was what it was all about and hence the objections are limited. If you are thinking about an education training programme, then those would be perhaps an aspect of your total educational program. How you fit that into your total educational program is your problem.

J Butler, Technical Change Centre:

Could Mr Looney expand on the teaching company schemes which are more educational than training packages, that might provide a partial answer to the enquiry here?

M W Looney:

I don't think they are because teaching company schemes are company-specific, so a company will always have a particular piece of equipment, or be going to implement a particular piece of equipment, so they are once again product-specific.

J Butler:

Yes, but surely over two years you are just teaching a teaching company associate to learn how to use a specific product? They are involved in a much more complex process.

M W Looney:

They are concerned with application, rather than the system for its own end, but it's always the application of a particular system.

S Bennett, Ford Motor Company:

In a large company like Ford, we are becoming increasingly aware that one of the main obstacles to change is the existing company structure, and in a technology like CADCAM it cuts across many existing empires. I wonder if other people have experienced the same thing and how, if at all, they have managed to cope with this very serious problem?

H Holland:

It is amazing that on this course, I have sat down to two meals and the people I have sat with have both raised this point about having got everything sorted out and then hitting company problems in terms of unions, etc. We even heard that something like CADCAM is from the floor cutting across existing company structure, so we should probably accept that there is a problem and do something about tackling it.

Mr Wilmett, Philips Electronics:

I think one of the important things is to organize round the technology and not fit the technology into an existing organization, because then you have got all the problems of cutting across the boundaries. It's far more efficient to organize around the technology. I am talking about the far broader issues than CAD, such as the total integrated business, because my view is that the discussions are mainly on CAD, whereas in fact many of our problems are across the whole front of CAD, computer manufacture, computer-aided testing, manufacturing requirements, planning which then brings the logistical approach of computing into the whole of the business operation.

D G Smith, Loughborough University:

Could Mr Looney enlarge on one thing that he just touched on briefly? To my mind education extends to a much wider field than just actually using computer equipment and training in this. What advice do you give to companies in regard to how they assess the suitability of computers for their particular work?

M W Looney:

When a company comes to us, we sit down at the outset and find out why they have come, and quite frequently this is one of the topics on the agenda. We are consultants, but in the practical experience centre type activity we are not providing a consultancy service. However, we will talk through these aspects of CADCAM, we will talk about the different types of systems that are on the market place, and how they should possibly go about selecting and evaluating systems. We will provide some general guidance on those areas.

D G Smith:

Can you direct them towards consultants, or people who can help them?

M W Looney:

Yes, we will do that. We will take it a stage further than that ourselves; we will undertake consultancy work for them along those lines if they want us to.

D G Smith:

Could you go a little more deeply and say what sort of criteria you suggest that they consider in the evaluation of the suitability of systems?

M W Looney:

I think there are the technical aspects of the system. First, you have to identify clearly what you want the system to do and you have to sit down and bear in mind that it's a tool that affects many areas of a company's operation. The system has to be specified from the outset against that background and not from within a particular department, so it's very important to get the specification right from the outset. You have to look very carefully at the suppliers you talk to – there are an awful lot coming on to the market at the moment. Having bought a particular system you are very much, at the present time anyway, locked in with them, so it's important to get that right. I think those are two of the main points we try to assess.

D G Smith:

When we are introducing any concept of new technology, in this instance CADCAM, the two key words are consultation and participation. The impact is going to be significant in other departments as well and it's very important that you undertake some form of audit to identify the areas of impact. In undertaking that audit you may find, as I have found, that forming project teams to undertake things like skill audits as well as look at the impact of the technology itself, is a great influence on the way people accept that change. If you can get key people from significant areas of impact to be members of that project team which then consults with the various departments, I think you will greater the element of success in the introduction and lessen opposition.

A Guy, Hatfield Polytechnic:

I would be interested to know from the speakers whether in the companies which come to them, it's the management who is trying to impose a system on to the work force or whether the users within the company are forcing the management to come forward and request these systems?

Unidentified contributor:

The best way to tackle a project of this nature is to involve the people at as high a level as possible. If we get them involved and aware that there is a need to do something and then be able to support that as it goes down the line, it then ends, hopefully, in the right sort of system for the company. On the other hand, of course, there will be a push from further down and as time goes on and as more and more people come through from schools (even now, from September with grants to primary schools) people are getting more and more understanding and experience of little micros and that push will become steadier and firmer as years go on. As people join companies and organizations there will be pressure to improve from both ends of the company. One would hope that in the normal event the managing director or the key people in the company ought to be the ones that are conversant with developments and this is the line that has taken in these awareness seminars on CADCAM throughout the country. First of all, you have a very brief introduction for half a day or three-quarters of the day with top people in the company, so that at the end of that time if they need convincing that CADCAM is something worth looking at, all they need to do then, presumably, is to go back to their company and call a meeting to set the thing in motion. Providing they have achieved that for a start, then that's something which is well worth while.

M W Looney:

We have found there are three, or at least three different areas, types of people coming to us. There are the very top people in the company who are taking the policy decisions. 'This is the way we ought to be going'. And then, pushing back downwards, there are then the functional managers who are trying to start off in their area and then trying to interest other people at their level in the technology. Then there are the users, as it were. The important thing is that everybody in the end pulls together and goes in the same direction as we have just heard.

Part 7:
International experience

43. International implementation of a CAAD project in schools of architecture

T W Maver and R Schijf

Abstract: With funding from the European Cultural Committee, work is in hand to develop a modular course structure which will promote CAAD education in schools of architecture throughout Europe. The paper identifies the need for the course structure and describes the pilot work jointly carried out by the University of Strathclyde and the Technical University of Delft. The course structure proposed in the paper will be the focus of discussion and elaboration at a meeting of European Schools of Architecture scheduled for October 1982, in Delft.

Introduction

Young people currently studying in schools of architecture will be at the height of their professional career around the year 2,000. The schools of architecture have a responsibility to introduce students to the concepts underlying the new generation of computer-based design aids and to their application in design project work, for, quite clearly, this new generation of modelling aids will have as dramatic an impact on the process and product of architectural design and on the relationship between architect and client, as did the introduction of the drawn plan and elevation, 5,000 years ago.

The aspects of architectural practice on which computers are already making an impact are:

- management : office organization and project control
- drafting : the production of working drawings
- design : analysis, synthesis, appraisal and decision

Each aspect is as important as the other two and ultimately the full potential of the computer will be realized only when all three are brought into systematic conjunction. In the limited time available in undergraduate courses, however, and given the difficulty of simulating office conditions, including relationships with the contractor, the starting point in schools of architecture should be with computer-aided design, or, more specifically, with computer-aided architectural design (CAAD).

Consideration of how best to introduce CAAD into the undergraduate curriculum brings one immediately to the differentiation between *using* and *developing* CAAD techniques. Primarily, in schools of architecture, we are concerned with fostering in the student an understanding of the causal relationships between design decisions and performance consequences: the *use* of CAAD programs, as we shall see in a later section of this paper, simply as 'black boxes' which predict the cost and performance characteristics of alternative design hypotheses, affords the student a special insight into the nature of the design activity and the complex interactions between economic, environmental, structural and aesthetic criteria. All students, then, should have the opportunity to *use* typical CAAD programs and, through their use, to develop a broad understanding of their ultimate potential and of their technical, economic, social and professional implications.

It would be a mistake, however, to limit our commitment to *use* only. As the technical complexity of building has grown, responsibility for significant sub-systems within the building have been passed by the architect to a range of specialist consultants – the structural engineer, the environmental engineer, the quality surveyor, the perspective artist, etc; comprehensive CAAD models offer the opportunity to regain the integrative responsibility for the design in its entirety. It is important, then, that the *development* of CAAD programs, ie the design of the design aids, is seen as an architectural responsibility and that at·least a selection of schools of architecture offer the opportunity to students to become expert in the theory and practice of CAAD. Those taking up the opportunity would become the authors of new software and the specifiers of hardware appropriate to the architectural profession and the building industry. There is, however, a real difficulty facing schools of architecture that wish to get started in CAAD: money is tight, expertise is scarce and the signal to noise ratio in the architectural press is low. This paper describes the first steps in an initiative, modestly funded by the European Cultural Committee, to develop a library of CAAD course units that could, in many combinations, be used at architectural schools throughout Europe for undergraduate, post-graduate and mid-career education (Schijf, 1982). The paper describes the pilot work to date, the set of proposals suggested by the pilot work and the likely next steps in the endeavour.

Pilot work

The Department of Architecture and Building Science at the University of Strathclyde and the Department of Architecture at the Technical University of Delft have been working jointly to pilot a range of course modules which might build into an appropriate modular course structure suitable for schools of architecture throughout Europe. With the help of a number of other schools, notably the Schools of Architecture in Antwerp, Tilburg, Sheffield and Aberdeen, experience has been gained in running studio projects using a range of computer programs. The following subsections described the programs and their use in three schools.

The computer programs

To date, some five or six computer programs have been piloted. The programs used were those readily available to the investigators and do not represent a definitive or complete list of the programs which will provide the back-up to the modular course structure. Those described here, are, however, typical of those which will be required.

GOAL:

GOAL (General Outline Appraisal of Layouts) was developed by ABACUS at the University of Strathclyde (Sussock, 1981) to allow the comprehensive appraisal of alternative design hypotheses at the earliest possible stage in design. The architect (ie the student) inputs (among other things) a proposed geometry and construction. Geometry (currently orthogonal, soon to include non-orthogonal) is input floor by floor using a digitizing tablet or at the terminal screen, either numerically or graphically. Constructional choices are made from an expandable constructions data base containing cost and thermal characteristics. Output from GOAL includes a range of cost and performance predictions, including:

- wall, floor and roof areas; volumes; wall to floor ratio, etc
- measures of functional planning efficiency
- environmental conditions – thermal, lighting, acoustic
- energy utilization – peak, annual; compliance with regulations
- costs – capital, recurring and costs-in-use

Design changes are easily made and their cost and performance consequences rapidly determined. Results from a large series of design searches can be filed and retrieved for comparison between students and between schools. The geometry held in file can be passed to the program BIBLE.

BIBLE:

BIBLE (Buildings with Invisible Back Lines Eliminated), also developed by ABACUS at the University of Strathclyde (Parkins, 1979), generates perspective views of any geometry passed to it from GOAL or input to it directly by the user. When initiated the program draws a small plan view at the centre of the terminal screen and provides a set of commands which allow the user, amongst other things, to:

- specify eyepoint, focus point, cone of vision, etc
- choose if the back lines are to be hidden, shown full or shown pecked
- choose a full three-point perspective or orthogonal projection
- match the perspective to an existing site photograph
- get the view drawn either on the screen or on a plotter

GABLE:

GABLE was developed at the School of Architecture at the University of Sheffield (Lawson, 1981). It runs on the Tektronix 4050 series of microprocessors and allows checking of daylight, sunlight, energy consumption and other appraisals to be carried out. Three-point perspectives of the geometry can be produced, from viewpoints inside and outside the building envelope and the production of conventional drawings suitable for the contractor is possible.

IVOGAR:

IVOGAR, under development in the Department of Architecture at the Technical University of Delft is intended to provide a general drawing aid and to effect appropriate interfaces to, and between, GOAL, BIBLE, GABLE and other software packages.

Use of the programs

Some or all of the computer programs are in use, exploratively, in a number of schools of architecture in the UK, Continental Europe and elsewhere in the world. The uses described here are specifically relevant to the development of the modular course structure.

Figure 1 *Comparative appraisal by a fourth year Strathclyde student of alternative hotel designs using the program GOAL (see page 9)*

Strathclyde:

The Department of Architecture and Building Science at the University of Strathclyde runs undergraduate and mid-career courses and will shortly introduce an MSc in CAAD. At undergraduate level, the differentiation is made between the needs of those who will *use* CAAD and those who wish, optionally, to *develop* CAAD. A number of second year and fourth year design projects have a CAAD bias including a five-week hotel design project in fourth year, which makes extensive use of GOAL and BIBLE. Figure 1 is a photo-reduction of an A1 board, one of six, submitted by a student taking the project; it shows four of the ten or so designs he appraised using GOAL. Figure 2 is a photo reduction of another of his boards, showing a range of perspective views based on the output from BIBLE.

Figure 2 *Perspective views of a Strathclyde student hotel design based on the output from the program BIBLE (see page 9)*

Delft:

The Department of Architecture at the Technische Hogeschool, Delft, offers third and fourth year students a number of CAAD options which provide a conceptual framework, give 'hands-on' experience of a range of available programs and provide the opportunity to use certain of the programs (GOAL, BIBLE, GABLE, IVOGAR and others) in a major studio project. Projects have included a 100-bed hotel and a central lecture room block for a University campus. Figures 3 and 4 show perspective views of the campus with the proposed lecture block as drawn by BIBLE.

Figure 3 *Perspective view of a Delft student design for a lecture block in the context of existing campus buildings, as drawn by the program BIBLE (see page 9)*

Figure 4 *Another perspective view of a Delft student design for a lecture block in the context of existing campus buildings, as drawn by the program BIBLE (see page 9)*

Antwerp and Tilberg:

Short courses and limited projects have been run, with inputs from Strathclyde and/or Delft, at the Nationale Hoger Instituut voor Bouwkunst en Stedebouw in Antwerp and the Akademie voor Bouwkunst in Tilburg. In the case of Tilburg, students travelled to Delft to use the computing facilities; from Antwerp, students dialled in over the international telephone network. As the time available for project work in both institutions was severely constrained, students did not generate design proposals from scratch; each was given a design proposal generated by a Delft students and asked to improve its cost/performance profile using GOAL and BIBLE. Figure 5 summarizes the comparative evaluation of design alternatives by Tilburg students.

Figure 5 *Comparative appraisal by a Tilburg student of four alternative hotel designs using the programs GOAL and BIBLE (see page 9)*

Exposition

This unit comprises lecture, seminars and demonstrations. The lecture material will include the history of computing, hardware types and trends, software concepts and artificial intelligence, the relevance of design methodology and systems concepts, a review of computer applications in architecture (management, drafting and design), economics, problems and prospects and future trends. Demonstrations may be in-house, in other schools and, best of all, in local practices.

Preparation

This unit gives the student 'hands-on' experience in a controlled way. It is intended to introduce the student to three aspects:

(i) the man-machine interface – keyboard, screen, tablet, plotter, operating systems, ie what button does what
(ii) software logic – program, files, algorithms, documentation, ie how the input gets transformed into an output in various example programs
(iii) data preparation – collecting, formating and inputting the data needed for a variety of example programs.

Proposed modular course structure

The experience to date at Strathclyde, Delft, Antwerp, Tilburg and elsewhere, points to a course structure with five types of unit, as follows:

1. Exposition

The concepts underlying CAAD, survey of the state of the art, and demonstrations

2. Preparation

'Hands-on' experience of a range of programs and discussion of their form, content and interfaces

3. Application

Using one or more programs in a studio design project

4. Instruction

Acquiring programming skills and knowlege of hardware and software systems

5. Development

Specifying, implementing and maintaining hardware and software systems

Units 1, 2 and 3 are sufficient to prepare students for CAAD use; units 4 and 5 are needed if the student wishes to go on to develop CAAD expertise. The units are described in turn in the following sub-sections.

Application

This is the central unit in the modular course structure and is intended to take its place alongside conventional design projects in the U/G studio. Students would be expected, in addition to submitting plans and elevations at the final crit, to show the sequence of their search for a solution with explicit comparative evaluation of the range of design alternatives which were appraised during the project. For projects of five or more weeks, students would, typically, design 'from scratch'; in shorter projects, students might be given a design (say from an earlier project or from another school) as a starting-point.

Instruction

This unit, offered only optionally, and perhaps only in certain schools, includes discussion of program languages (including instruction in one), data-bases and operating systems and of the comparative performance of micro, mini and main frame hardware.

Development

Again, this unit would be offered only optionally and only in certain schools were the degree of access to hardware and software and to expertise was high. This final unit, with associated project time, is intended to equip graduates to fulfil a specialist role in practice – specifying, implementing and maintaining CAAD systems.

The future

An invitation has been extended to the schools listed in section 1 of the Appendix to attend a meeting in Delft in October 1982, to discuss the draft proposals and to identify commitments which can be made to contribute to and/or to adopt part or all of the course structure. An appropriate response to the challenge of realizing the potential of the new generation of design aids will be possible, we believe, only by collaborative action.

Acknowledgements

The authors gratefully acknowledge the financial support of the European Cultural Committee and the collaborative effort already contributed by colleagues in a number of schools of architecture.

References

Lawson, B R (1981) GABLE: an integrated approach to interactive graphical techniques for modelling buildings. Proceedings, Computer Graphics 81, London

Parkins, R P (1979) BIBLE – a computer program for generating perspective views of buildings. University of Strathclyde ABACUS Occasional Paper no 75

Schijf, R (30 March – 1 April, 1982) Modular CAAD courses – a vehicle to discuss CAAD education. Butterworth Scientific Ltd. Proceedings CAD 82, Brighton

Sussock, H (1981) GOAL – General outline appraisal of layouts. University of Strathclyde ABACUS Occasional Paper no 62

Appendix

Modular CAAD courses project mailing list April 1982

Section 1		*English language*
UK	1.	Plymouth Polytechnic, Department of Architecture, Drake Circus, Plymouth, Devon, PL4 8AA Attn. Prof. Graeme M. Aylward
	2.	Robert Gordon's Institute of Technology, The Scott Sutherland School of Architecture, Garthdee, Aberdeen AB9 2QB Attn. Mr. Lamond, W. W. Laing
	3.	Brighton Polytechnic, Department of Architecture and Interior Design, Houlescoomb, Brighton BN2 4AT Attn. Mr. Philip Ranger
	4.	University of Sheffield, Department of Architecture, The Arts Tower, Sheffield S10 2TN Attn. Dr. Bryan Lawson
	5.	University of Edinburgh, Department of Architecture, 20 Chambers Street, Edinburgh EH1 1HZ Attn. Mr. Aart Bijl
	6.	Liverpool Polytechnic, School of Architecture, 53 Victoria Street, Liverpool Attn. Mr. Paul Coates
Ireland	7.	University College Dublin, School of Architecture, Richview Clonskeagh, Dublin 14 Attn. Prof. Dermot O'Connell
Denmark	8.	Arkitektskolen i Aarhus, Nørreport 20, 8000 Aarhus C Attn. Mr. Kristian Agger
Sweden	9.	Tekniska Högakolan i Lund, Sekt Lonen för Architektur, Box 725, 220 07 Lund Attn. Prof. Christer Bergenudd
	10.	id., Department of Structural Engeneering Attn. Prof. Per Christiansson
	11.	Institutet för Verkstadtsteknisk Forskning, Stockholmsenheten, 100 44 Stockholm Attn. Mr. Bendt Holmer
	12.	The Royal Institute of Technology, School of Architecture, Falk, S100 44 Stockholm 70 Attn. Prof. Olle Wahlström

Norway	13.	Norges Tekniska Høgskole, Department of Architecture, 7034 Trondheim Attn. Mr. L. Nordgärd
Finland	14.	The Finnish Association of Architects, Eteläesplanadt 22A, SF-00130 Helsinki 13 Attn. Mr. Matti Pöyry
Germany	15.	Technische Universität Berlin, Institut für Mechanik, Strasse des 17.Junl 135, D-1000 Berlin 12 Attn. Prof. Dr. -Ing. E. Kernchen
	16.	id., Institut für Bauplanung im Producktionsund Dienstleitungs Bereich Attn. Prof. Dr. -Ing. G. Nedeljkov
Switzer- land	17.	Eidgenössische Technische Hochschule Zürich, Architekturabteilung, Zürich Hönggerberg CH-8093 Attn. Prof. H. E. Kramel

Section 2   *French language*

Belgium	18.	Université de Liège, Laboratoire de Physique du Bâtiment, 15 Avenue des Tilleuls-BAT.D1, B 4000 Liège Attn. Prof. Albert Dupagne
	19.	Centre d'Etude, de Recherche et d'Action en Architecture, Rue Wilmotte 76, 1060 Bruxelles Attn. Mr. J. Massett
	20.	Institut Superieur d'Architecture, Saint Luc de Tournai, Chausée de Tournai 50, 7721 Ramegnies Chin. Attn. Mr. Alain Dequinze
	21.	Institut Supérieur d'Architecture, Rue Fabry 19, 400 Liège Attn. Charles Burton
	22.	Université Catholique de Louvain, Unité Architecture, Bâtiment Vinci, Place du Levant 1, B 1348 Louvain La Neuve Attn. Prof. E. Verhaegen
France	23.	Ecole d'Architecture de Toulouse, Chemin de Mirall, 31057 Toulouse Cedex Attn. Mr. Michel Léglise
	24.	Centre de Recherche Methodiques d'Architecture et d'Amenagement, Rue H. Picherit (CSTB), Nantes 44300 Attn. Prof. J. P. Peneau
	25.	Ecole d'Architecture de Grenoble, 10 Galerie des Baladins, F28100 Grenoble Attn. Prof. Bertrand David
	26.	Unité Pédagogigue d'Architecture de Marseille 70 Route Léon Lachamp, 13288 Marseille Luminy Cedex 9, Case 912 Attn. Mr. R. Billon
	27.	Unité Pédagogique d'Architecture no. 1, 11 Quai Malaquais, 75272 Paris Cedex 06 Attn. Mme. Catherine Urbain
	28.	Ecole d'Architecture de Bordeaux, Domaine de Raba 33405, Talence Cedex Attn. Mr. Y. Lormant
	29.	Ecole Spéciale d'Architecture, 254 Boulevard Raspail, 75014 Paris Attn. Mr. Francois Wehrlin, Directeur
	30.	Ecole National des Beaux Arts de Lyon, 10 Rue Neyret, 69001 Lyon Attn. Mr. Ph. Nahoum, Directeur

31. Unité Pédagogique d'Architecture de Lille, Le Forum – 43 Rue Gustave Delory, 59000 Lille
Attn. Mr. Pierre Eldin

Italy

32. Università degli Studi di Palermo, Facoltà di Architecttura
Attn. Prof. Arch. Margherita De Simone

33. Facultà di Architettura del Polytecnico di Milano, Laboratorio di Computer Graphics, Via Bonardi 3, 20133 Milano
Attn. Mr. Alessandro Polistina

34. Università di Bari, Instituto di Architectura e Urbanistica, Bari
Attn. Mr. Victor Nuzzolese

35. Facoltà di Architettura, Institute of Architectural Design, Via Monte Oliveto 3, 80100 Napoli
Attn. Mr. Claudio Cajati

36. Istituto Universitario di Architettura di Venezia
Attn. Prof. Valeriano Pastor

Spain

37. Escuela Tecnica Superior de Arquitectura de la Coruna
Attn. Mr. José-Antonio Franco Taboada

38. Universidad de Navarra, Escuela Tecnica Superior de Arquitectura, Pamplona
Attn. Mr. Juan Pedro Ros Martinez

Section 3

Belgium

Dutch language

39. Vrije Universiteit Brussel, Afd. Burgelijke Bouwkunde, Pleinlaan 2, B 1050 Brussel
T.a.v. Prof. Dr. Ir. W. P. de Wilde, Ms. M.

40. Provinciaal Hoger Architectuurinstituut, Gouverneur Verwilghensingel 3, Hasselt
T.a.v. Dhr. A. Neville, Dr. E. Vangeel

41. Stedelijk Hoger Instituut voor Architectuur en Stedebouw, Academiestraat 2, 9000 Gent
T.a.v. Dhr. Erik Balliu, Dhr. Marc Poriau, Dhr. Louis Hagen

Nether-
lands

42. Technische Hogeschool Eindhoven, Afd. Bouwkunde, Den Dolech 2, Postbus 513, 5600MB Eindhoven
T.a.v. Ir. Paul Dinjens, HG 5.09

43. Akademie van Bouwkunst Tilburg, Voltstraat 60, 5021SE Tilburg
T.a.v. Drs. Jan Vaessen

44. Akademie van Bouwkunst Rotterdam, Bospolderplein 16, Rotterdam
T.a.v. Ir. Frans Smits

45. Akademie van Bouwkunst Maastricht, Capucijnerstraat, 6211RT Maastricht
T.a.v. Ir. C. Kleinman

46. Akademie van Bouwkunst Arnhem, Sonsbeekweg 22, 6814BC Arnhem
T.a.v. Ir. A. Vos de Wael

Section 4

Organizers

47. University of Strathclyde, Department of Architecture, 131 Rottenrow, Glasgow G4 0NG, Scotland
Attn. Prof. T. W. Maver, Director ABACUS

48. National Hoger Instituut voor Bouwkunst en Stedebouw, Mutsaertstraat 31, 2000 Antwerpen
Attn. Dhr. Richard Foqué

49. Technische Hogeschool Delft, Afd. Bouwkunde, Postbus 5043, 2600GA Delft
T.a.v. Ir. R. Schijf, BK 8.12A

T Maver, ABACUS, University of Strathclyde, Department of Architecture, 131 Rottenrow, Glasgow G4 0NG

R Schijf, Department of Architecture, Technical University of Delft, Netherlands

Discussion

R W Howard, Construction Industry Computing Association:

I was just wondering how these courses tied up with the MSc course which I think you are starting shortly? Are they of equal importance or are they quite separate?

T W Maver, ABACUS University of Strathclyde:

The collaboration course I have been talking about is essentially for undergraduate students, although we always imagined that modules with it might be used in an MSc course or indeed for short mid-career courses which, as you well know, are very important. However, our own MSc course which we will be starting for the first time in October is somewhat different and at a somewhat more advanced level.

E A Warman, CADCAM publications:

How fully do you expose your students to the question of curious constraints on the design of a building that actually calls you into a circular-shaped square corner?

T W Maver:

The lessons to be learned from observing traditional design practice are limited; if somebody is trying to put up a set of shelves using only a nail file instead of a screwdriver, or if somebody is carrying out brain surgery with a plastic picnic knife, it is not very edifying to observe the process. On the other hand if you have access to a design which shows you readily, explicitly and reliably what happens if you take certain decisions, then you know you have got the tool which is appropriate, and no amount of study of design activity as currently practised helps that. Perhaps that's an arrogant view, but I am afraid that's the conclusion that I certainly have come to.

As to the constraints which the tool imposes on the design which I think is your question; it was the fashion originally in architectural software that if you were breaking the daylight regulations or if you were breaking the fire regulations the program would say, 'Stop, no, you can't do that'. What we are now discovering is that the explicit appraisal of design alternatives can shed a tremendous amount of light on whether or not the constraints are sensible. For instance it's obviously ridiculous to decide on that pattern of fenestration which maximizes daylight levels in the building without due consideration of the thermal consequences. I think we shall soon be in a better position to be able to say one ought to achieve certain targets across the full spectrum of design criteria, considered in combination.

S Monaghan, Paisley College of Technology:

This is the first time in today's lectures I have heard CAAD discussed at education level examining how it is implemented in a college or university. One of the constraints that we find in the mechanical engineering department at Paisley is the cost of the hardware and I wondered how many of your students are exposed to the exercise which you showed us?

T W Maver:

Obviously that's a very important question. We are rather lucky in our department of architecture in that there is a large personal development program going on. Our students have access to ten interactive terminals, most allowing graphical input and output.

We are better off than most, and I think somewhere or other in this collaborative exercise that we are engaged in, we ought to be able to find a way not only to make software available but of making hardware available as well, but I am bound to say that it is still a problem and I don't know how it is going to be resolved.

44. A cost-effective two-way active computer-aided tertiary education network for industrially developing countries

J P Paddock and I Harding-Barlow

Abstract: A new and unique system is proposed for conducting two-way active computer-aided tertiary education in all the geographical areas of industrially developing countries. This educational network would be of especial value in the training, retraining and upgrading of skills in all areas of technology. The computer-aided tertiary educational programs will include theoretical, technical, vocational and industrial training courses.

The system will be based on commercially available individual personal computers. The computers selected will be of moderate cost, be flexible, versatile and amenable to easy updating as new or improved components become available.

The network will actively link the teaching centre(s) with the students and the students with the teaching centre. Interactive two-way communications will use satellites, ground-to-ground radio, and land and microwave telephone systems. This network is designed to be adaptable to new techniques as they become available. The instructional centres can be linked to, and/or broadcast from anywhere in the world.

This computer-aided tertiary educational network system lends itself to cost-effective updating as instructional, computational and communications arts advance. This will permit industrially developing countries with modest means to keep pace with the need to train, retrain and improve the training of large numbers of persons in many new fields of technology using the latest teaching methods and state-of-the-art communications systems. The network is very flexible and is geared to serving relatively remote geographical areas all over the world, in addition to fulfilling the very diverse needs of industrially developing countries in all forms of tertiary education and training.

We propose a system of computer-aided academic and vocational education and training at the tertiary level in engineering and science, where students and teachers may be separated by appreciable distances. At the outset we will take CAD ED as meaning Computer-Aided and Designed Education.

Even though a new high-technology tool has become available to us, we may not ignore the educational aspects. Indeed, we must take especial pains to assure that the tool becomes and remains the servant of the educational process, instead of allowing it to take control and become the master. Thus we must make use of all the educational methods which have been practised, while developing to the fullest our use of the new technological aids now at our disposal. But we must acknowledge that just as no single method in education can be applicable or useful in every situation, no single computer-aided method will be ideal for every purpose. It follows, then, that the practitioner of computer-aided education should retain as much flexibility as possible. We will attempt to outline an approach which we believe may have potential in rapidly developing countries, where teachers are few, distances are great, and it is difficult to maintain high educational standards.

Correspondence programs in science and engineering have been criticized because it is felt that experimental courses can not be taught effectively to non-resident students, since there is limited face-to-face 'interactive' contact between student and teacher, and since experimental equipment is largely unavailable to the student. However, are the presently constituted practical courses the best method of teaching modern science and engineering to students, both on and off resident campuses? We need perhaps, to re-evaluate our present-day needs and determine what our teaching goals and requirements are and/or should be in the next decade. We may well find that in many instances we can teach many disciplines and practical courses more effectively than at present, with the aid of available and envisioned technology to both resident and correspondence students. It should be remembered that often the success of a course depends upon the personality and inventiveness of the teacher and how he or she exploits the available technological aids, including the computer. Tertiary engineering and scientific CAD ED should be a teacher-student interactive medium, not a computer-student interactive programmed learning one. As with all correspondence programs, telecommunicated ones are most successful with students who want to learn and are self-disciplined and motivated.

Telecommunicated CAD ED tertiary engineering and science programs can provide a substantial gain over what would be possible using traditional correspondence course methods because:

1. Computer-graphics displays and simulations can realisticaly train students in most types of experimental work, reducing the time required for on-site, 'hands-on' practical work by 70 to 90 per cent. (The training of commercial airlines pilots is now mainly accomplished using computer simulation equipment and comparatively little actual flying time.)
2. In addition to computer videotext, reference materials, simulations, many types of graphics, including animations, can be transmitted to the students.
3. The degree of interaction between student and teacher can be varied to produce a balance between educational achievement, availability of resources and cost effectiveness. It should always be remembered that various students groups require different levels of interaction with their instructors for successful completion of diverse courses.

We perceive some of the main advantages of telecommunicated CAD ED as being:

1. Quality education can be delivered to persons who otherwise would have little chance of receiving it.
2. Students who up to now have had to receive overseas training, could now receive it in their home countries.
3. Courses which up to now have not been cost-effective to teach because of limited numbers of students, can be taught because a large enough student pool can be drawn from wider geographical areas.
4. Competent teachers can teach students in the more remote geographical areas.
5. Where there is a short supply of teachers, courses can be taught via telecorrespondence.
6. Vocational training can be made available on site, even at small plants.
7. Once introduced to telecommunications, students will probably be more willing to use data-bases of all types, seek information via telecommunications networks and use computers freely in their personal and professional lives.
8. Students will become proficient in the day-to-day use of the least one computer, and learn what computers can and cannot do.
9. Because students will be recording their lectures as well as the interactive discussion sessions, they can review all materials as often as they wish.

It should be recognized that present classroom teaching methods will not always be successful in teleteaching and vice versa. At present there is a tendency to link computers with computer programming or computer literacy and CAD ED with programmed learning, automatic grading of test papers, multiple-choice testing, etc. We need, however, to consider CAD ED in wider terms. We feel CAD ED must not become a stereotyped method of teaching, where 'interactive' means computer-student interaction. We need to realize that there are ways of making correspondence CAD ED or teleteaching interactive between teacher and students and cost-effective at the same time. We must not be afraid to experiment with new creative CAD ED teleteaching methods because we do not yet know their full potential. As our understanding increases of the vast possibilities that are available to us via teleteaching, we will learn to use its many capabilities. For example: (i) It may be possible for both teachers and advanced students to take part in international computer-teleconferences, receiving the conference proceedings via their computers at the time of the conference, rather than many months later, and being able to ask questions of the various authors, also via their computers. (ii) Teleteaching courses can include guest lectures from experts anywhere in the world. And (iii) practical engineering applications which are common only in one part of the world, may be 'viewed' by students anywhere, using a colour digitizing camera to show the engineering details which would normally only be available during lengthy on-site visits.

The authors envisage a generalized teleteaching lecture program for tertiary education, in which the lecturer will prepare lectures in his/her normal manner with the following differences: instead of audio and blackboard presentation, the content will be typed into a computer, in a format similar to word processing, which will require no special computer language such as LOGO or PILOT. Then diagrams, graphs, illustrations and sometimes very simple simulations, will be digitized and added to the text. This total 'formal lecture' will be transmitted to the student at the end of an interactive teaching session and the student will be expected to study it before the next interactive teaching session.

An interactive teacher-student CAD ED teleteaching session will consist of at least four main parts:

1. Part one, in which the students ask questions about the previously transmitted lecture and also any other topics they may wish to cover. (From a technical point of view, the students would transmit their questions to central control, which would then 're-broadcast' the question and the lecturer's answers to the full teleclass.)
2. Part two, in which the lecturer will ask particular students to answer questions and/or discuss various aspects of the previous lecture or other course material.
3. Part three, where answers to tutorials and other course work problems are discussed. The lecturer may also ask certain or all students to transmit, and hence share with the rest of the class the results of individual projects.
4. The lecture which is to be discussed in the next interactive session will be transmitted along with tutorial work, etc, and stored on disc by the student.

If the telesession is a practical or laboratory rather than a lecture, most of the transmitted data will be in the form of pictorial representations, simulations, etc. It is often not realized that most practical engineering and science classes can be taught using the digitizing camera, scale models, student model kits and 35mm slides as the key pieces of equipment. As with all experimental course work, the overhead costs will be greater than for lectures only. However, it is felt that the ratio of the costs of a lecture to a practical course, would be similar for both the traditional on-site and the teleteaching programs. But, as stated above, the success of the telepracticals, as well as the telelectures will often depend upon the ingenuity of the lecturers. The authors are particularly interested in the teaching of telepracticals and feel that simulation programs should only be one aspect of these courses.

It has been asked, how will one test the students? Obviously the lecturer will be testing the students during every interactive period; however, it is suggested that formal testing take place once or at most, twice per year and take the form of short problems and/or calculations, followed by longer essay-type questions. The student will be given three hours and then asked to transmit, from disc his/her total 'examination paper'.

In addition to telelectures, telepracticals, and interactive teacher-student teleteaching sessions, the lecturer should be available to the individual students during certain hours of the day or evening via telephone, radio telephone, CB radio or other form of voice or data-line communication. In addition to the use of the computer, the lecturer can make use of audiotapes for individual remedial work, the answering of complex student questions and also as an adjunct to teleteaching. One of the authors taught a MSc course in toxicology for 70 students several years ago, using audiotapes and photocopied lesson materials, plus keeping in personal contact with the students by telephone. In this instance, the practical course work was conducted at a central location twice per year; however, it could have been very successfully taught as a telepractical, with the addition of colour 35mm slides for some of the histopathology (video-cassettes would not be cost-effective).

It has been asked whether students could share computers. We believe, however, that students should have their own, with 64K byte memory, a black and white video-monitor, a modem and at least 200K bytes of disc storage capacity. A printer, and where appropriate, colour monitors, a touch sensitive CRT and other peripheral attachments can be shared by several students, preferably no more than three or four.

Special add-on equipment such as a digitizing board, or light pen with the addition of high-resolution graphics equipment will be necessary in the teaching of computer-aided design to engineering students and for student thesis projects of many types. Much can be learnt from the computer-aided graphics-based engineering courses taught at the US Naval Academy, but lower-cost adaptations can be devised. Another useful source of ideas which can be adapted to the tertiary teaching environment is the film industry. Many of their computer-aided design ideas are costly, but if one examines the basic concepts behind them, one can develop low-cost, but effective methods applicable to telepractical tertiary science and engineering education.

Next we consider the criteria for the selection of computer equipment: what is available, whether it is cost-effective, and what will offer particular advantages in engineering and scientific training. There is no ideal computer which will give optimum results in all situations. However, what are feasible and realistic attributes of computers suitable for use in our proposed programs?

First, should the computer be 8, 16 or 32 bits? At the present time, 8 bit machines with the possibility of the addition of further CPUs, would seem to be the most cost-effective ones to use. This does not mean that in future a true 16 or 32 bit machine might not be more efficient and desirable. Tandy and Apple, for example, have an add-on 16 bit CPU for some of their computers, whereas Texas Instruments are using a 16 bit chip in one of their low-cost machines. The advantages of true 16 bit machines are their greater speed, larger memories, greater versatility, a simpler choice operating systems and increased graphics capability. The full capabilities of the 16 bit CPU are not being utilized in the presently available personal computers. One may ask whether it is better to use a single type of computer, say Apple or Tandy or Atari, or should one use a variety of types with attendant flexibility in the system. The question, of course, revolves around the type of CPU and operating system chosen, because the ease of inter-computer communications may depend upon these factors. The two most widely used 8 bit CPU chips are the Z80 (and Z80A) and the 6502. Among the better-known computers which use the 6502 are Apple, Atari, Commodore and Acorn, whereas the Z80 or Z80A is used in the Sinclair, Tangerine, Otrona, Tandy and Osborne. Some of the Tandy machines and the Fujitsu Micro-8 use the 6809 chip. The great advantage of the Z80 machines is that (provided enough memory is available) they can use the CP/M operating system. A number of the 6502 and 6809 based machines have optional Z80 CPUs so that they can use the CP/M operating system, for example, the Apple and Fujitsu Micro-8. The addition in this latter machine is particularly well integrated into the total system. The speeds of the 8 bit microprocessors vary from about 0.8 to 4.0 MHz.

How much memory should be available for use in engineering and scientific training? It seems that the absolute minimum required for this particular application would be 64K bytes, which would allow for the use of Pascal and C if and when these are necessary and also graphics applications.

For storage one may use cassettes, bubble memory, single or double density floppy discs, hard disc, or cartridge memory. At the present time 5¼-inch double density discs would seem to offer the maximum versatility, although ultra high-density 3-inch discs may soon partially replace them, and under certain circumstances, for example, in dusty conditions, bubble memories would offer very definite advantages, as would some of the new 3-inch discs. High-density floppy discs and cheaper bubble memories are still in the future.

Keyboard features are slowly becoming more uniform, although there is still quite a bit of variation. If one wishes to deal with a non-Romance language or generate complex single-character graphics one needs to study keyboard features with particular care. It should be noted that many manufacturers do not use the same keyboards on all their models or even within the same model type.

In choosing video-monitors, the major choices are between black-and-white and colour, and 80-column and 40-column text. Presentations in colour are often desirable in teaching and will be especially desirable in some tertiary teaching. However, the choice between black-and-white and colour is made difficult by the price difference. Moreover, having decided upon colour, one is faced with the choice between RGB monitors and TV receivers; RGB monitors are more expensive, but there is a variety of TV standards. Sometimes high resolution is of paramount importance, but often even the term 'high resolution' needs very specific definition. In many instances colour or high resolution may be required, whereas in comparatively few, both will be required. When both colour and resolution are required, thought should be given to the use of alternative methods of presentation, such as 35mm slides. A touch sensitive CRT screen can have great educational potential, particularly when one is involved in vocational tertiary engineering and scientific training to students in their second or third language, which happens frequently. Since a touch sensitive capability can now be added to many CRTs, this feature can be a realistic possibility.

Computer input and output should be accomplished only through standard ports. We recommend that at least one serial and one parallel port be available on each computer and suggest the RS423 or RS232 serial port and the IEEE488 parallel port.

Modems are of great importance in our world of inter-computer communications. It is to be earnestly hoped that greater standardization in both hardware and software will very rapidly take place. One of the authors has particularly strong feelings on the 'standard' low cost 1200 baud modems and 'standard' software, which are neither standard nor low-cost, and which lack proper descriptive literature. It is to be hoped that 1200 baud full-duplex modems will become standard, with 300 baud as an option, and that the necessary software for the inter-communication of all the major personal computer types be speedily developed.

When preparing course materials it is essential to have a variety of graphics capabilities available. Light pens, digitizing tablets, digitizing cameras and other schematic and graphics preparation aids are now becoming available at reasonable cost. Device dependence, however, can still be a problem although, we may hope, a diminishing one.

There is a great need for printers of high quality and low cost. For letter-quality printing, one may use electronic typewriters, which can be interfaced to computers relatively simply, but are still costly. Dot matrix printers are well suited for graphical output.

When considering computers in education it becomes vitally important to know the servicing record and availability of parts for the various computers. For many computers this is an unknown factor, because they have been on the market for a limited time. With the growth of a new 'rent a personal computer' industry, more information will be available shortly. However, it should be remembered that often the same make and model of a particular computer sold in one country may not be the same as that sold in another. Moreover the computer could be the same and the name different. When a program becomes large enough, then it will be desirable to buy a large number of computers of several types, to buy in quantity so as to take advantage of cost discounts, and to provide service and maintenance as part of the education program.

When buying computers, educators in economically developing countries should especially examine environmental factors, such as the effects of temperature, humidity, dust, altitude, quality and characteristics of the available power, vibration and shock. Listed below are some of the computer parts which are most likely to be affected by physical factors:

1. Low temperature – magnetic discs, integrated circuits
2. High temperature – magnetic discs, integrated circuit and plastic components
3. High humidity – circuit boards and inter-connections
4. High temperature plus high humidity, particularly in tropical and sub-tropical climates, often provide suitable conditions for the growth of fungus, especially on wiring, circuit boards and inter-connections which have been inadequately cleaned during manufacture. The use of contacts of high quality on integrated circuit and circuit board connectors is highly recommended.
5. High altitude – conductor spacing on circuit boards (this is of little concern except at high voltages)
6. Dust – disc drives and most floppy discs
7. Vibration and shock – plastic components, such as cases, circuit boards, mountings, disc drives, etc
8. Power surges – data errors are likely to occur
9. Wrong power rating – power supply failure in the computer and peripheral components

One facet we have not addressed as yet is size and number of basic computer components. Does one want an all-in-one computer or does one want a multi-component computer? Probably the answer is a bit of both – there is no fixed answer. At the present time and subject to change on further examination, the following is our rating of computer systems of especial use for teaching science and engineering in economically developing countries:

- Best-all-round general purpose computer – Fujitsu Micro-8 (Japanese version)
- Good general purpose computer – Hewlett-Packard 87
- Good single unit computer – Otrona Attache
- Good computer with some restrictions – Apple II
- To be re-evaluated in September 1982 – Osborne I and Digital Microsystems Fox
- Computers which look promising – Sinclair Spectrum and Acorn BBC Model B
- Evaluations planned – DAI personal computer, Tangerine Tigeress and LNW 80

The methods used to link the central teaching control to the students will depend largely upon the communications available. However, when suitable methods are not readily available or are of poor quality, a regional and/or site specific network could be set up. Since the factors involved in these circumstances are likely to be complex, the solutions will vary greatly, but must be cost effective. In general, the systems used will be based on state-of-the-art techniques and components. Any communications network developed may include some or all of the following: satellites, land and microwave telephone systems and ground-to-ground radio. It should be noted that the systems would be two-way interactive, but need not always consist of identical components. The total network will be designed to be adaptable to new techniques as they become available.

In conclusion, let us develop teleteaching to its fullest potential, by making the greatest possible use of all the modern technology available to us, realizing that it is the wave of the future. The educational network described above could be of especial value in the training, retraining and upgrading of skills in all areas of technology. The proposed method of interactive instruction is ideal in the fields of engineering (electronic engineering, computer-aided design etc); science (laser physics, analytical chemistry, environmental sciences, etc); agriculture (botanical genetics, biotechnology, etc) and health (paramedical training, occupational safety, etc). The computer-aided tertiary educational programs may include theoretical, technical, vocational and industrial training courses. This CAD ED network system lends itself to cost effective updating as instructional, computer and communications arts advance. This will permit industrially developing countries with modest means to keep pace with the need for qualified persons in many new fields of technology using the latest training methods and state-of-the-art equipment. The proposed system of education is very flexible and is geared to serving nearly all geographical areas anywhere in the world. In addition, it can fulfil the very diverse training needs of industrially developing countries in all forms of tertiary education.

And thus spake King Arthur, 'The old order changeth, yielding place to new, and God fulfils Himself in many ways, lest one good custom should corrupt the world.' (Tennyson).

I Harding-Barlow, Engineering and Scientific Consultant, 3717 Laguna Avenue, Palo Alto, CA 94306, USA

J Paddock, Engineering and Scientific Consultant, 3717 Laguna Avenue, Palo Alto, CA 94306, USA

Discussion:

T Otker, FDO Technisch Adviseurs BV:

I have two questions. First, what kind of standards are you using for industrially developing countries, and second, is there any need for this sort of education training in the sort of countries I have in mind anyway?

J P Paddock, Engineering and Scientific Consultants:

I am not sure that I can give you a really good answer to your first question. I think we have to adapt ourselves to the standards that are available and the nearest that we have are the IEE standards. These seem to be fairly universal and I won't go much further than that. I don't think we can deal with standards at this stage.

The second question, is this really needed? I think in the developing countries we are about to see a great need for it. I will point particularly to a recent government Commission in South Africa, which anticipated the need for training and retraining of five million people in the next ten years.

I Harding-Barlow, Engineering and Scientific Consultants:

Australia anticipates the need to train and retrain two million people within the next three to five years. These are just two countries that need to be considered. You can take just about any country that you name, particularly ones where you have mining or where other industry planned in the next couple of years and you will find CAI and CAD are needed.

T Otker:

I think there are important new developments in two developing countries at the moment, Africa and South America. In fact you can take any country, they are all coming up with new technology, new ideas and industries which I think is very interesting indeed. What you are doing is bringing our industry idea and philosophy to these countries, especially in CADCAM and showing them how to apply these tools in their refineries.

I Harding-Barlow:

Well, we don't want to impose shall we say, European, American, and Japanese culture on them but I think that a lot of them wish to use what we have and develop it according to their own needs. This is something we would like to see in various places in the world.

J P Paddock:

We would like to apply some true interactive techniques in different cultural settings in developing countries. As far as I can gather in the United States we have made enormous use of distance learning but not this type of interactive teaching.

I Harding-Barlow:

Some of the video type teaching that has been practical in the States hasn't been overly successful. One will really have to re-examine it. One of the things that we have been finding is that CAD ED in the States fulfils only a very limited usefulness and this is something that has disturbed us a great deal. When one attempts to apply the US-type CAD ED to Africa, Australia or even Europe one would have many problems.

Unidentified contributor:

My question is this. If we use a satellite basis and we are aiming at developing nations, who controls the switch? Interactive politics might create numerous problems.

I Harding-Barlow:

I don't think we want to impose our ideas on any country. I think we have had enough of that in colonial times but we want to try and be helpful, have interchange of technology, and try and make the most of rather limited educational resources throughout the world.

J P Paddock:

Just one other interesting point. In various parts of the world one may get different reactions to the same lessons, because the various cultures portray and visualize things differently.

45. Education and training for CAD – a comparative study of requirements for developing and developed nations

E A Warman and K Kautto-Koivula

Abstract: This paper presents comparisons, experiences and requirements for the introduction and transfer of technology relating to CADCAM between Finland and Great Britain. It then provides a further comparison with some technology transfer exercises undertaken with the socialist nations.

Introduction

Computer-aided design is now moving into its adolescence stage and experience and observations are indicating that some of the initial ideas on the approach and methodologies required for education and training in order to provide awareness, technology transfer and the development of revised approaches to problem-solving, are not correct. The pressures of increasing technological innovations and the need for industry to utilize the innovations in order to remain competitive make the traditional long-term approaches valueless. On the other hand, many of the technological innovations are beginnning to produce social problems. Such problems cannot be ignored in any national approach developed for dealing with CADCAM.

One aspect that started to highlight the need for a re-examination of the requirements of CADCAM education and training was discussions (Encarnacao, Torres and Warman, 1981) during the conference on CADCAM as a basis for the development of technology in developing nations. These discussions indicated, most clearly, that if a re-study of CAD educational needs was to have any added value, it must be considered from an international viewpoint. The reason being that CAD is a technology that is experiencing a world-wide requirement for a logical and structured technology transfer process. Thus it was essential that the viewpoints and requirements of different nations were considered.

Using contacts with other IFIP experts who had studied the educational and training requirements of CAD, a joint study was undertaken between Finland and the UK. At first sight one was dealing with two very different situations if industrial structure, type and tradition were key factors, or if the present educational processes, population distribution and density were major factors. As the study developed, a considerable amount of common ground was revealed even down to the problems of drawing office staffing and mix that had previously been discussed (Warman and Reader) for the British situation.

It was then considered worthwhile to use the data from this joint study and compare it with other information on two technology transfer exercises undertaken in Poland and Czechoslovakia with UNIDO assistance (Kozar, 1981; Zgoryelski, 1981).

This work, together with the results of discussions to attempt to determine a broad requirement for developing nation needs, is therefore the basis of this paper.

Computer-aided design provides the opportunity to rethink the fundamental aspects of engineering education and to establish an international consensus of the core requirements for CAD education and training.

The British scenario

Since 1977 (Warman and Reader), there have been massive economic forces acting upon the British economy and industry, particularly in the traditional engineering sectors.

Drawing and design offices in some areas of industry have suffered considerably because manpower reduction schemes, based upon voluntary redundancy and early retirement, with attractive severance payments, have encouraged the older, more experienced staff to leave employment.

The net results are that the ratios of higher academic to lower academic qualifications and the experience ratios have become more unfavourable.

What is encouraging is that some sectors of industry believe that CADCAM – or in its fuller sense – Computer-Aided Engineering – will be a key factor in helping industry to respond rapidly to the forthcoming economic upturn.

The belief in CAE with the very severe shortage of practitioners at all levels has led to intensive debates, but with very little action, as to the type, quantity and quality of CAE education that is required. Also the advances in technology that are continuing to take place are making some of the arguments null and void as to the type of education and training to be given and the quantity of people to be processed.

During the course of two confidential studies into various aspects of British industry related to the use of CAE methodologies, ancillary information was obtained that has indicated that the following factors are key points in deciding where the main thrust of effort should be:

1. the academic spread in design/drawing offices
2. the number of new engineering graduates entering industry
3. the time required for new graduates to be influential factors in industry
4. the mechanisms for setting course contents and standards for education
5. continued advances in technology

The points are further amplified below:

1. The academic spread in design/drawing offices

Mechanical engineering drawing offices were re-sampled to establish an up-to-date spread of academic qualifications of those staff working directly on the design/drawing functions. These new figures are presented in Table 1. The general pattern is similar to that previously reported (Warman and Reader) and the trend seems to be a slight worsening of the situation.

It is these people who are already in limited numbers, who will be expected to use the drafting and design systems that are being introduced.

The what, how, when and where they should be taught are most certainly key issues and without doubt, it is in these areas that the main thrust of education and training should be made. The education and training issues are not confined to the use of the equipment or a little FORTRAN programming. CAE provides new freedom of design and drafting expression and also has different constraints to classical methods. The education and training must therefore reflect these facts.

2. The number of new engineering graduates entering industry

Examinations of the statistical data, available from central government and other sources, concerning the production and deployment of engineering graduates raises many questions as to the quantity and quality of academic subjects to which they should be exposed during their undergraduate education.

The number who directly make a career in design is very limited as the previous data indicates. The cost problem (irrespective of the wisdom of so doing) of providing sufficient turnkey systems (that soon become obsolete), such that undergraduates obtain some practical exposure to CAD, is at present unacceptable.

Part of the solution seems to be the encouragement of more post-graduate study, either full- or part-time, with the assistance of industry. The current situation with respect to post-graduate studies and results over the broad spectrum of CAE is most encouraging and positive and the role of the post-graduate in industry must not be underrated.

3. The time required for graduates to become influential factors in industry

The very nature of industrial society with its pyramidial structures and the level at which key decisions are made, means that there is a considerable period of time before a graduate can become an influence factor. Thus, unless there is encouragement, with some mandatory requirements to enable graduates to be kept up-to-date, by the time that those who have remained in an engineering environment are in a position of influence, they are considerably out of date – this situation will become more crucial with each advance in technology.

4. The mechanism for setting course contents and standards for education

Without any question of doubt, there is considerable scope for putting some order into the number of bodies and their relationships, with regard to the setting of standards for the various levels of engineering training and education.

The professional engineering institutions set standards for membership, that may or may not be met by the universities, polytechnics and technical schools at various levels. The polytechnics do not have the autonomy of the universities, therefore their response time to course changes is longer. Without describing the web of relationships between the DoE & S, universities, NCAA, professional institutions etc, there is a far greater problem. That problem is the shortage of suitable people to teach the new technologies, *if* they are required to be taught.

5. Continued advances in technology

While the debates and lack of action continue, technology continues to advance and already there is much to question about the quantity and type of people – predicted some two/three years ago – required to make use of CAE techniques in industry. It is going to be more profitable to undertake a thorough study of technological advances in order to obtain some notion of future needs rather than implement ideas based upon the emotional pressures of the moment.

Finnish scenario

The use of CADCAM technology in Finnish industry is expanding fast. The most popular application areas have been the metal working industries, electrical and semi-conductor industries, textile and clothing industries and construction industry. The fast development of CADCAM technology has made it necessary for a small industrial country, like Finland, to expand its knowledge mostly by technology transfer.

The effective implementation and the use of CADCAM technology is possible only by efficient CADCAM education and training. It has been discovered that the need for education and training of CADCAM technology is very great in all application areas and at all organizational levels: managers, draftsmen, designers, development and maintenance personnel. This section of the paper describes some of the problems uncovered that prevent the effective use of education and training of CADCAM technology in Finland.

CADCAM education and training can be divided into two basic categories:

- basic technical education (undergraduate and graduate students)
- continuing education and training (graduate students and users of CADCAM systems)

Education and training problems in these two areas differ slightly.

Figure 1 *(see page 9)*

Basic education of CADCAM technology

Figure 2 describes the organizational relationships involved in the basic technical education in Finland. The objective of these organizations, with respect to CADCAM education, is either to give the basic knowledge of CADCAM technology in order to get the students orientated towards the new technology, or provide the possibility for post-graduate students to get higher degrees in CADCAM by fundamental research.

Figure 2   *CAD/CAM education and training in Finland (see page 9)*

The problems which have been discovered that prevent effective CADCAM education are:

1. The national strategies of CADCAM technology transfer are still unclear.

    Before the basic CADCAM education can be planned carefully, a policy should be decided at national level concerning the ratio between the imported and the national R&D of CADCAM technology.

2. Universities of technology and other technical educational organizations do not respond fast enough to industry's needs.

    Reasons for this can be partly the lack of motivation of professors and teachers and partly to the structure of the organizations; CADCAM technology is not the problem of only one department, it cuts across many departments.

3. Difficulties in getting tools for CADCAM education.

    Besides the theory, an important part of CADCAM education is practical exercises. The optimum solution for the tools in CADCAM education has not yet been found. Most of the commercial systems have been too expensive to get into all of the educational organizations. Buying only software, renting CADCAM systems or developing them by themselves are cheaper possibilities but are not always the best ones for CADCAM education.

4. Attitudes against the CADCAM education.

    In the beginning of CADCAM technology transfer, some of the professors and the teachers saw CADCAM technology as counter-productive to developing the skills of manual design, in particular the theoretical basis of design would soon become unknown. Some of them also thought that the use of CADCAM systems would prevent the students from seeing the extent of problems in design.

    Reasons for these attitudes could be based partly on the fact that many of the professors and the teachers are not as familiar with computers and methods of using them, as they should be.

5. Language problem.

    Most of the information concerning CADCAM technology is written in English. On higher levels of education, the use of a foreign language is not a problem. The lower down in educational organizations we go, the greater the problem is of using course material written in a foreign language.

6. Lack of CADCAM experts.

    One very important thing in preventing effective basic CADCAM education has been ther lack of CADCAM experts in different application areas. Without them and their experiences of CADCAM technology, it has been rather difficult to plan courses that give relevant information of the use and the development of CADCAM systems and technology.

Very often the industrial CADCAM experts should be used as teachers of these courses. Because of the lack of CADCAM experts this has been too seldom possible.

Continuing education and training of CADCAM technology

The purpose of continuing education and training of CADCAM technology has been to give the information of CADCAM technology mainly for the effective implementation and the use of CADCAM systems. A great deal of information is needed, especially when a company is seeking to start to consider the use of CADCAM technology.

Problems which have been discovered preventing the continuing education and training of CADCAM technology are:

1. Lack of organizations which are able to give education and training for the aforementioned purposes.

   The organization that is able to give such kind of education and training should have enough knowledge of the problems encountered during the selection, implementation and use of CADCAM technology.

   A large part of such kind of information may only be obtained by experience and it has been very difficult to find persons or organizations that have had this experience.

2. A large part of the information needed is application or company-oriented.

   An organization which chooses to give continuing CADCAM education training or consulting in CADCAM area, should have enough knowledge, besides the system technology, of the specialities concerning certain application areas or company oriented aspects.

   It has been discovered that it is rather difficult to plan a course that would be suitable or useful to many persons or organizations.

3. Language problem.

   Language has been found to be an even bigger problem in continuing CADCAM education and training than in basic education, especially in industry where very many of the older designers or draftsmen using CADCAM systems have a rather poor or miniscule background in English. Many surveys have shown how important a factor it can be in preventing the effective implementation and the use of CADCAM systems.

   It has been seen to be too time-consuming to teach English for this purpose. Also it is too time-wasting to try to translate all the system manuals. A solution which is a compromise of these two extremes is probably the best one.

4. Lack of traditions in CADCAM research and development.

   The use of CADCAM technology is a rather new phenomenon in Finland so there is still very little knowledge of the information needed in continuing CADCAM education and training. Very few researches and surveys have been made in order to get to know these needs.

Information needed in CADCAM education and training

In the above section we divided the field of CADCAM education and training into two categories, mainly by showing the differences and problems which are in these two areas. In this section we discuss the information content of CADCAM training and education and try to emphasize that the final recipient of this technology is a company.

By so doing, we can divide the needed information of CADCAM technology into three different categories (Figure 3):

 – CADCAM production technology
 – CADCAM organization technology
 – CADCAM product technology

Figure 3 *Types of CADCAM technologies (see page 9)*

CADCAM production technology tells how to use CADCAM systems. It can be divided further into three sub-categories:

 – System technology tells of the basic aspects of the software and hardware facilities which are going to be used.
 – Skilled labour technology includes mainly the methods and the ways to get the users to use a CADCAM system effectively.
 – Production organization technology contains the methods and the ways to implement a CADCAM system so that the productional objectives could be achieved.

CADCAM production technology contains typical information, such as:

 – basic information of CADCAM systems: uses, problems, possibilities, etc
 – analysis of the information process of design and connections to manufacturing
 – connections to material administration, production control systems, or other conventional data processing systems in a company
 – job scheduling and all kinds of preparation for implementation

CADCAM organization technology contains information telling ways of how to use CADCAM technology so that the financial objectives of a company may be achieved. Typical information is:

 – all the reorganization problems, re-education of personnel, etc
 – effects to the competition of a company
 – the share of responsibilities concerning CADCAM technology between different departments, etc
 – wide effects of new technology inside company: social aspects, ergonomy problems etc.

CADCAM product technology contains the information and knowledge of how to make CADCAM systems for new applications. Typically, it contains the following information:

 – computer science and information processing techniques, especially methods using interactive computer graphics
 – data-base technique
 – software engineering
 – mathematics, modelling, geometry
 – telecommunications technique
 – man-machine communication
 – application-oriented information

At the beginning of CADCAM system use, most of the needed technology normally concerns the production and organizational aspects. The more knowledge and experience of CADCAM technology that has been obtained, the more apparent is the need for CADCAM product technology.

First level comparison between Finland and Britain

Apart from the language problem, there is much common ground between the United Kingdom and Finland in the problems of education and training, at all levels, in CADCAM technology.

In some of the problems there is a slight phase difference because Britain was involved in CADCAM at a much earlier time.

However, the relative size difference of the two nations does not appear to have much relationship to the *rate of diffusion* of knowledge.

The shortage of experts in CADCAM technology is a significant factor in each case. Equally interesting is that although there are differences in the educational system, each nation has produced a tenuous web of relationships and responsibilities that mitigate against a cohesive national plan.

Again, although expressed slightly differently, there is the lack of response in each system, to the needs of industry. Also, the problem of getting equipment, in the educational institutions, to provide practical experience, is again a common problem. The question of academic reluctance is also common, it being kept more underground in Britain than in Finland. In hindsight, there is some sympathy with the fear that badly handled CADCAM education can destroy the fundamental and creative outlook of some users.

The age and experience spread in drawing and design offices appears also to be very similar. In addition to the age – reluctance to learn new technology relationship, Finland's problem is compounded by the reluctance to learn English; the language of CADCAM technology transfer.

Experiences in CADCAM education of the two nations will now be discussed to see how these technology transfer problems are being tackled.

Society structure and dissemination of knowledge

The older a nation is in the Western industrialized societies, the more elaborate is the structure and relationships that exist between the levels of the educational system, the activities of professional societies, industry and government. Whether or not Britain represents the ultimate in complexity is a matter of debate but it surely must be approaching this state.

Finland as an independent nation is much newer and, although it is developing a set of complex relationships, the situation is still fairly neat. The single technical research centre with its many laboratories and positive links to industry has a great deal of merit compared with multiple research centres having little co-ordination between each other and tenuous links with industry. Similarly, a single unit concerned with the continuing education of engineers and technologists, is of great importance.

Within the Eastern Europe socialist nations, starting with the Academy of Sciences at the top, there is a structure between research institutes for particular industrial sectors, universities and industry-based research activities. With this 'logical' structure has the dissemination of advanced technology been a simple and national process? Data collected would suggest most firmly that the answer is 'no'!

To summarize the results of work (Kozar, 1981; Zgorykelski, 1981), undertaken in Poland and Czechoslovakia the following key points emerge:

(a) There is a shortage of people skilled in the new techniques and their applications. Technical universities do not offer curricula in this field. Most of the existing knowhow is self-taught and usually proprietary.

(b) Basic education in informatics should contain sections that deal with engineering applications.

(c) Strategies and procedures to establish and develop CAE activities are by no means clear or uniformly accepted.

(d) There are severe problems of lack of management understanding.
(e) It is often better to select a single industry segment and concentrate effort on to this rather than operate on a broad point. Certainly the establishment of centres of excellence is a vital means of convincing by example.

Conclusions

This comparative study has shown most clearly that:

(a) There is a severe international shortage of CADCAM experts.
(b) To be able to solve all of the objectives for CADCAM education and training set by industry, a great deal of planning and co-operation between different educational organizations and industry is needed. This problem is common in spite of differing national infra-structures.
(c) The attention to CADCAM education at the lower levels – draftsmen, technicians etc – is very bad. The reasons seem to be lack of funding and a great difficulty in changing basic technical educational programmes.
(d) The presentation of the concepts of CADCAM and financial and social aspects need to be presented to all levels of management. They also need to form part of the curricula at business schools.
(e) CADCAM is an area where the pooling of international 'knowhow' is essential in order to obtain the maximum benefit from the small amount of expertise that exists, at present, within the world.
(f) Educational and training problems are different in small and large companies and thus need different and adaptive treatment.
(g) There must be established some methods and means of utilizing the international organizations, such as UNIDO, IIASA, IFIP, IFAC and ECE to assist in co-ordinating the very limited expert knowledge in order that a uniform and sensible spread of knowledge and experience is obtained through a unified education and training programme.

Appendix

An outline of CADCAM education and training in Finland

Basic CADCAM education:

Basic CADCAM education has been provided only at universities of technology. Two of them have bought a turnkey system as an education tool. Most of the experiences concerning the use of turnkey systems in CADCAM education have been positive. Students have been so interested in the CADCAM courses that it has been necessary to limit the amount of participants.

Besides turnkey systems, many software packages have been bought for CADCAM education purposes. This seems to be a much cheaper way to get tools for educational purposes, because all the universities and technical colleges have their own computers. At times problems have been encountered in converting software packages to the computer used.

Figure 2 shows the present situation in Finland concerning the basic CADCAM education situation in the different technical educational organizations. The next step in expanding basic CADCAM will be to provide sufficient knowledge for teachers at technical colleges and schools so that they will also be able to start the basic CADCAM education.

Technical colleges and schools are very important areas with respect to basic CADCAM education because most of the CADCAM users (designers, draftsmen) will have this basic educational level.

Continuing CADCAM education and training:

The most important organizations that provide the continuing CADCAM education and training in Finland at this moment are also shown in Figure 2.

The Technical Research Centre of Finland (VTT) is a state research organization and its main purpose is to transfer new technologies to Finland and make applied research for industry's needs.

VTT is the biggest technical research organization in Finland containing over 30 laboratories for different application areas and over 2,000 highly qualified personnel. CADCAM technology is considered one of the most important areas in VTT at this time. For several years VTT has undertaken research and surveys into CADCAM technology. It has very close contacts with Finnish industry, thus it has excellent opportunities to give continuing CADCAM education and training to industrial and governmental organizations.

INSKO – the Continuing Engineering Education Centre is the biggest organization in Finland for continuing education and training in technical subjects. It uses industrial experts as teachers in its courses, so the content of those courses is, in the main, very practical and based upon a wide range of experience.

Besides VTT and INSKO, there are some private consulting companies that have given continuing CADCAM training and consulting.

References

Committee of technology transfer, technology transfer and international division of labour (1980) Helsinki, in Finnish.

Encarnacao, Torres and Warman (1981) CADCAM as a basis for the development of technology in developing nations. North-Holland Proceedings of conference, Sao Paulo, Brazil.

Kautto-Koivula, K (1981) Strategies for CADCAM technology transfer: example from Finland. North-Holland Proceedings of conference, Sao Paulo, Brazil.

Kozar, Z (1981) Research and application of CAD and CAM systems in Czechoslovak mechanical engineering industries. North-Holland Proceedings of conference, Sao Paulo, Brazil.

Llewelyn, A (1982) CADCAM education and training. CAD 1982 Proceedings.

Ministry of Education, Helsinki (1981) *Educational development in Finland 1978–1981.* Reference publications 10

Uusitalo, M, Nykanen, M and Kautto-Koivula, K (December 1980) Survey of CADCAM situation in Finland. Technical Research Centre of Finland.

Warman, E A and Reader, F J Education and training for CADCAM – An industrial viewpoint.

Zgoryelski, M (1981) The concept of engineering computer applications promotion program. North-Holland Proceedings of conference, Sao Paulo, Brazil.

Discussion

R W Howard, CICA:

In my opinion Finnish architecture is probably the best in Europe. Their industrial design is very good and I think this is perhaps due to the appreciation of good design by the public at large, quite apart from having good designers. This means a good design becomes all-pervasive. Now if the sort of networks the previous speakers were talking about, obviously some time in the future, could actually link not only the teachers and the taught but also the consumers and the designers perhaps the sort of feedback from the dialogue about design could perhaps help other countries. Larger countries, like our own, where I think the average standard of appreciation of design is really very low, to rise to the level of countries which, like Finland being small, I think have a high level of communication. Larger countries seem to communicate less well and therefore the coincidence of ideas between consumers and designers is less good than in smaller countries like Finland, where it's very high.

E Warman, CADCAM Publications:

Maybe so. Yes, it is easier to communicate but if you sub-divide it, for example into small units there still seems to be some problem about communicating with respect to art and design. It may be that we are conditioned as a result of the horrendous period of architecture in the thirties, which I as a viewer, not an architect, think is abysmal. It could be of course that there is more space to indulge in a country like Finland or Sweden; it may be that we don't know how to use space.

It's an interesting point about communication. I am not quite sure whether this has come entirely out of today's discussion. If you are teaching and you could use the computer better than we do, for computer-aided design, it avoids the process of going through all the academic steps. You can actually do something on the computer, you can test the design, see what it looks like, then the questions you start asking, are not 'Is it strong enough?' but 'Is it a good design?', and compare it with other ones.

We have still an idea of the industrial revolution here in a different way, not just a question of space, but everything that was built then. It all seems to be cut and dried, but we could get back to some of the important debates that went on in the industrial revolution which no longer happen, for example the question, what is sublime and what is picturesque in terms of design.

T Otker, FDO, Holland:

We are dealing quite a lot in Scandinavian countries as well, especially in the CADCAM area. One of the things I found there is that there is training and education for students not only in working with hardware and software, but also for the control of information in organizations. I think most of the people in education and training are concentrating on the use of hardware and software and only 20 per cent represents the problems you would find in organizations in the industry. You use CADCAM as a definition in your study, but what's really behind it?

E Warman:

We had to use it as a definition to set some boundary points. There is in Scandinavia and Finland (because Finland isn't Scandinavia) certainly always has been in training and education of engineers a strong awareness that you have to provide the disciplines not only in terms of what I call the algorithms but also the management disciplines, so there has been a great deal of communication, there has been an awareness given to students right at the beginning such that they are part of the organization. If you have to communicate this is the way you do it. Something has to be made so that you have to communicate your information in a way that it can be made use of. There are small work forces, that were at one time extremely volatile with respect to social revolutions, more so than anywhere else in the world. Their Employment Acts are such that you cannot sack anyone. These are issues which can be problematical. The fact is that they do focus people's minds on to solving problems rather than sweeping them aside. I think this is what we have done. We in the UK have swept problems aside. It's not something that has happened overnight, it's been going on some period of time. When you start looking at the background of the history of some of these countries, the first thing you recognize is that they didn't happen overnight. They were doing some revolutionary things in respect of technology when we were still patting ourselves on the back.

T Okter:

Well, being a Scandinavian, I have some difficulty in philosophizing about this, but I think there are two or three factors which are important here. As an addition to explanations, I think the communications factor in a relatively small country, population-wise, was even more in the past than perhaps at present, extremely important. There were few people and everybody knew everybody else in industry for instance when I was there 30 years ago. There is a second factor which hits you when you compare British technical education institutions, particularly as they were, when the top management people in industry today were young. With very heavy emphasis on apprenticeship or sandwich courses or some other kind of long-term practice in industry maybe perpetuates obsolescent or obsolete skills to a new generation. The obsolescent skills may sometimes be important to learn but that's another question, *vis-à-vis* what I would call the continental tradition of technical education with this very heavy emphasis on the scientific problems. You pointed out the problem-solving which has played a role. In Finland there is a preference for young people. Finnish industry may suffer in ten to fifteen years time. Because one leaves in the ninth year of school in Sweden today, then it is necessary to get better grades to qualify later on for getting into engineering. It's become completely inverted in that particularly. In Finland industry is still a promising career and something which has some priority but not at all the high priority it was 20 years ago. It is, I think, a problem which is turning the other way round in Britain at present, for engineering education has much more attraction today than it had traditionally in Britain where at that time it was primarily administrative, colonial and other similar obligations which were most promising.

S T Monahan, Paisley College of Technology:

Our college is degree-awarding and as it's done under the auspices of CNAA we have to re-submit, not like the universities who can do it in-house and we are in the process of doing that. There are five of us here two from the electronics department, two from mechanical engineering department and one from Marsden computing and we are here specifically to get industry reactions to this CADCAM need. We are really here to find out, what does industry want? When we talk about education from the platform here are we talking about training by the vendors or are we talking about education in colleges, universities, what is it we as professional engineers in industry want the educationalists to provide?

E Warman:

It's my opinion that in industry, with which I have dealt with on a fairly broad level, they look for someone coming from the outside world who can operate the application system without having to train them. That's the first thing. It is very difficult to instil in their minds, particularly in the metal working industry, long-term requirements. You are going to need someone who is educated in a new method of thinking. You must listen to what he says, with a little patience because you won't necessarily understand it, so that you can get your industry to grow. People listen in those areas where by the very nature of the problem they are forced to listen, such as electronics and aerospace. These industries do not have a negative attitude. The first problem is to work on the management, to find out exactly what they do want, because what they really see at the moment when they talk about CAD staff is someone who has been trained elsewhere who knows all their problems.

I Harding-Barlow:

May I make a suggestion? If you can get your students to use their theoretical knowledge in practical ways, this would give them the best long-term education. I am speaking from my own experience since I decided very early that I would apply my university education to practical things. I am in my third career and I have always taken my original academic training, and applied it. I am wondering whether this wouldn't be one approach that might be useful.

46. Chairman's concluding remarks

A I Llewelyn

This conference has been of particular interest to those who attended the first one in 1977. The difference has been remarked upon and is indeed quite profound. At the first conference one was talking in a very diffuse way about possibilities and there was, I recall, much heated debate about the need. Now the need is obvious and one is dealing with specific problems and experience in the field of education and industrial training. People are actually doing what was previously speculation and that represents a tremendous advance.

There has, of course, been a five-year gap between conferences which was judged to be about the right timing. I think events have justified this decision and my impression is that this conference has been a success and that there should be a follow-up in about two years' time. There seems to be a general feeling that the next conference will take place in 1984 and planning will be on this basis. Comments and suggestions regarding the number of papers, problems with parallel sessions and the need for more time for discussion have been noted and will I am sure be taken into account when planning and organizing the 1984 conference. Indeed, as a result of suggestions already made, discussions are now in train for a series of workshops as one immediate outcome of this conference.

The proceedings of this conference will be edited and published as soon as possible and be available in adequate numbers. The printing of the 1977 proceedings was sold out very quickly and for some obscure reason was never reprinted. It is now, I believe, almost a collector's item! This time we will take care that more are printed and if necessary make reprints available.

I do not intend to attempt any summary of this conference but will venture two observations. One is prompted by the dialogue at the last session which I followed with great interest. It is an aspect which has percolated through many of the sessions, namely the improvement in product choice and quality, cost performance and reliability which is entirely due to the application of computer-aided design. The other related observation, which is of course the underlying theme of this conference, is the way in which this technology is changing the structure and pattern of industry and the consequent importance of getting the right balance in courses of education and training to meet future needs.

Finally, I would like to thank the organizers for the efficient arrangements which make all the difference to the success of a conference. Also to thank the authors and presenters of papers and the exhibitors who on this occasion contributed a special series of system presentations which provided an added value to the conference. It remains for me to thank you, the audience, for your participation and to express the hope that the time spent has been profitable, and that you will continue the evolutionary process which is now gathering pace and will return to take part in the 1984 conference and report further progress.

DATE DUE

NOV 1 2 1985			
MAR 30 87			
OCT 23 87 OCT 7			
MAR 1 4 1990 MAR 0 5 1990			
3 88 7 93			
JAN 4 '95 DEC 1 5 '94			
GAYLORD			PRINTED IN U.S.A.

WITHDRAWN FROM
OHIO NORTHERN
UNIVERSITY LIBRARY

HETERICK MEMORIAL LIBRARY
670.427 C11
/CADCAM in education and training : proc

onuu

3 5111 00121 5858

CADCAM
IN
EDUCATION
AND
TRAINING

Proceedings of the
CAD ED 83 Conference